Lecture Notes in Civil Engineering

Volume 129

Lecture Notes in Civil Engineering (LNCE) publishes the latest developments in Civil Engineering - quickly, informally and in top quality. Though original research reported in proceedings and post-proceedings represents the core of LNCE, edited volumes of exceptionally high quality and interest may also be considered for publication. Volumes published in LNCE embrace all aspects and subfields of, as well as new challenges in, Civil Engineering. Topics in the series include:

- Construction and Structural Mechanics
- Building Materials
- Concrete, Steel and Timber Structures
- Geotechnical Engineering
- Earthquake Engineering
- Coastal Engineering
- Ocean and Offshore Engineering; Ships and Floating Structures
- Hydraulics, Hydrology and Water Resources Engineering
- Environmental Engineering and Sustainability
- Structural Health and Monitoring
- Surveying and Geographical Information Systems
- Indoor Environments
- Transportation and Traffic
- Risk Analysis
- Safety and Security

To submit a proposal or request further information, please contact the appropriate Springer Editor:

– Mr. Pierpaolo Riva at pierpaolo.riva@springer.com (Europe and Americas);
– Ms. Swati Meherishi at swati.meherishi@springer.com (Asia - except China, and Australia, New Zealand);
– Wayne Hu at wayne.hu@springer.com (China).

All books in the series now indexed by Scopus and EI Compendex database!

More information about this series at http://www.springer.com/series/15087

Aeslina Abdul Kadir · Noor Amira Sarani ·
Shahiron Shahidan
Editors

Sustainable Waste Utilization in Bricks, Concrete, and Cementitious Materials

Characteristics, Properties, Performance, and Applications

Springer

Editors
Aeslina Abdul Kadir
Universiti Tun Hussein Onn Malaysia
Batu Pahat, Johor, Malaysia

Noor Amira Sarani
Universiti Tun Hussein Onn Malaysia
Batu Pahat, Johor, Malaysia

Shahiron Shahidan
Universiti Tun Hussein Onn Malaysia
Batu Pahat, Johor, Malaysia

ISSN 2366-2557 ISSN 2366-2565 (electronic)
Lecture Notes in Civil Engineering
ISBN 978-981-33-4917-9 ISBN 978-981-33-4918-6 (eBook)
https://doi.org/10.1007/978-981-33-4918-6

This Springer imprint is published by the registered company Springer Nature Singapore Pte Ltd.
The registered company address is: 152 Beach Road, #21-01/04 Gateway East, Singapore 189721, Singapore

Preface

This book entitled *Sustainable Waste Utilization in Bricks, Concrete and Cementitious Materials: Characteristics, Properties, Performance and Applications*. This book highlights current research, conceptual and practical utilization of waste in building materials. It examines various productions of industrial and agricultural wastes that have been generated worldwide that have a significant environmental impact towards the environment. The book discusses reliable guidance on how to incorporate these wastes effectively with greener technology and how to address its environmental impact to produce an environmentally friendly and sustainable green product in order to capitalize its practical application, properties' performance and economic advantages. The topics this book covered are on the physical, mechanical, environmental properties, leaching behavior, gas emissions and performance of sustainable construction materials.

The content of this book will assist and support the building sector as the construction and building materials sectors are the largest user of natural resources in the world; thus, they need substituting the raw material with renewable/recyclable materials. In terms of environmental aspects, this book emphasizes the importance of environmental conservation in efforts to conserve endangered natural resources for future generations. In line with the growing population growth, large energy consumption and rapid development of industrial processes have led to unbalanced use of natural resources and ultimately resulted to more wastes that significantly contribute to sources of pollutants and imbalance biodiversity. Therefore, the new formulations of building materials replaced with different types of wastes can offer a sustainable disposal method for waste materials, minimizing the environmental pollution and discovering potentials of alternative raw materials.

This book also aligned with Sustainable Development Goals (SDGs) that promote responsible consumption and production through the utilization of waste as resources that can be reused through recycling and recovery. Building sectors are also targeting to increase the number of green building materials in the market and to create policy amendments in allowing the use of recycled materials in constructions. In addition, the content of this book also fills up the National Priority

Areas (NPAs) in Environmental and Climate Change as the content promotes the reduction of waste disposal to the environment. As such, this book offers a valuable reference for researchers, industries and interested stakeholders in sustainable construction or any allied fields.

Johor, Malaysia

Aeslina Abdul Kadir
Noor Amira Sarani
Shahiron Shahidan
Editors

Contents

Contributors

Alida Abdullah Geopolymer and Green Technology, Centre of Excellence (CEGeoGTech), Universiti Malaysia Perlis, Kangar, Perlis, Malaysia

Mohd Mustafa Al Bakri Abdullah Geopolymer and Green Technology, Centre of Excellence (CEGeoGTech), Universiti Malaysia Perlis, Kangar, Perlis, Malaysia; Center of Excellence Geopolymer and Green Technology, School of Materials Engineering, Universiti Malaysia Perlis, Kangar, Perlis, Malaysia

Rafiza Abd Razak Faculty of Civil Engineering Technology, Universiti Malaysia Perlis, Kangar, Perlis, Malaysia;
Geopolymer and Green Technology, Centre of Excellence (CEGeoGTech), Universiti Malaysia Perlis, Kangar, Perlis, Malaysia

Aeslina Abdul Kadir Faculty of Civil Engineering and Built Environment, Universiti Tun Hussein Onn Malaysia, Parit Raja, Batu Pahat, Johor, Malaysia

Romisuhani Ahmad Center of Excellence Geopolymer and Green Technology, School of Materials Engineering, Universiti Malaysia Perlis, Kangar, Perlis, Malaysia

Sadeq Abdullah Abdo Al-Khadher Faculty of Civil Engineering and Built Environment, Universiti Tun Hussein Onn Malaysia, Parit Raja, Batu Pahat, Johor, Malaysia

Mohamad Yuzwan Aliff Aminuddin Faculty of Civil Engineering and Built Environment, Universiti Tun Hussein Onn Malaysia, Parit Raja, Batu Pahat, Johor, Malaysia

Noor Amira Sarani Faculty of Civil Engineering and Built Environment, Universiti Tun Hussein Onn Malaysia, Parit Raja, Batu Pahat, Johor, Malaysia

Syafiqa Ayob Faculty of Civil Engineering and Built Environment, Universiti Tun Hussein Onn Malaysia, Parit Raja, Batu Pahat, Johor, Malaysia

Ali Benlamoudi Faculty of Civil Engineering and Built Environment, Universiti Tun Hussein Onn Malaysia, Parit Raja, Batu Pahat, Johor, Malaysia

Rafikullah Deraman Jamilus Research Centre, Faculty of Civil Engineering and Built Environment, Universiti Tun Hussein Onn Malaysia, Parit Raja, Batu Pahat, Johor, Malaysia

Ooi Wan En Center of Excellence Geopolymer and Green Technology, School of Materials Engineering, Universiti Malaysia Perlis, Kangar, Perlis, Malaysia

Nurul Nabila Huda Hashar Faculty of Civil Engineering and Built Environment, Universiti Tun Hussein Onn Malaysia, Parit Raja, Batu Pahat, Johor, Malaysia

Azini Amiza Hashim Faculty of Civil Engineering and Built Environment, Universiti Tun Hussein Onn Malaysia, Parit Raja, Batu Pahat, Johor, Malaysia

Muhammad Fikri Hasmori Jamilus Research Centre, Faculty of Civil Engineering and Built Environment, Universiti Tun Hussein Onn Malaysia, Parit Raja, Batu Pahat, Johor, Malaysia

Mohd Ikhmal Haqeem Hassan Faculty of Civil Engineering and Built Environment, Universiti Tun Hussein Onn Malaysia, Parit Raja, Batu Pahat, Johor, Malaysia

Nur Fatin Nabila Hissham Faculty of Civil Engineering and Built Environment, Universiti Tun Hussein Onn Malaysia, Parit Raja, Batu Pahat, Johor, Malaysia

Kamarudin Hussin Geopolymer and Green Technology, Centre of Excellence (CEGeoGTech), Universiti Malaysia Perlis, Kangar, Perlis, Malaysia;
Center of Excellence Geopolymer and Green Technology, School of Materials Engineering, Universiti Malaysia Perlis, Kangar, Perlis, Malaysia

Wan Mastura Wan Ibrahim Center of Excellence Geopolymer and Green Technology, School of Materials Engineering, Universiti Malaysia Perlis, Kangar, Perlis, Malaysia

Mohd Hanif Ismail Jamilus Research Centre, Faculty of Civil Engineering and Built Environment, Universiti Tun Hussein Onn Malaysia, Parit Raja, Batu Pahat, Johor, Malaysia

Mohd Irwan Juki Faculty of Civil Engineering and Built Environment, Universiti Tun Hussein Onn Malaysia, Parit Raja, Batu Pahat, Johor, Malaysia

Nor Amani Filzah Mohd Kamil Faculty of Civil Engineering and Built Environment, Universiti Tun Hussein Onn Malaysia, Parit Raja, Batu Pahat, Johor, Malaysia

Faisal Sheikh Khalid Faculty of Civil Engineering and Built Environment, Universiti Tun Hussein Onn Malaysia, Parit Raja, Batu Pahat, Johor, Malaysia

Mohamed Khodja Division Research & Development SONATRACH, Central Direction of Research and Development, Algiers, Algeria

Andri Kusbiantoro Faculty of Engineering Technology, Universiti Tun Hussein Onn Malaysia, Parit Raja, Batu Pahat, Johor, Malaysia

Norsuria Mahmed Faculty of Chemistry Engineering Technology, Universiti Malaysia Perlis, Arau, Perlis, Jejawi, Malaysia

Liew Yun Ming Center of Excellence Geopolymer and Green Technology, School of Materials Engineering, Universiti Malaysia Perlis, Kangar, Perlis, Malaysia

Rosnita Mohamed Geopolymer and Green Technology, Centre of Excellence (CEGeoGTech), Universiti Malaysia Perlis, Kangar, Perlis, Malaysia

Rabiatul Syahindah Mustaffa Jamilus Research Centre, Faculty of Civil Engineering and Built Environment, Universiti Tun Hussein Onn Malaysia, Parit Raja, Batu Pahat, Johor, Malaysia

Sasitharan Nagapan Jamilus Research Centre, Faculty of Civil Engineering and Built Environment, Universiti Tun Hussein Onn Malaysia, Parit Raja, Batu Pahat, Johor, Malaysia

Noor Faiza Roslee Faculty of Civil Engineering and Built Environment, Universiti Tun Hussein Onn Malaysia, Parit Raja, Batu Pahat, Johor, Malaysia

Noor Fifinatasha Shahedan Faculty of Chemistry Engineering Technology, Universiti Malaysia Perlis, Arau, Perlis, Jejawi, Malaysia

Shahiron Shahidan Faculty of Civil Engineering and Built Environment, Universiti Tun Hussein Onn Malaysia, Parit Raja, Batu Pahat, Johor, Malaysia

Subaer Geopolymers and Green Material Group, Physics Department, FMIPA, Universitas Negeri Makassar, Makassar, Indonesia

Ong Shee Ween Center of Excellence Geopolymer and Green Technology, School of Materials Engineering, Universiti Malaysia Perlis, Kangar, Perlis, Malaysia

Zarina Yahya Geopolymer and Green Technology, Centre of Excellence (CEGeoGTech), Universiti Malaysia Perlis, Kangar, Perlis, Malaysia

Heah Cheng Yong Center of Excellence Geopolymer and Green Technology, School of Materials Engineering, Universiti Malaysia Perlis, Kangar, Perlis, Malaysia

List of Figures

Technological Properties of Fly Ash-Based Lightweight Geopolymer Brick

Development of Ash-Based and Slag-Based Pressed Geopolymer

Characteristic of One-Part Geopolymer as Building Materials

Performance on Physical and Mechanical Properties of Fired Clay Brick Incorporated with Palm Kernel Shell for Lightweight Building Materials

Characterization of Petroleum Sludge Prior to Be Treated in Cement Plants

Fabrication of Lightweight Ceramic Materials Using Geopolymer Technology

Performance of Sintered Pozzolanic Artificial Aggregates as Coarse Aggregate Replacement in Concrete

Density, Compressive Strength and Water Absorption Properties of Sand Cement Brick Containing Recycled Concrete Aggregate (RCA) and Crumb Rubber (CR) as Partial Sand Replacement Materials

Performance of Concrete Containing Synthetic Wire Waste as Fiber Materials

The Effect of Metakaolin in Production of Low Thermal Conductivity Cement Sand Brick

List of Tables

Development of Fly Ash Concrete Using Glass Bubble for Thermal Insulation Building Application

Characteristic of One-Part Geopolymer as Building Materials

Performance on Physical and Mechanical Properties of Fired Clay Brick Incorporated with Palm Kernel Shell for Lightweight Building Materials

Characterization of Petroleum Sludge Prior to Be Treated in Cement Plants

Fabrication of Lightweight Ceramic Materials Using Geopolymer Technology

Performance of Sintered Pozzolanic Artificial Aggregates as Coarse Aggregate Replacement in Concrete

Density, Compressive Strength and Water Absorption Properties of Sand Cement Brick Containing Recycled Concrete Aggregate (RCA) and Crumb Rubber (CR) as Partial Sand Replacement Materials

Performance of Concrete Containing Synthetic Wire Waste as Fiber Materials

Environment Acceptability as Building Material from Organic Contaminants Leaching Behavior in Sludge

The Effect of Metakaolin in Production of Low Thermal Conductivity Cement Sand Brick

Configuration of Physical and Mechanical Characteristics of Fly Ash and Bottom Ash Replacement in Self-compacting Concrete (SCC)

Mohd Ikhmal Haqeem Hassan, Aeslina Abdul Kadir, Noor Amira Sarani, and Sadeq Abdullah Abdo Al-Khadher

Abstract Generally, fly ash (FA) and bottom ash (BA) are produced by the process of combustion in coal-fired power plants that considered as waste which includes heavy metals in its composition. Moreover, these metals are harmful to both the atmosphere and human health. Apparently, for these ashes, treatment methods are available, and one of them is the replacement of FA and BA as concrete materials. Therefore, the primary purposed of this paper was to investigate the properties and characteristics of self-compacted concrete (SCC) incorporated with fly ash and bottom ash. SCC was developed with different percentages of FA and BA as a replacement for cement and sand with ratios of 10, 20 and 30% respectively. The properties tested were slump flow and t500 test, sieve segregation, and J-ring test for physical characteristic. Next, density, compressive power, and water absorption were examined to determine its mechanical characteristic. The results revealed that both physical and mechanical characteristics with 20% of FA and BA in SCC were suitable to be defined as SCC. A comparable set of data in compressive strength were observed, and densities have been established as standard weight concrete. In conclusion, this research demonstrates that FA and BA material are possibly be used as a new material for concrete production.

Keywords SCC · Fly ash · Bottom ash · OPC · Density · Compressive strength

M. I. H. Hassan · A. A. Kadir (✉) · N. A. Sarani · S. A. A. Al-Khadher
Faculty of Civil Engineering and Built Environment, Universiti Tun Hussein Onn Malaysia, 86400 Parit Raja, Batu Pahat, Johor, Malaysia
e-mail: aeslina@uthm.edu.my

M. I. H. Hassan
e-mail: mohd.ikhmal.haqeem@gmail.com

N. A. Sarani
e-mail: nramira1987@gmail.com

S. A. A. Al-Khadher
e-mail: sadeq@uthm.edu.my

A. Abdul Kadir et al. (eds.), *Sustainable Waste Utilization in Bricks, Concrete, and Cementitious Materials*, Lecture Notes in Civil Engineering 129,
https://doi.org/10.1007/978-981-33-4918-6_1

1 Introduction

The combustion phase in coal-fired power plants, some waste is generated called fly ash (FA) and bottom ash (BA). The ability of these wastes to be replaced by construction materials, depending on their form, size and chemical properties. Furthermore, the use of these wastes required the replacement of these items based on the efficiency guidelines set out by the United States Environmental Protection Agency [1].

The FA or fine ash fraction is raised by exhaust gases and captured by highly effective air filters before reaching an atmosphere. That consists mainly of subtle, glassy, cement-like spheres. Besides, BA is a portion of the ash that drops down the grids just under the boilers and be combined with water and drained into the ponds [2]. The type for FA and BA is generally based on the type of furnace and coal origin [3, 4]. Thus, it reported that 80% of the output would become fly ash through the coal burning process and that 20% of the production would remain low ash [5]. The form of FA and BA is generally depended on the method of a furnace and coal origin [3, 4]. Other than this, FA and BA are classified as scheduled waste containing a large amount of toxic and heavy metals [6].

Besides, FA has been commonly used as a substitute for Portland cement with a weight of approximately 30% without materially affecting the strength and efficiency of concrete [7–9]. FA concrete typically consists of silica, alumina and iron, and pozzolanic material is a substance containing alumina and silica that forms cement in the presence of moisture [10]. FA is being used in recent years, primarily attributable to the utility of waste materials as well as the production of functional materials, in particular, due to improved consistency and durability of fly ash [11–13]. In comparison, a few researchers studied the mechanical properties of BA concrete produced in the concrete mix [4]. The results show that BA offered a compressive strength comparable to standard concrete and found that the concrete strength improved with a rise in the bottom ash. Overall, this research emphasizes on the processing of SCC by adding the required proportion of FA to cement replacement and BA waste to completely exploit the advantages of waste in terms of its physical and mechanical properties.

2 Materials and Method

2.1 Material

Fly ash and bottom ash were obtained from one of the coal-fired power plants on the Malaysian Peninsula (Fig. 1a, b). FA and BA were incorporated into SCC mix by replacing cement and sand in the percentage of 0, 10, 20 and 30% respectively. Details of the mix configuration for this research, as shown in Table 1. There were ten design mix proportions with a total binder content of 530 to 551 kg/m^3 with water to binder ratio of 0.41.

Fig. 1 Samples of: a Fly Ash (FA), b Bottom Ash (BA)

Table 1 Mix proportion incorporated with FA and BA

Mixture	Total Binder (kg/m^3)	Cement (kg/m^3)	Fly ash (kg/m^3)	Coarse aggregate (kg/m^3)	Sand (kg/m^3)	Bottom ash (kg/m^3)	Water (kg/m^3)	SP (kg/m^3)
FA0BA0	551	551	0	594	914	0	228	5
FA10BA0	551	495	55	594	914	0	220	4
FA20BA0	540	432	108	594	914	0	216	4
FA30BA0	530	371	159	594	914	0	212	4
FA0BA10	551	550	0	594	822	67	228	5
FA0BA20	551	550	0	594	731	133	228	5
FA0BA30	551	550	0	594	640	200	228	5
FA10BA10	551	495	55	594	822	67	220	4
FA20BA20	540	432	108	594	731	133	216	4
FA30BA30	530	371	159	594	640	200	212	4

2.2 *Method*

Experimental tests were carried out for physical characteristics were filling ability using slump flows and t500 tests, passing ability using J-ring test and segregation test. The slump flow test for SCC is similar to the slump flow test for traditional concrete. The difference is that apart from calculating the vertical height, the horizontal spreads were measured to obtain a slump flow. The t500 test is a complementary indicator slump flow of SCC where t500 was a time between when the cone was elevated from the base plate to the point (500 mm). The slump flow test was evaluated according to BS EN 12350-8 [14], as shown in Fig. 2.

Next, the J-Ring flow spread was performed that indicates limited deformity of SCC due to the blocking effect of the reinforcement bars and the flow time T50 indicates the rate of deformation within a given flow distance (Fig. 3). This method

Fig. 2 Slump flow of FA and BA incorporated in SCC using t500 test

Fig. 3 Passing ability of FA and BA incorporated in SCC using the J-Ring test

should comply with BS EN 12350-12 [15].

Meanwhile, the Sieve Segregation Resistance test is designed to determine the resistance of the SCC to segregation by referred to BS EN 12350-11 [16] (Fig. 4). Consequently, the mechanical characteristics testing for FA and BA incorporated with SCC include density, water absorption and compressive strength tests were conducted for hardened concrete. Curing was done following BS EN 12390-2 [17] and density, compressive strength and water absorption at 28 days by referring BS EN 12390 [18] and BS 1881-122 [19].

Fig. 4 Segregation resistance of FA and BA incorporated in SCC using Sieve Segregation Test

3 Result and Discussion

3.1 Physical Characteristic of FA and BA Incorporated in SCC

The slump flow test was done to achieve a target average range diameter of more than 500 mm but less than 700 mm. The time of spread to t500 shall also not be less than 2 s. Spread diameters by all mixtures were reported higher than the required level of the SCC and spreading time as shown in Figs. 5 and 6 was consistent with the

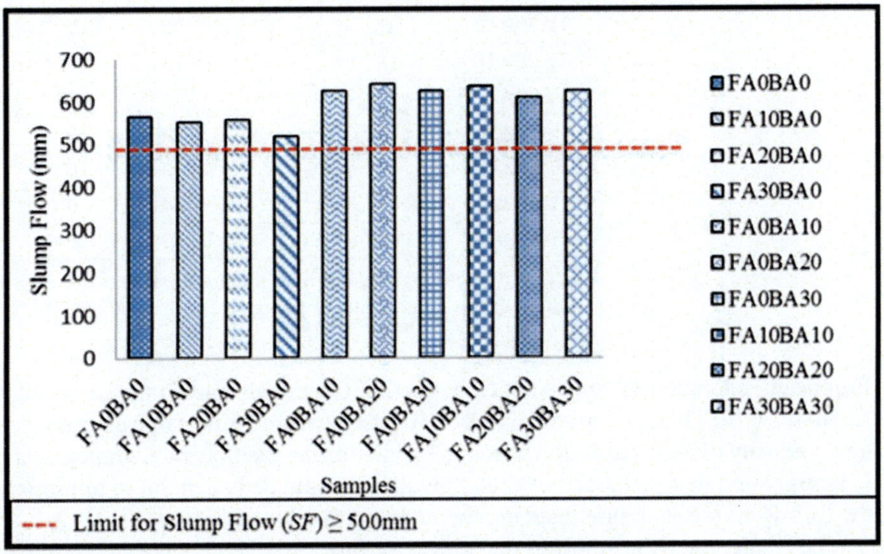

Fig. 5 Results of slump flow of FA and Ba incorporated in SCC

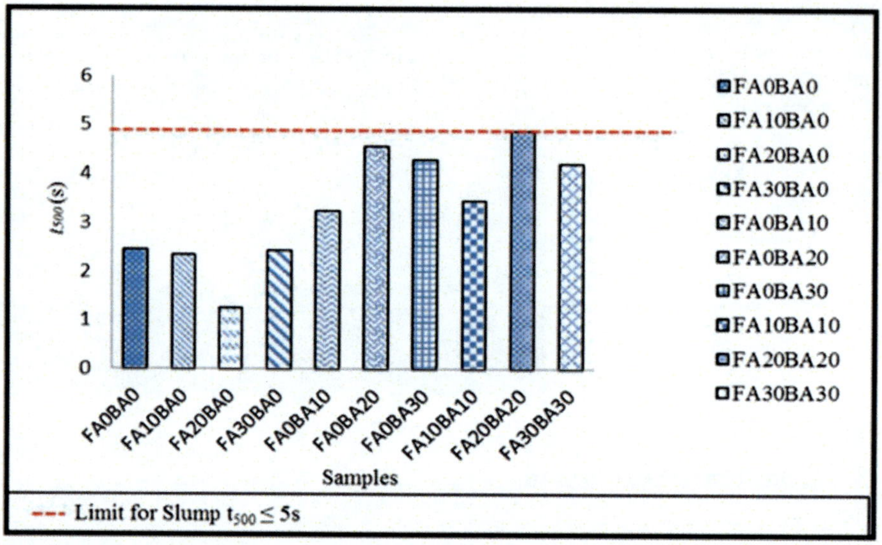

Fig. 6 Results of t500 (s) of FA and Ba incorporated in SCC

Table 2 Passing ability of FA and BA incorporated in SCC by using the J-ring test

Samples	Height of spread	Passing ability	
		h_{avg}	h_o
FA0BA0	20	40	0
FA10BA0	16	30	0
FA20BA0	14	32	0
FA30BA0	21	35	0
FA0BA10	18	25	0
FA0BA20	19	28	0
FA0BA30	15	19	0
FA10BA10	10	10	0
FA20BA20	14	14	0
FA30BA30	19	19	0

h_{avg}—average height between height inner and outer of J-ring set; H_o—a height in the centre of J-ring set

European Federation of Specialist Construction Chemicals and Concrete Systems, EFNARC [20]. Thus, according to EFNARC, the measured time value does not evaluate viscosity of SCC but is correlated to by identifying the flow rate. Guneyisi et al. [21] suggested that variation between flow times would be beneficial in minimizing the formwork pressure or enhancing the resistant to segregation

The results for J-ring showed that all the targeted mix percentages were met the performance criteria and for concrete passing ability (Table 2). The study pertains

Table 3 Result of segregation resistance using sieve segregation test

Samples (%)	Sieve segregation test					
	Mass of pan, mp (kg)	Mass of pan + sieve receiver (kg)	Mass of pan + sieve receiver + concrete (kg)	(Mass of pan + passed concrete), mps (kg)	Mass of concrete, mc (kg)	Segregated proportion, SR (%) ≤ 20%
FA0BA0	0.90	2.00	6.10	1.55	4.10	15.85
FA10BA0	0.92	2.04	6.24	1.26	4.20	8.10
FA20BA0	0.92	2.02	5.20	1.18	3.18	8.18
FA30BA0	0.92	2.00	6.46	1.30	4.46	8.520
FA0BA10	0.92	2.00	4.90	1.36	2.90	15.17
FA0BA20	0.92	2.00	5.70	1.40	3.70	12.97
FA0BA30	0.92	1.70	5.30	1.04	3.60	3.33
FA10BA10	0.90	2.02	5.88	1.18	3.86	7.25
FA20BA20	0.92	2.00	5.82	1.32	3.82	10.47
FA30BA30	0.92	2.00	6.20	1.24	4.20	7.62

the passing flexibility of the spreading height between 0 and 10 mm as complying to standard and were following the proposed requirement of SCC workability as stated out in EFNARC [22].

Findings of resistance towards segregation are listed in Table 3. Depending on the result, the research goal was fulfilled by all the design mix proportions. Segregated resistance within 20% was needed to be obtained by the sample referred to in BS EN 12350:11 [16]. The lowest segregated portion value, 3.33% of FA0BA30 sample and the highest segregated portion value, 15.85% of the control sample (FA0BA0) were reported.

3.2 *Mechanical Characteristic of FA and BA Incorporated in SCC*

On the other hand, the analysis of the mechanical characteristic of FA and BA incorporated in SCC were presented. The density of the SCC integrated with FA and BA were calculated by weighing the cube samples before the compression test. For consistency, the average density is calculated for each concrete mixes, and the results are shown in Table 4. According to that, the FA30BA30 samples do have the lowest density with 2170 kg/m^3. Instead, FA10BA0 and FA20BA0 samples with 2304 kg/m^3 reported a higher density of SCC samples. It shows that with FA substitution of up to 20% in samples, the density was higher than with BA alone. Even so, all samples were comparable with BS EN 206:2013 as standard weight concrete with a range between 2000–2600 kg/m^3. Other than that, Wongkeo et al. [23] proposed that density of

Table 4 Density of SCC samples

Sample	Density (kg/m^3)
FA0BA0	2189
FA10BA0	2304
FA20BA0	2304
FA30BA0	2274
FA0BA10	2244
FA0BA20	2289
FA0BA30	2244
FA10BA10	2296
FA20BA20	2207
FA30BA30	2170

concrete may vary depending on the density of aggregate, the amount of air space that is compressed or trapped, and the content of water and cement.

Figure 7 depicts the water absorption results of FA and BA of SCC samples. The findings indicate that with the increase of BA, water absorption decreased significantly for FA0BA10 (1.08%), FA0BA20 (1.02%) and FA0BA30 sample with 0.67% respectively. It is proposed by Siddique [25] that the integration of BA into concrete

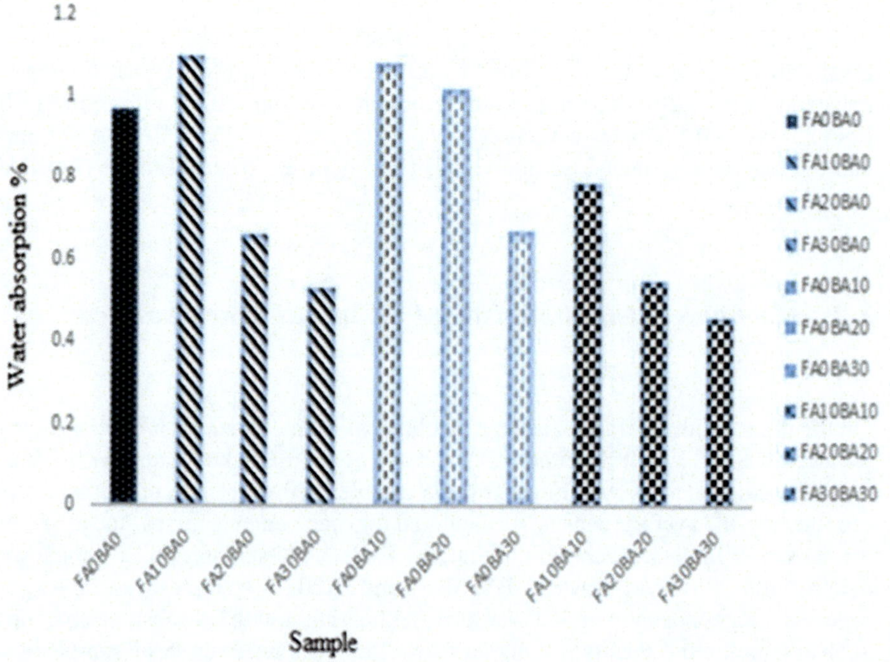

Fig. 7 Water absorption of SCC samples

contributed to low absorption characteristics. The very same tendencies are also shown for the cooperation of FA and the mix of FA and BA. The findings for FA samples are therefore 1.10, 0.66 and 0.53%, with the replacement of 10–30% respectively.

Meanwhile, the results for both the tests incorporated of FA and BA are 0.79, 0.55 and 0.46% respectively for FA10BA10, FA20BA20 and FA30BA30. As per the result, the greater rate of FA and BA added the lesser water absorption rate recorded. Liu [13] argued that this might potentially be attributable to the fact that its FA is more refined than cement. Therefore, it filled the empty spaces in the samples and leads to lower porosity. Significantly, all SCC mixture samples had a low water absorption characteristic (only about 10%). It is consistent with the report published by Singh and Siddique [24].

The final section examines the compressive strength of cube samples at 7, 14 and 28 days of age. The compressive intensity was measured in compliance with BS EN 12390-3 [26]. All SCCs blended with FA and BA attained compressive strengths of a higher value than standard concrete. By comparison to BS EN 206 [27], the sample strength classes ranged from class C45 to class C70 at 28 days and fit the criteria to be classified as normal-weight and heavyweight concrete. Figure 8 display a comparative analysis of the compressive strengths between the control sample (FA0BA0) and the samples with different percentages of FA and BA substitution in the SCC. A study by Kasemchaisiri and Tangtermsirikul [28] revealed that the compressive strength of SCC decreased with the increase of BA percentages.

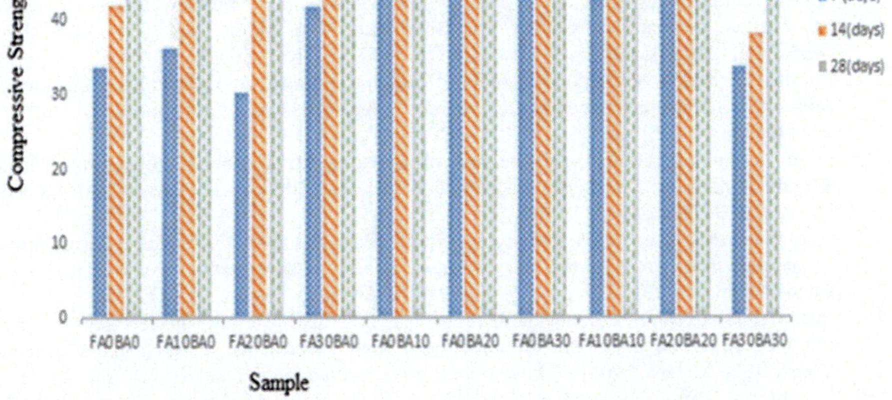

Fig. 8 Compressive strength at various percentages of FA, BA and FA BA

4 Conclusion

In conclusion, it has been noted that FA and BA can be used as an alternative for SCC. All tests performed from all these waste materials showed properties within its ranges to be defined as SCC. As for compressive strength, it reveals that compressive strength of the concrete from day 7 to day 28. By reference to BS EN 206, the sample strength grades ranged from class C45 to class C70 at 28 days and met the criteria to be graded as normal-weight and heavyweight concrete. Water absorption shows with the increase of BA would result in the low water absorption of the samples. It was proposed that the SCC developed of FA and BA (FA20BA20) yield the excellent physical and mechanical efficiency of the SCC. The use of FA and BA is not only capable of minimizing the amount of these waste materials sent to the landfill site, but also fit better in terms of economic viability in concrete technology as low-cost replacements.

References

1. USEPA Environmental Protection Agency United States (1996) Hazardous waste characteristics scoping study. US Environmental Protection Agency, Office of Solid Waste
2. Ahmaruzzaman M (2010) A review on the utilization of fly ash. Prog Energy Combust Sci 36(3):327–363. https://doi.org/10.1016/j.pecs.2009.11.003
3. Jaturapitakkul C, Cheerarot R (2003) Development of bottom ash as pozzolanic material. J Mater Civ Eng 15:1–48. https://doi.org/10.1061/(ASCE)0899-1561
4. Muhardi A, Marto A, Kassim K, Makhtar AM, Wei LF, Lim YS (2010) Engineering characteristics of Tanjung Bin coal ash. Electron J Geotech Eng 15:1117–1129
5. Syahrul M, Sani M, Muftah F, Muda Z (2010) The properties of special concrete using washed bottom ash (WBA) as partial sand replacement. Int J Sustain Constr Eng Technol 1(2):65–76
6. Izquierdo M, Querol X (2012) Leaching behaviour of elements from coal combustion fly ash: an overview. Int J Coal Geol 94:54–66. https://doi.org/10.1016/j.coal.2011.10.006
7. Behnood A, Modiri Gharehveran M, Gozali Asl F, Ameri M (2015) Effects of copper slag and recycled concrete aggregate on the properties of CIR mixes with bitumen emulsion, rice husk ash, Portland cement and fly ash. Constr Build Mater 96:172–180. https://doi.org/10.1016/j.conbuildmat.2015.08.021
8. Sua-Iam G, Makul N (2015) Utilization of coal and biomass fired ash in the production of self-consolidating concrete: a literature review. J Clean Prod. https://doi.org/10.1016/j.jclepro.2015.03.038
9. Uysal M, Sumer M (2011) Performance of self-compacting concrete containing different mineral admixtures. Constr Build Mater 25(11):4112–4120. https://doi.org/10.1016/j.conbuildmat.2011.04.032
10. Garrabrants AC, Kosson DS, DeLapp R, van der Sloot HA (2014) Effect of coal combustion fly ash use in concrete on the mass transport release of constituents of potential concern. Chemosphere 103:131–139. https://doi.org/10.1016/j.chemosphere.2013.11.048
11. Jalal M, Pouladkhan A, Harandi OF, Jafari D (2015) Comparative study on effects of Class F fly ash, nano silica and silica fume on properties of high performance self-compacting concrete. Constr Build Mater 94:90–104. https://doi.org/10.1016/j.conbuildmat.2015.07.001
12. Naganathan S, Subramaniam N, Mustapha K (2012) Development of brick using thermal power plant bottomash and fly ash. Asian J Civ Eng 13:275–287

13. Liu M (2010) Self-compacting concrete with different levels of pulverized fuel ash. Constr Build Mater 24(7):1245–1252. https://doi.org/10.1016/j.conbuildmat.2009.12.012
14. British Standards Institution (2010) Testing fresh concrete—Part 8: self-compacting concrete—Slump-flow test, London. BS EN 12350-8
15. British Standards Institution (2010) Testing fresh concrete—Part 12: self-compacting concrete. J-ring test, London. BS EN 12350-12
16. British Standards Institution (2010) Testing fresh concrete—Part 11: self compacting concrete—Sieve segregation test, London. BS EN 12350-11
17. British Standards Institution (2010) Testing hardened concrete—Part 2: making and curing specimens for strength tests, London. BS EN 12390-11
18. British Standards Institution (2010) Testing hardened concrete. BS EN, London, p 12390
19. British Standards Institution (2010) Testing concrete—Part 122: method for determination of water absorption, London. BS 1881-122
20. EFNARC (2005) Specifications and guidelines for self-compacting concrete. www.efnarc.org, pp 1–32
21. Güneyisi E, Gesoglu M, Al-Goody A, İpek S (2015) Fresh and rheological behavior of nano-silica and fly ash blended self-compacting concrete. Constr Build Mater 95:29–44. https://doi.org/10.1016/j.conbuildmat.2015.07.142
22. EFNARC (European Federation of National Associations Representing producers and applicators of specialist building products for Concrete) (2002) Specification and guidelines for self-compacting concrete. Hampshire UK
23. Wongkeo W, Thongsanitgarn P, Ngamjarurojana A, Chaipanich A (2014) Compressive strength and chloride resistance of self-compacting concrete containing high level fly ash and silica fume. Mater Des 64:261–269. https://doi.org/10.1016/j.matdes.2014.07.042
24. Singh M, Siddique R (2014) strength properties and micro-structural properties of concrete containing coal bottom ash as partial replacement of fine aggregate. Constr Build Mater 50:246–256. Constr Build Mater 50. https://doi.org/10.1016/j.conbuildmat.2013.09.026
25. Siddique R (2013) Properties of self-compacting concrete containing class F fly ash. Mater Des 32:1501–1507. https://doi.org/10.1016/j.matdes.2010.08.043
26. British Standards Institution (2019) Testing hardened concrete—Part 3: compressive strength of test specimens, London. BS EN 12350–33
27. British Standards Institution (2013) Concrete—specification, performance, production and conformity. BS EN, London, p 206
28. Kasemchaisiri R, Tangtermsirikul S (2008) Properties of self-compacting concrete in corporating bottom ash as a partial replacement of fine aggregate. ScienceAsia 34:1. https://doi.org/10.2306/scienceasia1513-1874.2008.34.087

Toxicity Characteristics and Heavy Metals Leachability of Self Compacted Concrete Containing Fly Ash and Bottom Ash as Partial Cement and Sand Replacement

Mohd Ikhmal Haqeem Hassan, Aeslina Abdul Kadir, Noor Amira Sarani, and Azini Amiza Hashim

Abstract This research aimed to examine the toxicity characteristics of self-compacted concrete (SCC) containing fly ash (FA) and bottom ash (BA) as a partial replacement of ordinary Portland cement and fine aggregate. In particular, the other objective was to identify the heavy metals leaching nature of FA and BA in SCC as well as to determine their possible use as construction materials. FA and BA derived from the phase of combustion in coal-fired power plants. It contains heavy metals within their compositions. SCC mixtures were prepared to have various percentages of FA (substitution of cement) and BA (substitution of sand) of 0, 10, 20, and 30% respectively for subsequent experiments. Several investigations were performed out, such as the characterization of the main composition and heavy metals of the materials through X-Ray Fluorescence (main composition and heavy metal characterization of the raw materials), the compressive strength test, the Toxicity Characteristics Leaching Procedure (TCLP) and the Synthetic Precipitation Leaching Procedure (SPLP). Results showed that SCC containing FA and BA replacement had obtained compressive strengths of a similar range or higher than the control SCC (without any replacement of FA and BA). Sample FA10BA10 or 10% substitution of FA and BA recorded the highest compressive strength value at 58.07 ± 0.50 MPa. From the results of TCLP and SPLP, it founded that the inclusion of ashes up to 30% was safe as the concentration of heavy metal leaching did not surpass the concentration of

M. I. H. Hassan · A. A. Kadir (✉) · N. A. Sarani · A. A. Hashim
Faculty of Civil Engineering and Built Environment, Universiti Tun Hussein Onn Malaysia, 86400 Parit Raja, Batu Pahat, Johor, Malaysia
e-mail: aeslina@uthm.edu.my

M. I. H. Hassan
e-mail: mohd.ikhmal.haqeem@gmail.com

N. A. Sarani
e-mail: nramira1987@gmail.com

A. A. Hashim
e-mail: aziniamizaa@gmail.com

A. Abdul Kadir et al. (eds.), *Sustainable Waste Utilization in Bricks, Concrete, and Cementitious Materials*, Lecture Notes in Civil Engineering 129,
https://doi.org/10.1007/978-981-33-4918-6_2

pollutants for toxicity characteristics. In conclusion, this research suggests that the disposal and use of FA and BA as a promising replacement of construction materials may be used to minimize their environmental issues, improve efficiency and reduce the cost of production of SCC in the future.

Keywords Toxicity characteristic · Leaching · Heavy metals · Construction materials · Fly ash · Bottom ash · Self-compacted concrete

1 Introduction

Fly ash (FA) and bottom ash (BA) are usually the excesses of waste created by the combustion process in coal-fired power plants. The total quantity of FA and BA produced represents 80% of the total residues produced from incineration [1]. Residual weights range from 5 to 30% before combustion, depending on their composition [2]. FA is more like a type of fine particulate matter with many heavy metals, organic compounds and chlorides [3]. Besides, BA can explain as particles consisting of different elements such as metals, minerals, ceramics and unburned materials [4, 5]. Both FA and BA contain several hazardous elements that must be recognized as a threat to the environment and its life-threatening effects.

Researchers Chang and Wey [6] and Aubert et al. [7] have previously identified the decreasing impact of the FA and BA disposal applications in Taiwan and France, respectively [6, 7]. Consequently, incineration of solid waste has contributed to a reduction in the amount of disposal, which, in turn, has resulted in the formation of ash as combusted by-products. BA had already been reused primarily as road ingredients and as aggregate or sand material in concrete. Meanwhile, due to the high content of heavy metals in the FA, ashes had very little use and were often disposed of in landfills. However, in recent years, FA is powerfully applicable as a substitute for cement in concrete industries due to its cement properties.

The purpose of this research was, therefore, to integrate scheduled wastes such as FA and BA into the use of SCC as building materials. Previous researchers have been committed to the exploration of the direct effects of the introduction of waste into their subject matter of study based on improvements in mechanical and chemical properties. However, environmental impacts were either ignored or less elaborated. Therefore, both the compressive strength and leachability effects of the integration were calculated, analyzed and presented in this paper.

2 Materials and Method

2.1 Study Location

The FA and BA were collected from a thermal processing plant in Peninsula, Malaysia. The material composition of FA, BA, and OPC was analyzed using the X-ray fluorescence (XRF) test performed using the Bruker AXS S4 Pioneer. The FA and BA samples used in the test were prepared in the form of pellets, with a sample at a wax ratio of 8:2 using the Pressed Pellet Technique. Between all the elements and compounds in the ash detected by XRF, only the metal elements measurable by the Atomic Absorption Spectrometry (AAS) as shown in Table 1.

Besides, aggregates and sand used in the manufacture of SCC usually complied with the specifications of BS EN 206-1 [8]. For aggregates, sizes of 14 to 20 mm prepared using a sieving process. OPC was used in compliance with BS EN 197-1 [9], although mixtures were tested to conform with EN 943-2 [10].

A series of mixtures were prepared accordingly for the control sample with 0% of FA and BA and mixture with different percentages FA and BA (10, 20, 30% of FA and BA). The overall binder for all samples ranged from 530 to 550 kg/m^3, respectively. Standard moulds of 150 mm^3 in compliance with BS EN 12390-1 [11] were used to contain fresh SCC items. The configuration of the mix for this research, as shown in Table 2.

A compression test assessed the compressive strength at 28 days in BS EN 12390-3 [12]. Next, the crushed cubes are smashed using a steel hammer to produce smaller pieces of the samples. The fragments were then further crushed using the Aggregate Impact Value (AIV) equipment to reduce the size of the solid particles to less than 9.5 mm. The crushed fragments were sieved and retrieved as samples for the Toxicity Characteristic Leaching Procedure according to Method 1311 [13].

TCLP was used as a set of guidelines for preparing collected samples of concrete specimens for the leachate analysis to be completed. Since the SCC cubes were

Table 1 Heavy metals of FA, BA and Ordinary Portland Cement (OPC)

Heavy metals	Formula	Concentration (mg/L)		
		FA	BA	OPC
Chromium	Cr	228	176	54
Manganese oxide	MnO	900	800	800
Iron (III) oxide	Fe_2O_3	41,600	52,600	30,200
Nickel	Ni	107	88	19
Copper	Cu	101	38	26
Zinc	Zn	52	31	164
Arsenic	As	38	12	37
Lead	Pb	62	18	60

Table 2 Proportions of SCC mix incorporated with FA and BA

Mix design	Total binder (kg/m^3)	Cement (kg/m^3)	FA (kg/m^3)	CA (kg/m^3)	Sand (kg/m^3)	BA (kg/m^3)	Water (kg/m^3)	SP (kg/m^3)
FA0BA0 (control)	550	550	0	593	914	0	228	5
FA10BA0	550	495	55	593	914	0	220	4
FA20BA0	540	432	108	593	914	0	216	4
FA30BA0	530	371	159	593	914	0	212	4
FA0BA10	550	550	0	593	822	67	228	5
FA0BA20	550	550	0	593	731	133	228	5
FA0BA30	550	550	0	593	640	200	228	5
FA10BA10	550	495	55	593	822	67	220	4
FA20BA20	540	432	108	593	731	133	216	4
FA30BA30	530	371	159	593	640	200	212	4

concrete specimens, the extraction method had to be performed under the Sample group containing more than 0.5% of the dry solids. The extraction fluid was prepared by diluting 5.7 mL of glacial acetic acid (CH_3COOH) with pure water to a volume of 1 L. A total of 50 g of the sample was prepared and placed in a 2 L extraction bottle, and the extraction fluid was then poured in the rotary agitation apparatus, and it was left to spin for 18 h from end to end. The solution in the extraction bottle was then diluted to dispel the solid particles. The fluid portion of the sample was held at a pH of less than 2.0 and stored in the refrigerator at 5 °C for a leachate determination analysis using AAS, which was performed using Perkin Elmer Analyst 800.

Also, the Synthetic Precipitation Leaching Technique (SPLP) according to Method 1312 [14] was used to provide information on the mobility (leachability) of organic and inorganic components from liquids, soils, and waste. Extraction fluids with pH 4.2 solution have been used for this study. The scale of the samples used was less than 9.5 mm.

3 Result and Discussion

Table 3 indicates that compressive intensity increased from day 7 to day 28. All SCCs incorporated with FA and BA attained compressive strengths with a similar range but higher than standard concrete. By comparison to BS EN 206 [15], the sample strength classes ranged from class C45 to class C70 at 28 days and met the criteria to be graded as normal-weight and heavy-weight concrete.

At various replacements of OPC with 0, 10, 20, and 30% of FA, strengths were observed to be in the range of 30–42 MPa at 7 days, 41–59 MPa at 14 days and 49–69 MPa respectively. All the samples incorporated with FA were higher than

Table 3 Compressive strength of SCC samples incorporated with FA and BA

Samples	Compressive strength (MPa)		
	7 (days)	14 (days)	28 (days)
FA0BA0	33.60 ± 0.68	41.88 ± 0.39	49.30 ± 0.61
FA10BA0	36.21 ± 0.67	58.04 ± 0.33	68.79 ± 0.24
FA20BA0	30.04 ± 0.32	49.46 ± 0.14	67.48 ± 0.56
FA30BA0	41.77 ± 0.72	45.42 ± 0.31	59.91 ± 0.46
FA0BA10	49.5 ± 0.64	56.40 ± 0.47	58.01 ± 0.17
FA0BA20	49.24 ± 0.91	50.90 ± 0.71	56.76 ± 0.91
FA0BA30	47.02 ± 0.14	50.64 ± 0.97	57.94 ± 0.50
FA10BA10	43.04 ± 0.26	51.71 ± 0.56	58.07 ± 0.50
FA20BA20	45.49 ± 0.51	52.04 ± 0.58	57.33 ± 0.11
FA30BA30	33.58 ± 0.67	38.1 ± 0.54	46.81 ± 0.33

the control sample, which is FA0BA0. It shows that the replacement of FA has improved the compressive strength of the SCC. The highest strength was recorded from FA10BA0 sample with 68.79 MPa at 28 days. The increase in strength of fly ash concrete may be attributed to continuous hydration and the filling of pores with Calcium Silicate Hydrate gel formed due to pozzolanic action of coal fly ash [16].

In the meantime, for compressive strength with a different sand replacement of 10, 20, and 30% of BA shown that strengths were observed range of 33–50 MPa, 41–57 MPa, and 49–59 MPa at 14 and 28 days respectively. SCC incorporated with BA gained higher intensity at the early age of the SCC, varying from 47 to 49 MPa, but steadily increased at 14 and 28 days with the highest reported value from the FA0BA10 sample at 58.01 MPa.

3.1 *Effect on pH for Toxicity Characteristic Leaching Procedure (TCLP) and Synthetic Precipitation Leaching Procedure (SPLP) Tests*

One of the crucial factors that affect the leaching of heavy metals concentrations is the pH value. Yu et al. [17] suggested that the leachability of heavy metals is much dependent on the pH of leaching. Therefore, all the samples were subjected to TCLP leachate (pH 2.88 ± 0.1) and SPLP leachate (pH 4.2 ± 0.1). The pH results were recorded after 18 h of agitation using a rotary agitation apparatus. The results were shown in Fig. 1.

The results showed that the pH value for TCLP leachant for all the samples was much lower than the pH value for SPLP. In TCLP leachants, the lowest value was recorded with pH 5.2 from the FA30BA30 sample. Meanwhile, the highest value for pH in TCLP leachants was recorded from FA0BA0 sample with pH 11.4. Samples with replacement in FA only were demonstrating the decreasing value of pH with

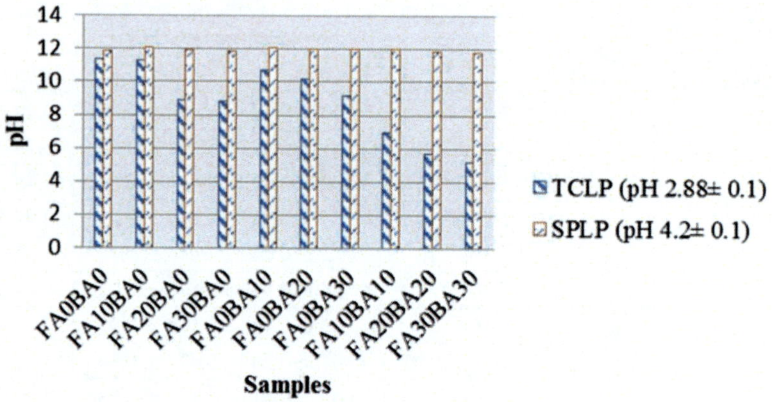

Fig. 1 Comparison of pH for TCLP and SPLP samples

the increment of the percentages of FA replacement. Similar trends were observed for samples with the combination of FA and BA.

The pH values of SPLP leachant in the samples were recorded in the range of 11.78 to pH 12.09. It is probably due to the alkaline nature of the materials that could significantly change the initial pH of the leachant [18]. It resulted in lower heavy metals were leached in SPLP compared to TCLP samples. Besides, Kim et al. [19] also suggested that most of the elements in FA and BA only slightly soluble. Heavy metals are most soluble in acidic leachant while those elements that form oxyanions are more soluble at high pH.

3.2 *Comparison of Leachability of Heavy Metals Using TCLP and SPLP Method*

The findings of the leachate analyzed were compared with the concentration limits for heavy metals set by the USEPA [20] and the value recommendations for chemicals of health significance in drinking water [21]. Also, both methods used identical particle sizes, which are smaller than 9.5 mm. However, there was a gap in the leaching fluid used in the SPLP and TCLP experiments. It was confirmed that most of the TCLP heavy metal concentrations were higher than the SPLP results.

Figures 2, 3, 4 and 5 shows a comparison of SCC heavy metal concentrations using SPLP and TCLP for all samples. Both test methods reported significant leachate concentrations of target metals for FA and BA content. However, due to the disparity in the leaching fluid used in the TCLP and SPLP experiments, the leaching concentrations with marginal variations can be observed. Concentration, volatility and solubility are several variables that assess the capacity for leaching [22]. Most of the TCLP heavy metal concentrations were slightly higher than the metal concentrations found in the SPLP test, particularly for As. It is because metal solubility usually decreases

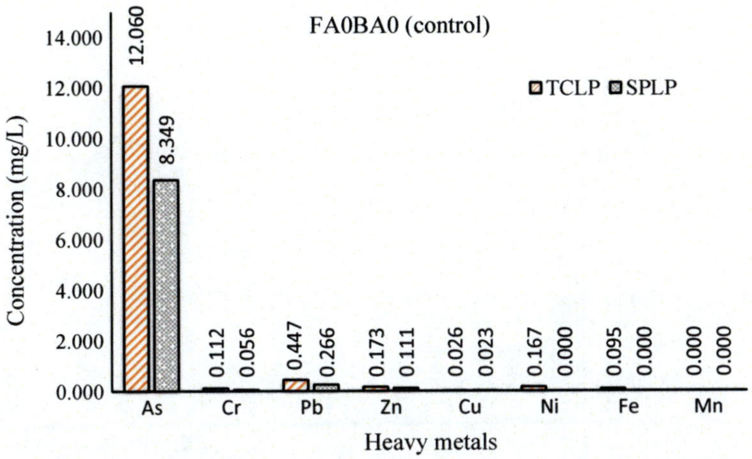

Fig. 2 Comparison of heavy metals concentrations for SCC using TCLP and SPLP for FA0BA0 (control) sample

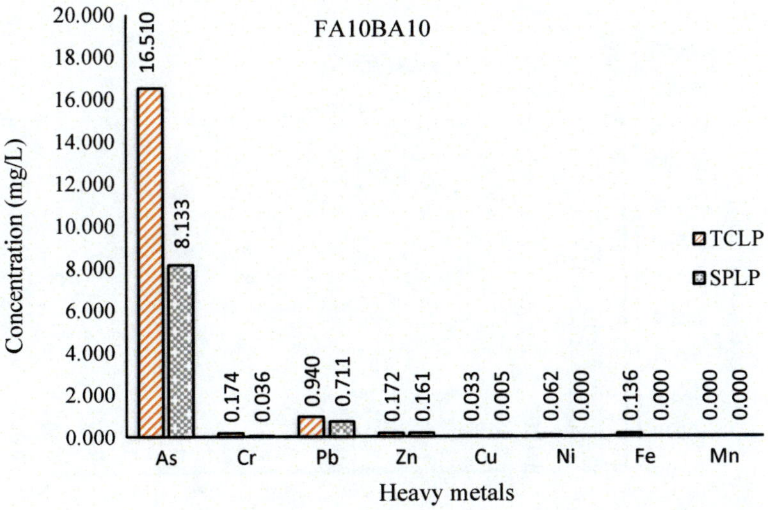

Fig. 3 Comparison of heavy metals concentrations for SCC using TCLP and SPLP for FA10BA10 sample

with a rise in pH. TCLP involved leaching under slightly acidic buffered conditions with pH 2.88 and pH was 4.2 in the SPLP test. Other heavy metals consisting of Pb, Zn, Ni, Fe and Mn have been leached at concentrations that are exceptionally low and do not reach the limit of the permissible concentration in leachate. That is because FA consists of aluminium oxide and iron hydroxide, which are common sorbents for the removal of Arsenic from water.

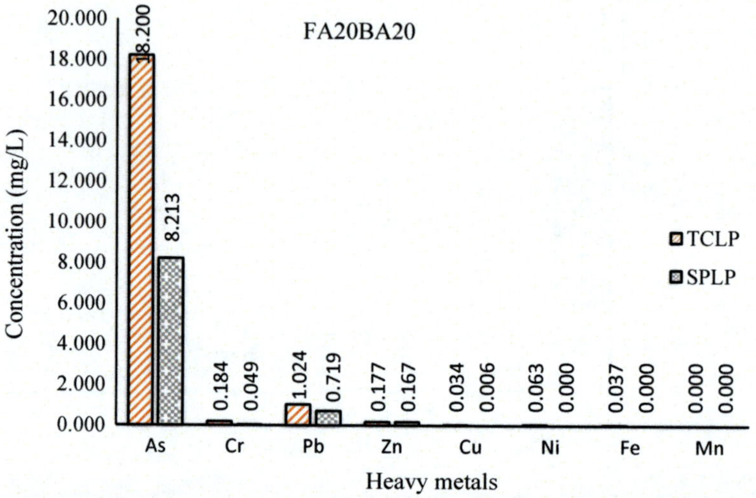

Fig. 4 Comparison of heavy metals concentrations for SCC using TCLP and SPLP for FA20BA20 sample

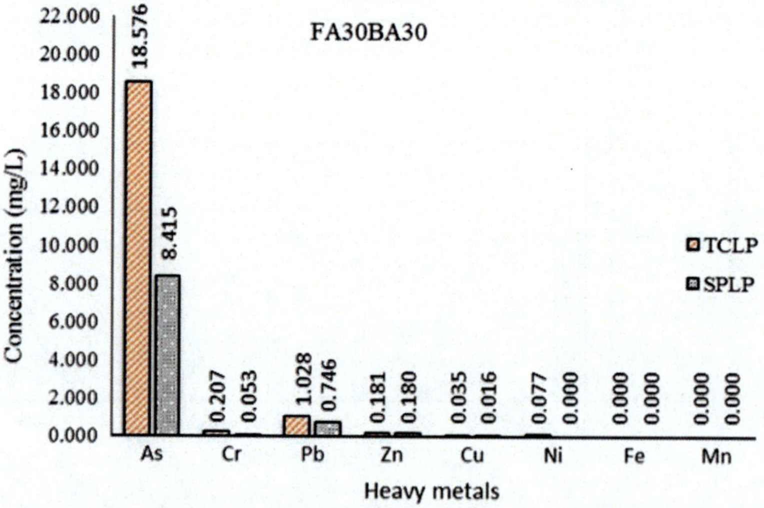

Fig. 5 Comparison of heavy metals concentrations for SCC using TCLP and SPLP for FA30BA30 sample

The element concentrations in all TCLP samples were below the acceptable limit set by the USEPA, except for Arsenic. In comparison, samples with the substitution of FA alone and samples with a combination of FA and BA resulted in higher leaching of As compared to samples with BA alone. The highest leaching of As was reported in the FA30BA30 sample at 18.576 mg/L. Arsenic has gained significant popularity as it is mobile over a broad pH range. Arsenic acid solution releases increase with pH, although this pattern is reversed in alkaline solutions [23]. The difference in leaching is caused by the pH dependency of most heavy metal elements [22].

A similar pattern observed in SPLP test which all elements were below the USEPA limit except for As. SPLP using a leachant with a pH similar to that of groundwater or surface water where the coals are used, stored, or disposed of, rather than the acetic acid of TCLP could be a more representative test for whether the waste materials might be hazardous based on its toxicity. Therefore, the concentrations of the elements in SPLP leachates, the results are compared to the World Health Organization limit for drinking water [21]. The results demonstrated that elements such as As and Pb exceed the WHO guidelines for drinking water quality. The highest concentration for As and Pb elements were recorded higher from sample FA30BA30 with 8.415 and 0.746 mg/L, respectively. Other than that, the FA0BA0 sample was the only sample that exceeds the WHO guidelines for Cr with 0.006 mg/L higher than the guidelines that consider extremely low concentrations. Elements such as Cu, Ni, and Mn were below the guidelines or not detected in SPLP. Meanwhile, Zn and Fe are not of health concern at concentration normally observed in drinking water as has been stated in the WHO guidelines.

4 Conclusion

Based on the characteristics of FA and BA, it was found that the concentrations of elements in FA were usually higher than in BA. FA thus displays higher concentrations of heavy metals compared to BA and OPC. As for compressive strength, it indicates that compressive strength increased from day 7 to day 28. By comparison to BS EN 206:2013, the sample strength classes ranged from class C45 to class C70 at 28 days and met the criteria to be graded as normal-weight and heavy-weight concrete. The other goal was to assess the leachability of SCC heavy metals incorporated with FA and BA. From TCLP, Arsenic leaching was the only heavy metal that leached out of the samples and exceeded the limit set by USEPA. The highest value for As was reported in FA30BA30 sample with 18.576 mg/L and indicated the highest value of all samples. Meanwhile, for SPLP, the findings again indicate that the highest concentration of heavy metals leached from the samples was As. The concentrations of Arsenic in the control sample are 8.349 mg/L. For FA and BA samples, which are FA10BA10 (8.133 mg/L), FA20BA20 (8.213 mg/L) and FA30BA30 (8.415 mg/L) of As concentration. The pH importance of the leaching agent is an important factor influencing the leaching of heavy metals. TCLP findings indicate higher value relative to SPLP results because the leaching agent is acidic compared to SPLP. In conclusion, a sustainable approach to the reuse of FA and BA in this research is useful for environmental and construction purposes.

References

1. Ginés O, Chimenos JM, Vizcarro A, Formosa J, Rosell JR (2009) Combined use of MSWI bottom ash and fly ash as aggregate in concrete formulation: environmental and mechanical considerations. J Hazard Mater 169(1):643–650. https://doi.org/10.1016/j.jhazmat.2009.03.141
2. Shi HS, Kan LL (2009) Leaching behavior of heavy metals from municipal solid wastes incineration (MSWI) fly ash used in concrete. J Hazard Mater 164(2):750–754. https://doi.org/10.1016/j.jhazmat.2008.08.077
3. Sajwan KS, Punshon T, Seaman JC (2006) Production of combustion products and their potential uses. In: Sajwan KS, Twardowska I, Punshon TAAK (eds) Coal combustion byproducts and environmental issues. Springer, New York
4. Kikuchi R (2006) Alternative by-products of coal combustion and simultaneous $SO_2/SO_3/NO_X$ treatment of coal-fired flue gas: approach to environmentally friendly use of low-rank coal. In: Sajwan KS, Twardowska I, Punshon TAAK (eds) Coal combustion byproducts and environmental issues. Springer, New York, pp 4507–4511
5. Kadir AA, Hassan MIH (2014) An overview of fly ash and bottom ash replacement in self-compaction concrete. Key Eng Mater 594–595:465–470. https://doi.org/10.4028/www.scientific.net/KEM.594-595.465
6. Chang FY, Wey MY (2007) Comparison of the characteristics of bottom and fly ashes generated from various incineration processes. J Hazard Mater 138:594–603. https://doi.org/10.1016/j.jhazmat.2006.05.099
7. Aubert J-E, Husson B, Vaquier A (2004) Use of municipal solid waste incineration fly ash in concrete. Cem Concr Res 34:957–963. https://doi.org/10.1016/j.cemconres.2003.11.002
8. British Standards Institution (2000) Concrete—Part 1: specification, performance, production and conformity. London. BS EN 206-1
9. British Standards Institution (2000) Cement—Part 1: Composition, specifications and conformity criteria for common cements. London. BS EN 197-1
10. European Norm (2019) Protective clothing against dangerous solid, liquid and gaseous chemicals, including liquid and solid aerosols—Part 2: performance requirements for Type 1 (gas-tight) chemical protective suits for emergency teams (ET). London. EN 943-2
11. British Standards Institution (2000) Testing hardened concrete—Part 1: shape, dimensions and other requirements for specimens and moulds. London. BS EN 12390-1
12. British Standards Institution (2002) Testing hardened concrete—Part 3: compressive strength of test specimens, London. BS EN 12390-3
13. United States Environmental Protection Agency (1992) Test Method 1311—Toxicity Characteristic Leaching Procedure (TCLP). USEPA 1311
14. United States Environmental Protection Agency (1994) Test Method 1312—Synthetic Precipitation Leaching Procedure (SPLP). USEPA 1312
15. British Standards Institution (2013) Concrete—Specification, performance, production and conformity. BS EN, London, p 206
16. Siddique R, Aggarwal P, Aggarwal Y (2012) Influence of water/powder ratio on strength properties of self-compacting concrete containing coal fly ash and bottom ash. Constr Build Mater 29:73–81. https://doi.org/10.1016/j.conbuildmat.2011.10.035
17. Yu Q, Nagataki S, Lin J, Saeki T, Hisada M (2005) The leachability of heavy metals in hardened fly ash cement and cement-solidified fly ash. Cem Concr Res 35(6):1056–1063. https://doi.org/10.1016/j.cemconres.2004.03.031
18. Li XG, Lv Y, Ma BG, Chen QB, Yin XB, Jian SW (2012) Utilization of municipal solid waste incineration bottom ash in blended cement. J Clean Prod 32:96–100. https://doi.org/10.1016/j.jclepro.2012.03.038
19. Kim HK, Lee HK (2011) Use of power plant bottom ash as fine and coarse aggregates in high-strength concrete. Constr Build Mater 25(2):1115–1122. https://doi.org/10.1016/j.conbuildmat.2010.06.065

20. USEPA Environmental Protection Agency United States (1996) Hazardous waste characteristics scoping study. Office of Solid Waste. US Environmental Protection Agency, Washington, D.C.
21. World Health Organization (WHO) (2006) Guidelines for drinking-water quality
22. Kim AG, Kazonich G, Dahlberg M (2003) Relative solubility of cations in class F fly ash. Environ Sci Technol 37(19):4507–4511. https://doi.org/10.1021/es0263691
23. Kosson DS, Garrabrants AC, DeLapp R, van der Sloot HA (2014) pH-dependent leaching of constituents of potential concern from concrete materials containing coal combustion fly ash. Chemosphere 103:140–147. https://doi.org/10.1016/j.chemosphere.2013.11.049

Technological Properties of Fly Ash-Based Lightweight Geopolymer Brick

Wan Mastura Wan Ibrahim, Mohd Mustafa Al Bakri Abdullah, Kamarudin Hussin, Aeslina Abdul Kadir, and Romisuhani Ahmad

Abstract Lightweight brick reduces the overall self-weight of the structures resulting in the reduction of the foundation size, cost, and other specifications. However, conventional lightweight brick production causes several environmental impacts and produces low mechanical properties, so there is a clear need to search and replace for more efficient and durable alternatives beyond the limitations of the conventional lightweight brick. Geopolymer represents an excellent opportunity to ensure greater sustainability in the construction industry especially for the use of industrial waste such as fly ash. This research focuses on the production of fly ash-based lightweight geopolymer brick using superplasticizer as a foaming agent. The effects of geopolymeric synthesis parameters such as the sodium hydroxide (NaOH) concentration (6, 8, 10, 12 and 14 M), ratio of foaming agent to water (1/10, 1/20, 1/30 and 1/40) by volume, ratio of foam to geopolymer paste (0.5, 1.0, 1.5 and 2.0) by volume, on the lightweight geopolymer paste that affect the mechanical and microstructure properties were studied in detailed. The compressive strength, water absorption and density were analyzed to determine the mechanical properties of lightweight geopolymer brick. The microstructure properties of lightweight geopolymer brick were tested by using Scanning Electron Microscope. The results

W. M. W. Ibrahim (✉) · M. M. A. B. Abdullah · K. Hussin · R. Ahmad
Center of Excellence Geopolymer and Green Technology, School of Materials Engineering, Universiti Malaysia Perlis, 01000 Kangar, Perlis, Malaysia
e-mail: wanmastura@unimap.edu.my

M. M. A. B. Abdullah
e-mail: mustafa_albakri@unimap.edu.my

K. Hussin
e-mail: kamarudin@unimap.edu.my

R. Ahmad
e-mail: romisuhani@unimap.edu.my

A. A. Kadir
Faculty of Civil Engineering and Built Environment, Universiti Tun Hussein Onn Malaysia, Parit Raja, Malaysia
e-mail: aeslina@uthm.edu.my

A. Abdul Kadir et al. (eds.), *Sustainable Waste Utilization in Bricks, Concrete, and Cementitious Materials*, Lecture Notes in Civil Engineering 129,
https://doi.org/10.1007/978-981-33-4918-6_3

indicated that the lightweight geopolymer brick has an optimum NaOH concentration of 12 M, with highest compressive strength of 15.2 MPa at 7 days, an optimum ratio of foaming agent to water (1/10) and ratio of foam to geopolymer paste (1.0) with the highest strength of 16.6 MPa (7 days). A potential new lightweight brick can be produced by using a low-cost foaming agent and easy to process for addition to geopolymer paste.

Keywords Lightweight brick · Geopolymer · Fly ash · Mechanical properties · Microstructure properties

1 Introduction

Concrete Masonry Units (CMUs) are one of the most widely used elements in building facades and widely used in building construction as load-bearing and non-loadbearing walls [1]. Demand is increasing for affordable and lightweight construction materials with superior mechanical properties. Lightweight concrete can be classed according to its unit weight or density, which normally ranges from 320 to 1920 kg/m^3 according to the ACI Committee 213 Guide for Structural Lightweight Aggregate Concrete [2]. The challenge in making a lightweight concrete is decreasing the density while maintaining the strength of lightweight concrete. Lightweight concrete masonry could be produced either by using lightweight aggregates (natural lightweight aggregates or artificial lightweight aggregates) or admixtures (air entraining agents or admixtures that develop gases) [3].

There are two types of commercialized lightweight concrete masonry available in markets which are Cellular Lightweight Concrete (CLC) bricks and Aerated Autoclaved Concrete (AAC) bricks. The AAC brick production suffers from some disadvantages such as increased plant precautions due to the explosive nature of hydrogen gas, difficulties in the production control and high energy consumptions [4]. Besides, production of this lightweight concrete also involves high-pressure steam curing (autoclaved) with high curing temperature (180–210 °C) to and needs of superplasticizer to improve the compressive strength of lightweight concrete [5, 6]. The CLC brick was made using cement, water, fly ash and foaming agent and undergoes curing process in water curing procedure for 12–14 days and in steam curing procedure 12 h are required. The performance of CLC brick in terms of compressive strength is low for the density of CLC brick range of 400–1800 kg/m^3 [7].

The geopolymer technology is one of the green technologies. It utilizes low cost of production, which uses a variety of solid aluminosilicate source materials to produce a promising material with excellent properties [8]. Wu and Sun [9] stated that any aluminosilicate (Al–Si) materials could be the precursors for manufacturing of geopolymer such as blast furnace slag, fly ash, kaolin, mine tailings, pozzolan, silica fume, natural minerals and building residues. Most of these precursors, fly ash merits special consideration since it is a waste material produced from coal combustion and available in abundance.

Geopolymeric bricks are considered as a new technology in which are eco-sustainable masonry units because they possess good mechanical and thermal properties as well as widen the possibilities to recycle waste material to useful products especially for building material [10]. Geopolymer lightweight bricks more sustainable type of the lightweight concrete because of utilization of waste as source materials, manufactured at temperatures below 100 °C, reduce the carbon dioxide (CO_2) emissions and has better resistance to the chemical and fire [11].

Currently, there is minimal information about the effect of usage of superplasticizer only as a foaming agent on the physical, mechanical and thermal properties of lightweight geopolymer for masonry applications. Therefore, the important properties such as compressive strength, density, water absorption and thermal insulation of lightweight geopolymer using superplasticizer as foaming agent were investigated in this research. In this investigation, the lightweight geopolymer was produced by using superplasticizer only as a foaming agent because the effect of sulfate (SO_3) group improves the flow characteristics and could produce stable foam due to the long side chain (organic molecules) and high charge density. Besides, the dispersion of cement particles occurs due to the steric hindrance which is better in air entrainment. Therefore, this study would produce a lightweight geopolymer in the simplest way by using one type of foaming agent only without the addition of stabilizer.

2 Brick

Bricks are extensively used in building and construction material worldwide. The worldwide annual production of bricks is currently about 1391 billion units and the demand for bricks is expected to be continuously increasing [12]. Conventional bricks were manufactured from clay with a high temperature of kiln firing (900–1000 °C) or from ordinary Portland cement (OPC) concrete [13].

Clays are resulted from the disintegration of rocks such as pegmatite and granite, and for the manufacturing of brick, they commonly come from sedimentary or aquatic deposits. The overall process basically consists of screening, grinding, washing and moulding the clay into bricks shape either by hand or machine [14]. The clay produced from the quarrying operations is energy exhaustive, generates high level of wastes and unfavorably affects the landscape. The high temperature kiln firing consumes a significant amount of energy and also releases large quantities of greenhouse gases. Clay bricks integrate the energy of approximately 2.0 kWh and release about 0.41 kg of CO_2 per brick [15–17]. The uses of other raw materials apart from clay like waste materials are something that needs to be emphasized to protect the clay resource and the environment.

Concrete bricks are produced from ordinary Portland cement (OPC) and aggregates. Production of OPC has led to the releasing of a significant amount of CO_2 and other greenhouse gases (GHGs) causing global warming. With every ton of cement produced, almost a tonne of CO_2 is emitted [18]. The world currently uses 3.6 billion tonnes of cement per year, which contribute to 8% of CO_2 emissions according to the most recent available data.

Development of an alternative binder with advanced and efficient technology is urgently required for saving the cement industry and reduces CO_2 emissions. Besides, the conventional bricks also have a high density which results in high construction cost. Raw materials and manufacturing processes would affect the density of bricks which could be different between 1300 and 2200 kg/m^3. The density of bricks affects the weight of walls and the variations in weight have consequences on structural design, thermal and acoustical design of the wall [19]. Addition of foaming agent or uses of lightweight aggregate seems to be an alternative to the production of lightweight bricks.

3 Lightweight Brick

The main reason for introducing foaming agent into bricks is to improve the physical and mechanical properties of bricks to be lightweight, high workability, good thermal insulation and good fire resistance. One of the crucial characteristics of lightweight bricks is low density. Haider [20] mentioned that the lower density of bricks would create opportunities for cost-saving in both design and construction parts. In the construction parts, the lower density contributes to the cost-saving due to easier handling and the potential for a decrease in transport costs.

Ling and Teo [21] investigated the potential use of industrial and agricultural wastes that are Expanded Polystyrene (EPS) and Rice Husk Ash (RHA) respectively as source material for the production of 'Green' lightweight concrete bricks. The investigation revealed that the thermal conductivity decreases with increasing the RHA content and reduces the density of the sample. The thermal conductivity of lightweight concrete bricks at 28 days was in the range of 0.254–0.682 W/mK and the density of bricks was in the range of 1750–1838 kg/m^3.

Research done by Kumar et al. [22] demonstrated that the compressive strength of lightweight fly ash brick increases up to 63% as compared to the conventional clay brick and the water absorption also decreases to 48% when compared to the conventional clay brick. Balamurugan et al. [23] studied the formation of lightweight bricks using ordinary Portland cement (OPC) and protein-based foaming agents. The results obtained shows that maximum water absorption of foam concrete brick below the allowable limit of 15% and the compressive strength reduced due to the thermal effect.

Mahendran et al. [24] has made a comparison between Chamber Clay bricks, Fly ash bricks, Autoclaved Aerated Concrete (AAC) blocks, Cellular Lightweight Concrete (CLC) blocks and Porotherm blocks based on their engineering properties. The results obtained were simplified in Table 1. Based on the data analysis, it was concluded that even clay bricks are used for so many years in the construction field; it still has its limitations too. In comparison with clay brick, it shows better results in strength and heating load, but it does not come under lightweight blocks and thermal

Table 1 Data analysis of different types of bricks/blocks [24]

Types of brick/block	Compressive strength (MPa)	Water absorption (%)	Density (kg/m^3)	Thermal conductivity (W/mK)
Clay bricks	7.93	10.54	1885.60	1.0773
Fly ash bricks	9.60	13.52	1805.56	–
Porotherm blocks	1.46	14.00	741.67	0.48
AAC blocks	3.29	59.67	562.67	0.7585
CLC blocks	2.49	8.72	972.26	0.8340

efficiency. The other blocks like Porotherm block, AAC block and CLC block come under lightweight and thermal efficient blocks but do not perform load-bearing. Hence, the quality of bricks or blocks may vary depending on manufacturing units [24].

According to the previous report by Ling and Teo [21], Siram [25] and Xu et al. [26], there are still uses an OPC cement and clay as a source material in making lightweight bricks. Even in the production of commercial lightweight masonry units, which are cellular lightweight (CLC) blocks and autoclaved aerated concrete (AAC) had utilized OPC cement and required high energy consumptions. Due to this reason, an alternative method for lightweight concrete masonry units should be produced with improvement in terms of the mechanical properties, manufacturing and process with low energy required, and also utilization of waste materials as source materials.

4 Geopolymer

Generally, 'geopolymer' is the chemical reaction between aluminosilicate reactive material with strong alkaline solutions such as sodium hydroxide solution and sodium silicate solution yield amorphous aluminosilicate material [8]. Geopolymers have been increasing in interest, study and utilization worldwide for some decades. The further inspiration for discovering this alternative is attributed to its good fire and acid resistance, high compressive strength, low shrinkage, and high flexural strength [27]. Geopolymers are usually produced using either metakaolin or fly ash as a precursor have framework structures initiating from condensation process of tetrahedral aluminosilicate units of different Si/Al ratio such as (Al–O–Si–O–Si–O–)M^+, (Al–O–Si–O–)M^+, (Si–O–Al–O–Si–O–Si–O–)M^+ etcetera. where M^+ is an alkali ion, usually come from Na^+ or K^+ ion, which balances the charge of the tetrahedral Al [28].

Figure 1 represents the process formation of fly ash-based geopolymer. The simplified and fundamental principle of the formation of fly ash-based geopolymer is the

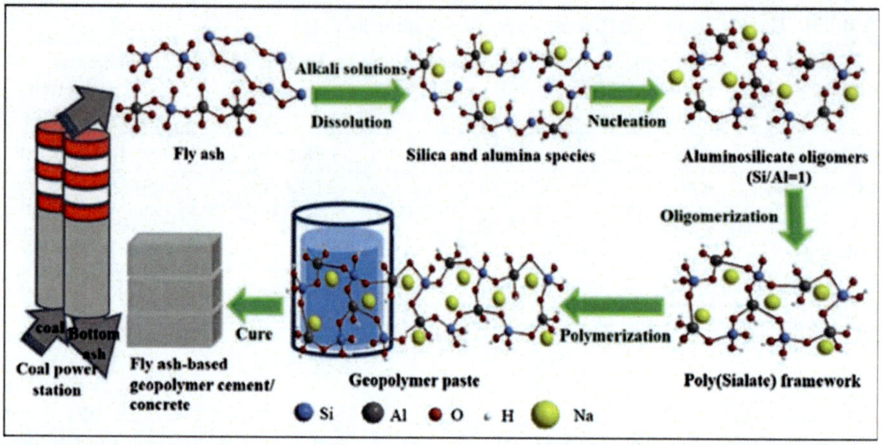

Fig. 1 The schematic drawings are showing the process from fly ash to fly ash-based geopolymer cement/concrete [29]

decomposition of aluminosilicate in the fly ash and then proceeds for the polycondensation process. The process begins with the reactions between fly ash and alkaline solution and condensation between the resultant Si^{4+} and Al^{3+} species, followed by other complicated nucleation of monomer, followed by the oligomerization, and last is polymerization process, which finally principal to a new aluminosilicate-based polymer with new amorphous three-dimensional network structure [29].

Figure 2 shows the different patterns of aluminosilicate chains formed in aluminosilicate oligomers depending on the Si/Al ratio. Duxson et al. [30] mentioned that the geopolymer compositions are much related to the mechanical strength of geopolymer which could be identified by the Si/Al ratio. It would be predictable that

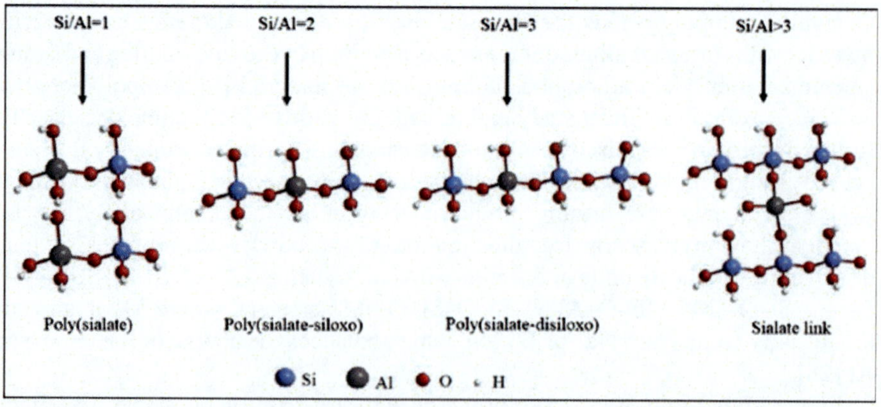

Fig. 2 Aluminosilicate chains in the aluminosilicate oligomers with different Si/Al molar ratio [29]

Table 2 Development of geopolymer application [32]

Geopolymer applications	Year
Fire-resistant wood-chipboards	1973–1976
Ceramic applications	1977–1978
Low temperature geopolymeric setting of ceramic, L.T.G.S.	1977–1982
Geopolymeric binders	1979–1995
Aviation applications	1994–2000

the strength of completely condensed tetrahedral aluminosilicate network structures should increase continuity with silica content, caused by the increased strength of Si-O-Si bonds in comparison to Al-O-Al and Si-O-Al bonds. Saidi et al. [31] reported that the compressive strength of geopolymer increases with increasing Si/Al ratio varying from 1.8 to 2.33 molar ratios. The strength value of geopolymer with Si/Al = 2.33 is almost three times that of geopolymer with Si/Al = 1.8. This performance can be primarily attributed to the formation of stronger Si-O-Si bonds at the expense of Si-O-Al bonds.

Geopolymer has developed several useful applications since 1972 listed by Davidovits [32]. The applications show genuine geopolymer products patented from several countries had brilliantly withstood 25 years of use and that are continuously commercialized. The geopolymer applications were summarized in Table 2.

4.1 Geopolymerization Process

The chemical composition of geopolymer is rather similar to zeolite but has an amorphous microstructure [33]. During the geopolymerization process, when the aluminosilicate powder of source material is mixed with the alkaline solution, a paste produced and rapidly converts into a hard geopolymer form. Therefore, there is not enough space and time for the paste or gel to develop into a well-crystallized structure; this is the important difference between geopolymers and zeolites. After shorter setting and hardening time, geopolymers with the strongly packed polycrystalline structure are formed presenting better mechanical and physical properties compared to zeolites which have lower density and a cage-like crystalline structure [18].

Figure 3 depicts the reaction mechanism that occurred in geopolymerization process. The reaction mechanism clearly shows the transformation process of solid aluminosilicate material into a synthetic alkali aluminosilicate [34]. The dissolution of the solid aluminosilicate source by alkaline solution produces aluminate and silicate species. The dissolution of amorphous aluminosilicates is rapid at high pH, and this quickly creates a supersaturated aluminosilicate solution. The gel formation occurred in the concentrated solution during the condensation process and formed large networks of oligomers. This process releases the water consumed during the dissolution process. After the gelation process, the system continues to rearrange

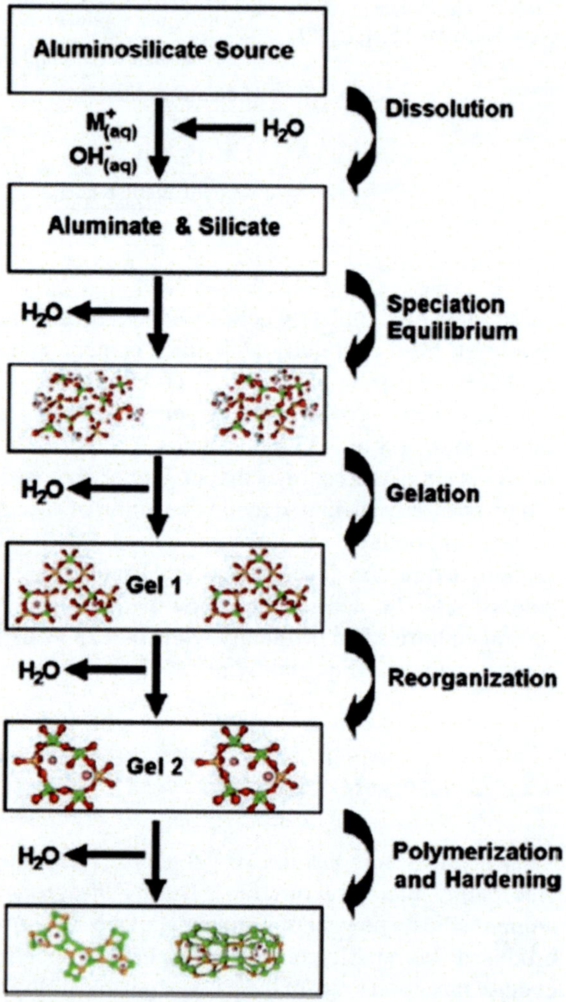

Fig. 3 Conceptual model for geopolymerization [34]

and reorganize, as the connectivity of the gel network increases, resulting in the three-dimensional aluminosilicate network commonly attributed to the geopolymers [34].

Geopolymerization is an exothermic reaction and the formations indicate that any materials containing Silicon (Si) and Aluminum (Al) could act as precursors to produce geopolymer materials [35, 36]. Overall et al. [18] concluded that the basic steps of geopolymerization process involve the dissolution of solid aluminosilicate in alkaline solution, transportation or diffusion of dissolved Al and Si complexes from the particle surface to the inter-particle space, development of a gel phase subsequent from the polymerization of monomer between Al and Si complexes and added silicate solution and finally hardening of the gel phase.

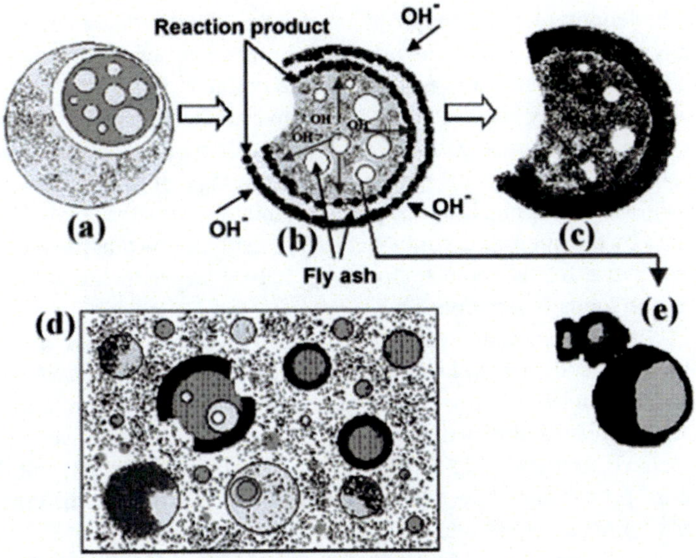

Fig. 4 Descriptive model of the alkali activation of fly ash [37]

Fernández-Jiménez et al. [37] also proposed a modelling structure of the alkali activation of fly ash for more understanding of the mechanism that occurred during the geopolymerization process as seen in Fig. 4. The mechanism reactions involved at this stage are the dissolution process and it was clearly shown that the reaction product generated both inside and outside the shell of the sphere. The initial chemical attack to the surface of particles and continues until expands into a larger hole and exposing smaller particles. At the same time, the alkaline solution will penetrate the smaller particles inside the larger spheres and starts to fill up the reaction product to form a solid matrix.

4.2 *Lightweight Geopolymer*

Lightweight geopolymer could be produced either using air-entraining agents such as chemical foaming agent, organic foaming agent, surfactants or lightweight aggregate. Different types of foaming agents resulted in different properties and performance of lightweight geopolymer. Some efforts have been made to present lightweight geopolymer materials through diverse methods and procedures.

Skvara et al. [38] studied the preparation and properties of fly ash-based geopolymer foams using silica fume, aluminium powder, FeSi and SiC powder as gas-foaming agents. Lightweight geopolymers were cured at a temperature of 80 °C for 12 h and the macrospores are developed by releasing hydrogen (H_2) gaseous results from the Al (or, respectively, Si) reaction in the strong alkaline medium.

Lightweight geopolymers are fireproof and can resist temperatures of up to 1000 °C without any indication of breakdown. After exposure at temperatures ranging from 400 to 800 °C, their shape does not change and collapse.

Previous research by Nyale et al. [39] also investigated the potential use of fly ash in the production of foamed geopolymeric materials by using sodium hypochlorite (NaOCl) solution as a foaming agent. The morphology of the foamed geopolymer contained agglomerated and irregular particles, thus showing the level of cenospheres was attacked by the hydroxide component. The results show that the time of foamed geopolymer paste left at room temperature before cured in the oven affects the porosity of the foamed geopolymer.

Wu and Sun [9] determined the properties of lightweight geopolymer made from metakaolin, class F fly ash and two types of lightweight aggregate which are Cenospheres and Expanded polystyrene (EPS) beads. The samples are cured at 80 °C for 24 h and the results obtained shows that with Cenosphere lightweight aggregate, the compressive strength of 22.5 MPa, water absorption of 22.2% and dry density of 980 kg/m^3 have been achieved while with EPS plastic beads, the corresponding values are 12.4 MPa, 19.7% and 1000 kg/m^3 were reported.

As per mentioned in the literature review, the geopolymer method has high potential to be applied in lightweight application due to their excellent mechanical properties with low density achieved.

5 Factors Affecting Properties of Lightweight Geopolymer Brick

The performance and properties of lightweight geopolymer brick were influenced by several factors including concentration of sodium hydroxide solution, mixed design, curing temperature and curing time.

5.1 Concentration of Sodium Hydroxide (NaOH) Solution

The most important factor for the production of geopolymer material is the concentration of sodium hydroxide (NaOH) solution used [40]. The concentrations of the alkaline solution affect the release of Si^{4+} and Al^{3+} from fly ash during the geopolymerization process. A high concentration of NaOH solution is usually useful for gaining high compressive strength but there is an optimum range [29].

Rattanasak and Chindaprasirt [41] studied the effect of NaOH solution on the synthesis of geopolymer using fly ash as raw material. Results revealed that the concentration of NaOH solution affected the solubility of fly ash. High strength geopolymer up to 70 MPa was obtained when the mixture was formulated with 10 M NaOH and sodium silicate to NaOH ratio of 1.0.

Hanjitsuwan et al. [42] studied the effects of sodium hydroxide (NaOH) concentration on setting time, compressive strength and electrical properties of high calcium fly ash geopolymer pastes. Five NaOH concentrations (8, 10, 12, 15 and 18 M) were studied. The pastes with high NaOH concentrations showed increased setting time and compressive strength due to a high degree of geopolymerization as a result of the increased leaching of silica and alumina from fly ash.

Hardjito [43] studied the effects of NaOH concentration on the compressive strength of fly ash-based geopolymer mortar. The results obtained reported that NaOH concentration was proportional to the compressive strength of geopolymer mortar. They have concluded that higher concentration of NaOH solution results in a higher compressive strength of geopolymer mortar.

Tsaousi et al. [4] reported that the increase of NaOH concentration caused the Si dissolution increases due to a substantial excess amount of free NaOH and generated lower SiO_2/Na_2O mass ratios in the aqueous phase. The lower SiO_2/Na_2O mass ratios assist the formation of monomer and oligomer Si species and consequently prevent the polycondensation process which extends the setting time of geopolymeric perlitic pastes. Thus, it was proved that molarity of NaOH solution influenced the mechanical properties for geopolymer material due to the improvement of their properties, especially in the compressive strength properties.

5.2 *Mix Design*

The low density and porous structure of lightweight geopolymer material have been shown to enhance some of its physical properties such as thermal and sound insulation and provide higher fire resistance when compared to the solid material [44]. Despite possessing many attractive properties, the low strength of lightweight geopolymer material often limits its structural implementations. In order to improve the lightweight properties of geopolymer material, it is critical to understand the effects of the foaming agent ratio to be added to the geopolymer paste for developing the properties of the lightweight geopolymer.

Liu et al. [45] found that with the increase in foaming agent hydrogen peroxide (H_2O_2) content from 0 to 5%, the density and the compressive strength of foamed geopolymer decrease gradually from 1593.7 to 276.6 kg/m^3 and from 30.27 to 0.35 MPa, respectively. A study by Kargin et al. [46] also demonstrated that the strength and average density of geopolymer foamed concrete decreased when the amount of the foaming agent increased.

Feng et al. [47] showed that different amounts of foaming agent (H_2O_2) had effects on the properties of the geopolymer sample. As the amount of H_2O_2 increased, the numbers of voids with air contained inside the material could be produced, thus resulting in the reduction of thermal conductivity and density and an increase in apparent porosity. The strength of the geopolymer also decreased with the increases of H_2O_2 amount due to the increased porosity of the sample.

The knowledge found from literature proved that by increasing the amount of foaming agent resulting in weakening the properties of foamed geopolymer such as density and compressive strength due to the formation of large pore size. Yang et al. [48] studied the production of foamed geopolymer by using active metal powder as the foaming agent and organosilane water repellent agent was used as a modifier to control the pore size distribution. The increase in the amount of metal (Al or Zn) powders lead to a lower density and lower thermal conductivity of the foamed geopolymer obtained due to the higher porosity.

Stable lightweight concrete masonry production depends on many factors such as type of foaming agent, method of preparation of foaming agent to initiate a uniform or homogeneous distribution of air voids (bubbles), design calculation accuracy of the mixture, and lightweight concrete production [49]. Due to this reason, an appropriate amount of foaming agent added to the geopolymer paste should be controlled with a selected ratio to reduce the density of lightweight geopolymer without affecting much to their strength behavior.

6 Mechanical and Microstructure Properties of Fly Ash-Based Lightweight Geopolymer Bricks at Different Conditions

Lightweight geopolymer brick proposes some important properties to the construction industry such as low density, low thermal conductivity, acceptable compressive strength and water absorption as well as greater fire resistance.

6.1 Compressive Strength of Fly Ash-Based Lightweight Geopolymer Brick with Different NaOH Concentration

The results of compressive strength for fly ash-based lightweight geopolymer with different concentrations of NaOH varying from 6, 8, 10, 12 and 14 M were illustrated in Fig. 5. The highest strength is observed at 12 M NaOH concentration with 15.2 MPa while the lowest strength (11.1 MPa) was obtained at 6 M NaOH concentration. The compressive strength increased gradually from 11.1 MPa (for 6 M) to 12.4 MPa, 13.6 MPa and 15.2 MPa for 8 M, 10 M and 12 M, respectively. However, beyond 14 M of NaOH concentration, the compressive strength was slightly decreased from 15.2 to 14.4 MPa as depicted in Fig. 5.

Figure 5 showed an increase in strength when the concentration of NaOH increased from 6 to 12 M. This results can be clarified by the activation of the internal Si and Al components caused by the increased breakage of the glassy chain of fly ash, which is influenced by the high alkalinity resulting from the increase of

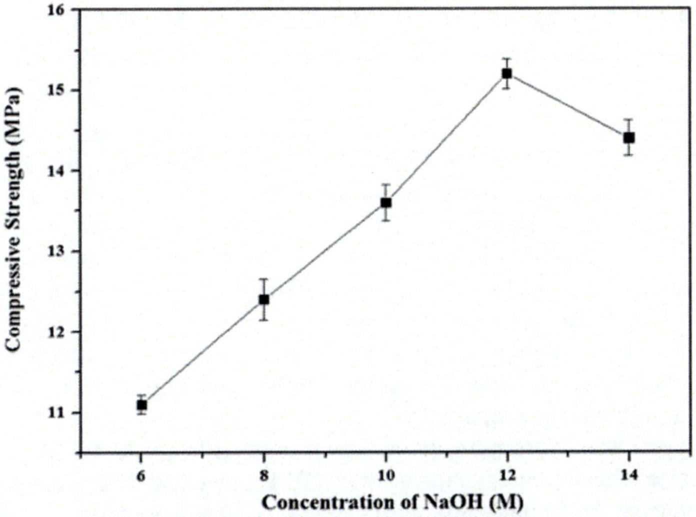

Fig. 5 The compressive strength of lightweight geopolymer with different concentration of NaOH solution

the molarity of NaOH [50]. As high NaOH concentration enhances the dissolution of raw materials and increases the solubility of silicates and aluminates [51].

However, the results obtained show a decrease in strength when the concentration of NaOH increased from 12 to 14 M. This is mostly because high concentration of NaOH (14 M) leads to the excess amount of OH^- anions and Na^+ cations. The existence of Na^+ cations in geopolymer material is to balance the negative charges produced by the formation of Si-O-Al bonding during the dissolution process, meanwhile the OH^- anions involved during hydrolysis process of fly ash [52]. Huseien et al. [53] also reported similar results where the compressive strength decreased when the concentration of NaOH was varied from 12 to 16 M and increase in alkali concentration enhanced geopolymerization process resulting in an increase in the compressive strength of geopolymer.

A low concentration (6 M) of NaOH in the activator solution produces lowest strength. This is because the dissolution of fly ash is very limited and the interparticle volume could not be fully filled by the precipitated gels [54]. The increase of NaOH concentration in the aqueous phase of the geopolymeric system results in the increased leaching of Si and Al ions of fly ash and therefore increases the strength of lightweight geopolymer. However, under extremely high NaOH concentrations in excess of OH^- concentration, dissolution of fly ash was accelerated and precipitations of aluminosilicate gel rapidly, which inhibits another geopolymeric precursor [55].

6.2 *Water Absorption of Fly Ash-Based Lightweight Geopolymer Brick with Different NaOH Concentration*

The water absorption of lightweight geopolymer was depicted in Fig. 6. The lowest water absorption was found at 12 M of NaOH concentration with 7.7% and the highest water absorption was shown at 6 M of NaOH concentration with 11.8%. A decrease in water absorption from 11.8 to 7.7% was observed by the increase concentration of NaOH solution from 6 to 12 M. However, water absorption was slightly increased at 14 M concentration of NaOH from 7.7 to 8.0%. The results shown in Fig. 6 are consistent with the results shown in Fig. 5, where the water absorption of lightweight geopolymer has a linear relationship with the compressive strength. The effect of water absorption can be seen in the reduced compressive strength of lightweight geopolymer.

The lowest water absorption percentage is observed at 12 M of NaOH concentration because when optimum amount of NaOH is presented during the geopolymerization process, the samples produced undergo complete geopolymerization during curing period due to the formation of compacted structure as depicted in Fig. 8, forming strong geopolymer material. Jitchaiyaphum et al. [56] mentioned that water absorption mainly depends on capillary pore volume and the volume of artificial pores governs the compressive strength and density for lightweight materials. Sufficient amount of NaOH concentration caused the leaching of Si, Al and other minor ions increased and produced lower water absorption and increased compressive strength.

The NaOH concentration has a great influence on the water absorption percentage of fly ash-based lightweight geopolymer. Higher concentration of NaOH produced lowest water absorption due to the increasing rate of geopolymerization which

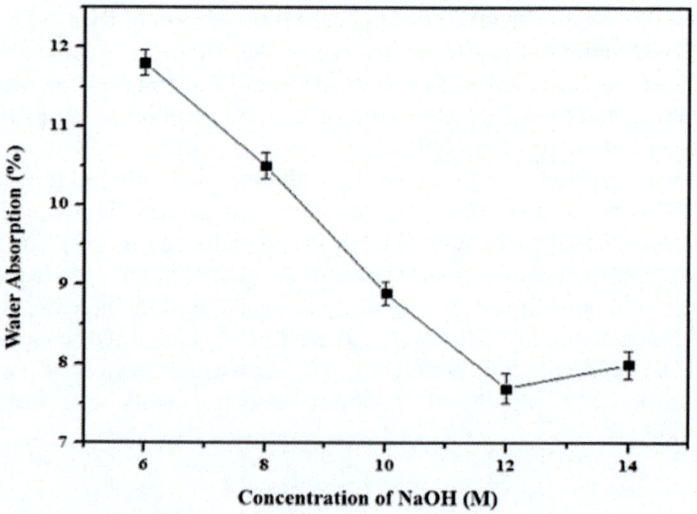

Fig. 6 The water absorption of lightweight geopolymer with different molarity of NaOH solution

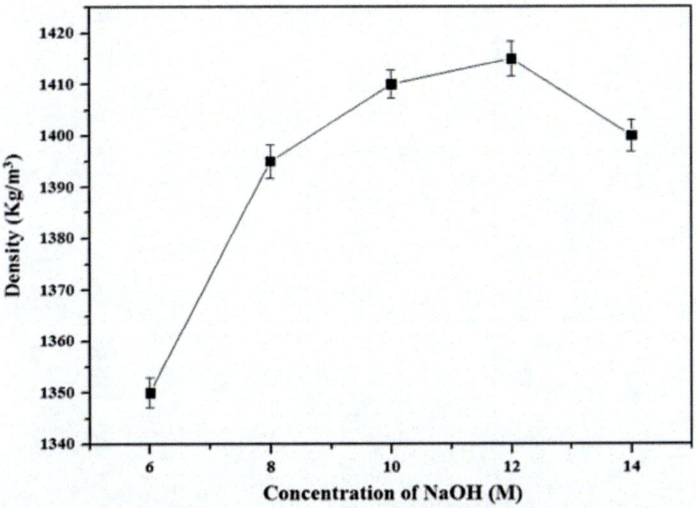

Fig. 7 The density of lightweight geopolymer with different molarity of NaOH solution

produce a denser structure hence the potential for capillary action is reduced. The water absorption results could be used to represent an open porosity of geopolymer [57]. The increases in the concentration of NaOH solution contributed to reducing the water content and reducing the voids content. Görhan and Kürklü [58] stated that the lower water absorption and apparent porosity were observed when higher alkali content of activators was used to produce geopolymer.

6.3 Density of Fly Ash-Based Lightweight Geopolymer Brick with Different NaOH Concentration

The densities of lightweight geopolymer with different concentrations of NaOH are shown in Fig. 7. The densities vary in the range of 1350–1415 kg/m^3. Increases of NaOH concentration from 6 to 12 M lead to enhance the density of lightweight geopolymer. The lowest density (1350 kg/m^3) of lightweight geopolymer was depicted at 6 M of NaOH concentration while the highest density (1415 kg/m^3) was observed at 12 M of NaOH concentration. The density of lightweight geopolymer at 14 M was slightly decreased from 1415 to 1400 kg/m^3.

Generally, the density of light geopolymer depends on the strength of lightweight geopolymer where an increase in strength accompanies the increase in lightweight geopolymer density. This is because high concentration of NaOH solution caused the greater dissolution process due to the leaching of silica and alumina, contributing to the increases in geopolymerization reaction [59]. Nyale et al. [60] also found that the

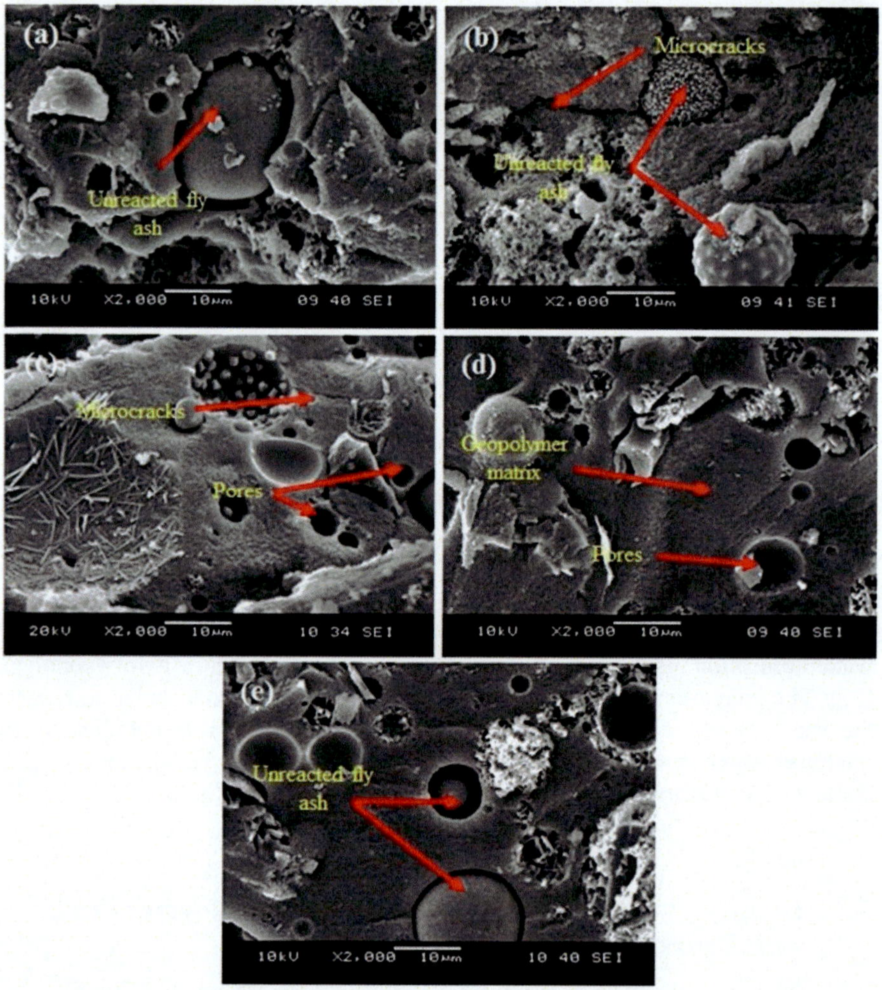

Fig. 8 Microstructure of lightweight geopolymer with different concentration of NaOH solution: **a** 6 M, **b** 8 M, **c** 10 M, **d** 12 M and **e** 14 M

synthesized foamed geopolymer's density increased with decreasing water absorption. The reduction of density is due to moisture loss during curing in geopolymer samples to the atmosphere. This observation was supported by Bakkali et al. [61].

The use of higher concentration of NaOH resulting in the increase of geopolymer paste viscosity and this condition prevents the leaching process of silica and alumina of fly ash and thus reduces the rate of geopolymerization [52]. Besides, too high NaOH concentration (14 M) causes disruption of stable and crystalline structure of the geopolymer matrix due to the excessive quantity of free Na^{+} ions which leads to cramping therefore causes density of lightweight geopolymer decreases.

6.4 Microstructure of Fly Ash-Based Lightweight Geopolymer Brick with Different NaOH Concentration

Figure 8 shows the microstructure of lightweight geopolymer with different concentrations of NaOH solutions. The microstructure of lightweight geopolymer showed some unreacted or partially reacted fly ash particles and formed a continuous alumina-silicate gel. Cracks and pores which could limit the strength of geopolymer were also observed in the geopolymer matrix (Fig. 8). The lightweight geopolymer samples with 12 M of NaOH concentration show more compacted structure with less amount of pores and cracks presented.

The microstructure revealed that several unreacted particles, presence of pores and microcracks, exist in lightweight geopolymer matrix. This is supported by the previous study done by Abdullah et al. [62] reported that the increasing concentration of NaOH solution produced less unreacted fly ash microspheres proportions. Leaching of silica and alumina was started when fly ash reacts with NaOH and the amount of leaching was dependent on NaOH concentration [41]. However, the white precipitate on the surface of lightweight geopolymer samples was produced by the Na^+ that was not to bond with Si-O and Al-O (non-bridge) and the remaining Na^+ reacted with CO_2 in the atmosphere to form the sodium carbonate [48]. Increases in concentration of NaOH solution show the microstructure of lightweight geopolymer denser due to the complete reaction between fly ash and activator solution hence causing a good compressive strength being obtained.

From this study, the optimum NaOH concentration of fly ash-based lightweight geopolymer was obtained at 12 M. Analysis has proven that the NaOH concentration affected the compressive strength, water absorption, density and microstructure of lightweight geopolymer.

6.5 Compressive Strength of Fly Ash-Based Lightweight Geopolymer Brick with Different Foaming Agent/Water Ratio and Foam/Geopolymer Paste Ratio

The compressive strength results of fly ash-based lightweight geopolymer with various ratios of foaming agent to water (by volume) and foam to geopolymer paste (by volume) are shown in Fig. 9. The strength of lightweight geopolymer is directly affected by the foaming agent content. The increases in ratio of foaming agent to water (by volume) and ratio of foam to geopolymer paste (by volume) decreased the compressive strength of lightweight geopolymer. The lowest strength (6.2 MPa) was observed at a ratio of foaming agent to water (by volume) of 1/40 with ratio of foam to geopolymer paste 2.0, by volume. The highest strength (17.8 MPa) achieved at ratio of foam to geopolymer paste of 0.5 and ratio of foaming agent to water of 1/10, by volume.

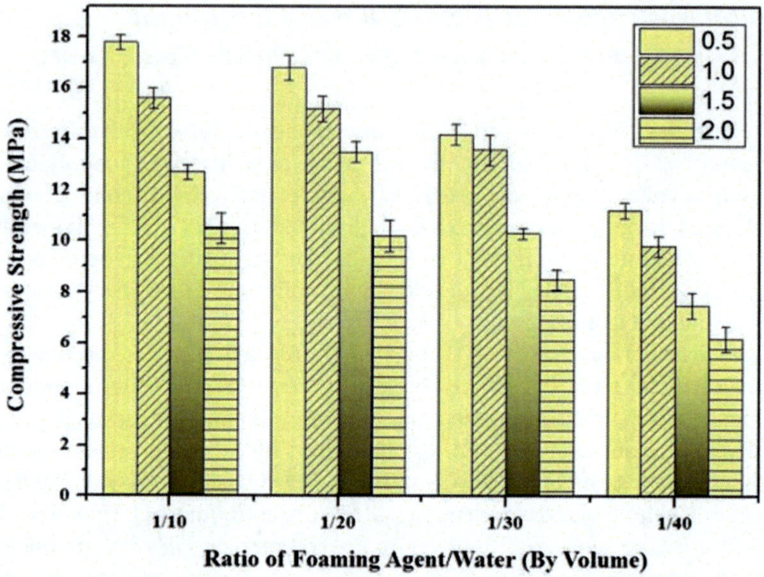

Fig. 9 The compressive strength of lightweight geopolymer with different ratio of foaming agent/water ratio and foam/geopolymer paste ratio

The strength obtained from this study was similar to the research done by Risdanareni et al. [63] which also found that the strength decreased with increasing the doses of foaming agents. The compressive strength of lightweight geopolymer is closely related to its density, both decreasing with the addition of foaming agent. At a higher foam volume, the merging of bubbles seemed to produce larger voids that result in a wide distribution of void size and lower the strength [64]. The amount of foaming agent added to the geopolymer paste influenced the pore produced and consequently caused the strength to decrease. This is due to the nucleation and stability of the pore structure for lightweight geopolymer paste strongly depends on the viscosity of the initial mix [63].

6.6 Water Absorption of Fly Ash-Based Lightweight Geopolymer Brick with Different Foaming Agent/Water Ratio and Foam/Geopolymer Paste Ratio

The water absorption of lightweight geopolymer with a different ratio of foaming agent to water and foam to geopolymer paste ratio was presented in Fig. 10. The lowest water absorption (5.6%) was depicted at foaming agent to water ratio of 1/10 and foam to geopolymer paste ratio of 0.5, by volume. The highest water absorption (20.4%) was found in a sample with a foaming agent to water ratio of 1/40 and

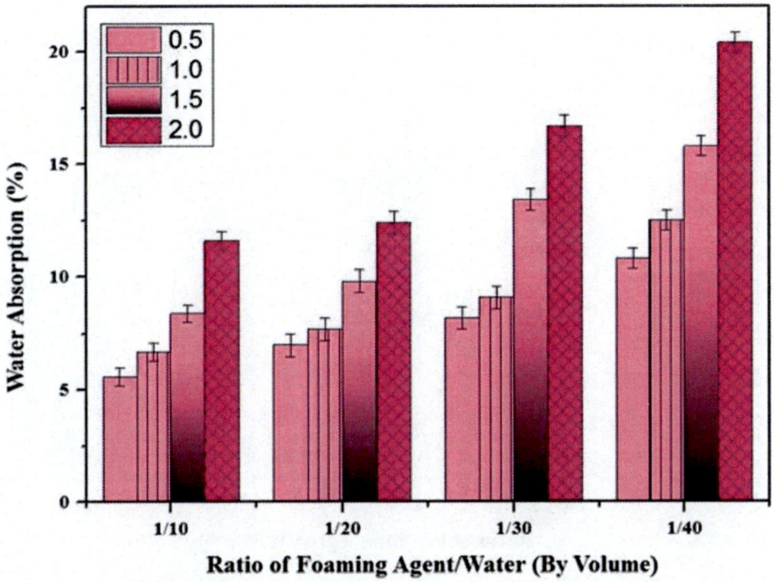

Fig. 10 The water absorption of lightweight geopolymer with different ratio of foaming agent/water ratio and foam/geopolymer paste ratio

foam to geopolymer paste ratio of 2.0, by volume. This result shows that the increase of foaming agent content caused increased pore amount, which contributed to the increasing in water absorption value.

Increasing volumes of foam and amount of pores which are beneficial for flows of water appeared to cause the increase of water absorption. The results are supported by the research done by Zhao et al. [65] which also discovered that increasing the amount of foam would increase the water absorption percentage. This is due to the fact that the sample has a high specific surface area so that it can absorb huge amounts of water. Masi et al. [66] also have presented an increasing trend of water absorption values with increasing the amount of foaming agent in geopolymer paste. It was influenced by the extent of large pores that are not completely filled by water due to air present in the cavities. It is also detected that with a lower density, an increase in water permeability was observed, possibly due to the capillary effect and the interconnection between air pores [65].

6.7 Density of Fly Ash-Based Lightweight Geopolymer Brick with Different Foaming Agent/Water Ratio and Foam/Geopolymer Paste Ratio

Figure 11 showed the density values of lightweight geopolymer with various ratio

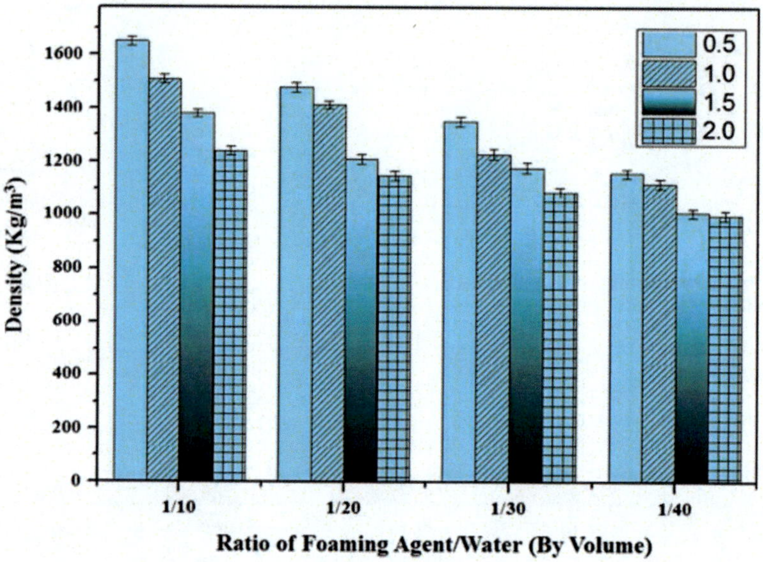

Fig. 11 The density of lightweight geopolymer with different ratio of foaming agent/water ratio and foam/geopolymer paste ratio

of foaming agent to water and foam to geopolymer paste. The density data shown in Fig. 11 are the average value of three samples and the density values are in the range between 900 to 1650 kg/m^3. As can be seen from the results obtained, the highest (1650 kg/m^3) density value was presented at the samples with the ratio of foaming agent to water of 1/10, by volume and ratio of foam to geopolymer paste of 0.5, by volume, respectively. While, the lowest (920 kg/m^3) density value appeared at the sample with the ratio of foaming agent to water of 1/40, by volume and ratio of foam to geopolymer paste of 2.0, by volume, respectively.

This is because the density of lightweight geopolymer is much related to the compressive strength of the sample. Low density provides low strength with high water absorption. It was observed that the use of this type of foaming agent (polyoxyethylene alkyether sulfate) had relatively uniform pores as seen in Fig. 12 and high viscosity of the mixture even in a low concentration of foam used.

It was clearly shown that the increase in the volume of foaming agent would decrease the density of lightweight geopolymer. The increase of foaming agent caused the amount of voids inside the material with air contained could be generated, thus resulting in the reduction of density. As mentioned by Feng et al. [47] the reduction in density of lightweight geopolymer is due to the increasing of pores inside the materials which also caused the increase in porosity. Liu et al. [64] reported that the addition of foam reduced the density due to the creation of tiny air bubbles in the geopolymer samples, which resulted in higher porosity and reduced compressive strength of foamed geopolymer. The addition of foaming agent reduced the density

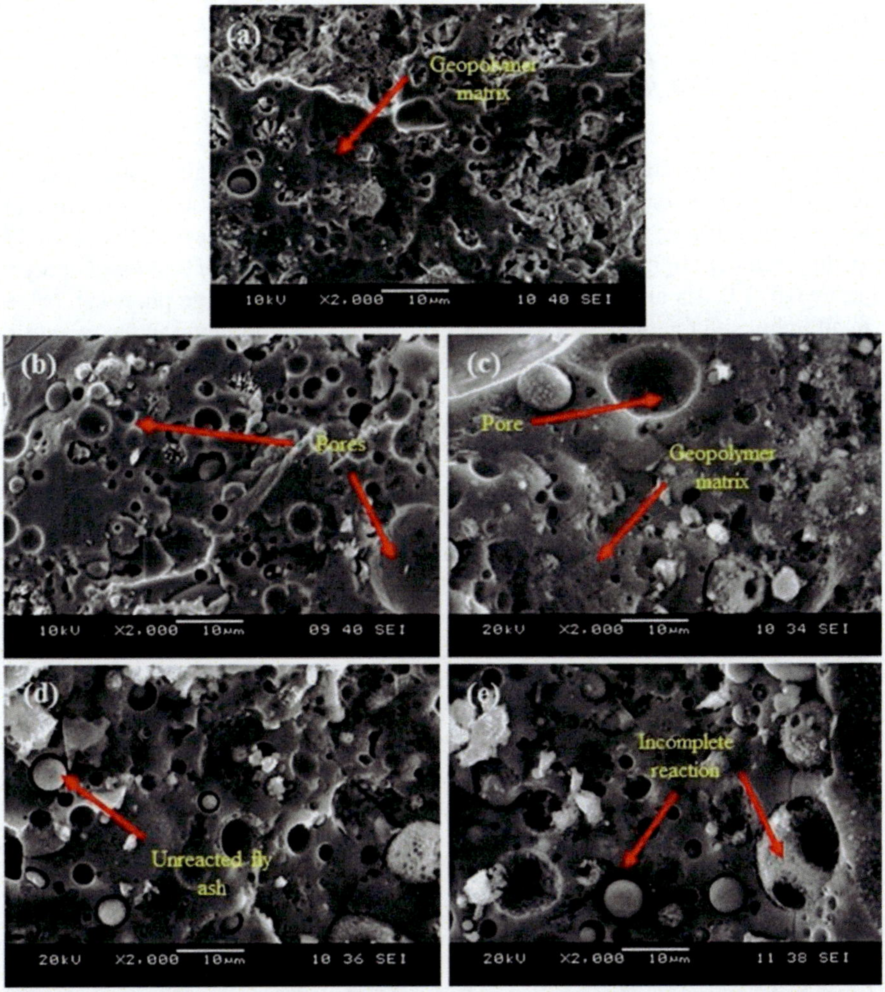

Fig. 12 The microstructure of lightweight geopolymer with different ratio of foam/geopolymer paste and control sample: **a** control sample, **b** 0.5, **c** 1.0, **d** 1.5 and **e** 2.0, by volume

of geopolymer paste during the hardening process where the air voids produced are confined within the binding skeleton.

6.8 Microstructure of Fly Ash-Based Lightweight Geopolymer Brick with Different Foaming Agent/Water Ratio and Foam/Geopolymer Paste Ratio

Figure 12 shows the morphology structure of the lightweight geopolymer with different ratio of foam to geopolymer paste with constant ratio of foaming agent to water of 1/10, by volume. The unreacted fly ash particles have still appeared for all the samples of lightweight geopolymer with different ratios of foam to geopolymer paste; 0.5, 1.0, 1.5 and 2.0, by volume. Minimal small and large pores can be seen in all lightweight geopolymer samples. The microstructure reported the quantity of the pores appeared from the samples investigated in this study, showing that, for a ratio of foam to geopolymer paste of 2.0, by volume, the quantity of pores generated seems increased as compared to the samples with the ratio of foam to geopolymer paste of 0.5, by volume.

The results show that the greater the foam's volume, the greater the amount of pore produces [67]. These results confirm that the difference amount of foaming agent added to the geopolymer paste can influence the microstructure of lightweight geopolymer. According to Liu et al. [64], the relative smaller pores in the matrix decrease and relative larger pores in the matrix increase with increasing foaming agent content. The introduction of foaming agent contributes to increasing the number and size of the pores. From the results obtained, the optimum ratio of foaming agent to water and ratio of foam to geopolymer paste were selected at 1/10, by volume and 1.0, by volume, respectively. The ratio was chosen based on the best value of compressive strength, water absorption and density required in ASTM Standard [68] for lightweight bricks application.

7 Conclusion

It could be concluded that the densities of lightweight geopolymer at different molarities of NaOH solution was substantially lower and could be applied for non-load bearing brick as stated in Standard specification for non-load bearing concrete masonry units [69]. The optimum compressive strength, density and water absorption obtained with the different ratio of foaming agent to water, and effects of foam to geopolymer paste, the lightweight geopolymer has been fulfilled the standard requirement for ASTM Standard and could be applied for lightweight concrete building bricks as stated in ASTM C 55. The samples obtained in present work show potential in the application of wall insulation materials. The fly ash-based lightweight geopolymer prepared by using superplasticizer as foaming agent has a good potential to be applied for load-bearing and non-load-bearing concrete masonry units or concrete bricks.

References

1. El-Hassan H, Shao Y, Ghouleh Z (2013) Effect of initial curing on carbonation of lightweight concrete masonry units. ACI Mater J 110:441–450. https://doi.org/10.14359/51685791
2. ACI 213 (2010) Guide for structural lightweight-aggregate concrete reported by ACI Committee 213. ACI COMMITTEE REPORT, 1–38
3. Kan A, Demirboğa R (2009) A novel material for lightweight concrete production. Cem Concr Compos 31(7):489–495. https://doi.org/10.1016/j.cemconcomp.2009.05.002
4. Tsaousi GM, Douni I, Taxiarchou M, Panias D, Paspaliaris I (2016) Development of foamed inorganic polymeric materials based on perlite. IOP Conf Ser Mater Sci Eng 123(1):12062. https://doi.org/10.1088/1757-899X/123/1/012062
5. Wongkeo W, Thongsanitgarn P, Ngamjarurojana A, Chaipanich A (2014) Compressive strength and chloride resistance of self-compacting concrete containing high level fly ash and silica fume. Mater Des 64:261–269. https://doi.org/10.1016/j.matdes.2014.07.042
6. Aminudin E, Md. Din MF, Hussin MW, Iwao K, Mohanadoss P, Ichikawa Y (2015) Properties of industrial boiler ash as sand replacement and thermal improvement in aerated concrete. Malaysian J Civ Eng 27(1):155–169
7. Awana M, Kumar C (2017) Cellular lightweight concrete. In: International conference on emerging trends in engineering, technology, sciences and management, Adithya Institute of Technology, pp 241–246
8. Davidovits J (1989) Geopolymers and geopolymeric materials. J Therm Anal 35:429–441. https://doi.org/10.1007/BF01904446
9. Wu HC, Sun P (2007) New building materials from fly ash-based lightweight inorganic polymer. Constr Build Mater 21(1):211–217. https://doi.org/10.1016/j.conbuildmat.2005.06.052
10. Petrillo A, Cioffi R, Ferone C, Colangelo F, Borrelli C (2016) Eco-sustainable geopolymer concrete blocks production process. Agric Agric Sci Procedia 8:408–418. https://doi.org/10.1016/j.aaspro.2016.02.037
11. Hajimohammadi A, Ngo T, Mendis P (2017) How does aluminium foaming agent impact the geopolymer formation mechanism? Cem Concr Compos 80:277–286. https://doi.org/10.1016/j.cemconcomp.2017.03.022
12. Arshad MS, Pawade PY (2014) Reuse of natural waste material for making lightweight bricks. Int J Sci Technol Res 3(6):49–53
13. Mineral Products association (2013) Brick and block production. Mortar Ind Assoc Learn Texts 05(20):1–10
14. Kadir AA, Mohajerani A, Roddick F, Buckeridge J (2009) Density, strength, thermal conductivity and leachate characteristics of lightweight fired clay bricks incorporating cigarette butts. World Acad Sci Eng Technol 53(5):179–184
15. Ahmari S, Zhang L (2012) Production of eco-friendly bricks from copper mine tailings through geopolymerization. Constr Build Mater 29:323–331. https://doi.org/10.1016/j.conbuildmat.2011.10.048
16. Zhang L (2013) Production of bricks from waste materials—a review. Constr Build Mater 47:643–655. https://doi.org/10.1016/j.conbuildmat.2013.05.043
17. Murekar NR, Satpute RS, Chaudhari MM (2017) Using waste material for making light weight bricks. In: International conference on recent trends in engineering sciences and technology (ICRTEST 2017). pp 467–470
18. Komnitsas K, Zaharaki D (2007) Geopolymerisation: a review and prospects for the minerals industry. Miner Eng 20(14):1261–1277. https://doi.org/10.1016/j.mineng.2007.07.011
19. Deboucha S, Hashim R (2011) A review on bricks and stabilized compressed earth blocks. Sci Res Essays 6(3):499–506. https://doi.org/10.5897/SRE09.356
20. Haider W (2007) In-plane response of wide spaced reinforced masonry shear walls. Dissertation, Central Queensland University, Queensland
21. Ling IH, Teo DCL (2012) Compressive strength and durability properties of lightweight concrete bricks under full water curing and air-dry curing. Int J Sustain Energy Dev 1, pp 14–20. https://doi.org/10.20533/ijsed.2046.3707.2012.0003

22. Kumar R, Patyal V, Lallotra B, Kumar D, Brick C (2014) Study of properties of light weight fly ash. Int J Eng Res Appl (IJERA) 49–53
23. Balamurugan G, Chockalingam K, Chidambaram M, Aravindha kumar M, Balasundaram M (2017) Experimental study on light weight foamed concrete. Int Res J Eng Technol 4(4):677–686
24. Mahendran K, Sivaram T, Shahulhameed MRL (2016) A comparative study on various building blocks as an alternative to conventional bricks. In: International conference on emerging trends in engineering and management research (ICETEMR-16). pp 1097–1109
25. Siram KKB (2012) Cellular light-weight concrete blocks as a replacement of burnt clay bricks. Int J Eng Adv Technol 2(2):149–151
26. Xu Y, Jiang L, Xu J, Li Y (2012) Mechanical properties of expanded polystyrene lightweight aggregate concrete and brick. Constr Build Mater 27(1):32–38. https://doi.org/10.1016/j.conbuildmat.2011.08.030
27. Li X, Ma X, Zhang S, Zheng E (2013) Mechanical properties and microstructure of class C fly ash-based geopolymer paste and mortar. Materials (Basel) 6(4):1485–1495. https://doi.org/10.3390/ma6041485
28. Bakharev T (2006) Thermal behaviour of geopolymers prepared using class F fly ash and elevated temperature curing. Cem Concr Res 36(6):1134–1147. https://doi.org/10.1016/j.cemconres.2006.03.022
29. Zhuang XY, Chen L, Komarneni S, Zhou CH, Tong DS, Yang HM, Yu WH, Wang H (2016) Fly ash-based geopolymer: clean production, properties and applications. J Clean Prod 125:253–267. https://doi.org/10.1016/j.jclepro.2016.03.019
30. Duxson P, Provis JL, Lukey GC, Mallicoat SW, Kriven WM, Van Deventer JSJ (2005) Understanding the relationship between geopolymer composition, microstructure and mechanical properties. Colloids Surf Physicochem Eng Asp 269:47–58. https://doi.org/10.1016/j.colsurfa.2005.06.060
31. Saidi N, Samet B, Baklouti S (2013) Effect of composition on structure and mechanical properties of metakaolin based PSS-Geopolymer. Int J Mater Sci 3(4):145. https://doi.org/10.14355/ijmsci.2013.0304.03
32. Davidovits PJ (2002) 30 years of successes and failures in geopolymer applications. Market trends and potential breakthroughs. In: Geopolymer 2002 Conference. pp 1–16
33. Chindaprasirt P, Chareerat T, Hatanaka S, Cao T (2010) High-strength geopolymer using fine high-calcium fly ash. J Mater Civ Eng 23:246–271. https://doi.org/10.1061/(ASCE)MT.1943-5533.0000161
34. Duxson P, Mallicoat SW, Lukey GC, Kriven WM, van Deventer JSJ (2007) The effect of alkali and Si/Al ratio on the development of mechanical properties of metakaolin-based geopolymers. Colloids Surf Phys Eng Asp 292(1):8–20. https://doi.org/10.1016/j.colsurfa.2006.05.044
35. Xu H, Van Deventer JSJ (2000) The geopolymerisation of alumino-silicate minerals. Int J Miner Process 59(3):247–266. https://doi.org/10.1016/S0301-7516(99)00074-5
36. Hardjito D, Cheak CC, Lee Ing CH (2009) Strength and setting times of low calcium fly ash-based geopolymer mortar. Mod Appl Sci 2(4):2–11. https://doi.org/10.5539/mas.v2n4p3
37. Fernández-Jiménez A, Palomo A, Criado M (2005) Microstructure development of alkali-activated fly ash cement: a descriptive model. Cem Concr Res 35(6):1204–1209. https://doi.org/10.1016/j.cemconres.2004.08.021
38. Skvara F, Šulc R, Tišler Z, Skricik P, Šmilauer V, Zlámalová Cílová Z (2014) Preparation and properties of fly ash-based geopolymer foams. Ceram Silikaty 58(3):188–197
39. Nyale SM, Babajide OO, Birch GD, Böke N, Petrik LF (2013) Synthesis and characterization of coal fly ash-based foamed geopolymer. Procedia Environ Sci 18:722–730. https://doi.org/10.1016/j.proenv.2013.04.098
40. Memon FA, Nuruddin MF, Khan S, Shafiq N, Ayub T (2013) Effect of sodium hydroxide concentration on fresh properties and compressive strength of self-compacting geopolymer concrete. J Eng Sci Technol 8(1):44–56
41. Rattanasak U, Chindaprasirt P (2009) Influence of NaOH solution on the synthesis of fly ash geopolymer. Miner Eng 22(12):1073–1078. https://doi.org/10.1016/j.mineng.2009.03.022

42. Hanjitsuwan S, Hunpratub S, Thongbai P, Maensiri S, Sata V, Chindaprasirt P (2014) Effects of NaOH concentrations on physical and electrical properties of high calcium fly ash geopolymer paste. Cem Concr Compos 45:9–14. https://doi.org/10.1016/j.cemconcomp.2013.09.012
43. Hardjito D (2005) Studies on fly ash-based geopolymer concrete. Dissertation, Petra Christian University
44. Brooks R, Bahadory M, Tovia F, Rostami H (2010) Properties of alkali-activated fly ash: high performance to lightweight. Int J Sustain Eng 3(3):211–218. https://doi.org/10.1080/19397038.2010.487162
45. Liu Z, Shao NN, Qin JF, Kong FL, Wang CX, Wang DM (2015) Strength and thermal behavior of low weight foam geopolymer using circulating fluidized bed combustion fly ash. J Cent South Univ 22(9):3633–3640. https://doi.org/10.1007/s11771-015-2904-0
46. Kargin A, Baev V, Mashkin N (2017) Fly-ash geo-polymer foamed concrete. In: AIP conference proceedings, p 020005
47. Feng J, Zhang R, Gong L, Li Y, Cao W, Cheng X (2015) Development of porous fly ash-based geopolymer with low thermal conductivity. Mater Des 65:529–533. https://doi.org/10.1016/j.matdes.2014.09.024
48. Yang T, Chou C, Chien C (2012) The effects of foaming agents and modifiers on a foamed-geopolymer. In: The world congress on advances in civil, environmental and materials research (ACEM), pp 905–914
49. Kallunkal G, John E (2016) Optimization of foam concrete masonry blocks. Int J Eng Res Gen Sci 4(5):85–106
50. Ryu GS, Lee YB, Koh KT, Chung YS (2013) The mechanical properties of fly ash-based geopolymer concrete with alkaline activators. Constr Build Mater 47:409–418. https://doi.org/10.1016/j.conbuildmat.2013.05.069
51. Pavithra P, Srinivasula Reddy M, Dinakar P, Hanumantha Rao B, Satpathy BK, Mohanty AN (2016) Effect of the Na_2SiO_3/NaOH ratio and NaOH molarity on the synthesis of fly ash-based geopolymer mortar. Geo-Chicago 2016:336–344
52. Chindaprasirt P, Jaturapitakkul C, Chalee W, Rattanasak U (2009) Comparative study on the characteristics of fly ash and bottom ash geopolymers. Waste Manag 29(2):539–543. https://doi.org/10.1016/j.wasman.2008.06.023
53. Huseien GF, Ismail M, Khalid NHA, Hussin MW, Mirza J (2018) Compressive strength and microstructure of assorted wastes incorporated geopolymer mortars: Effect of solution molarity. Alexandra Eng J 57(4):3375–3386. https://doi.org/10.1016/j.aej.2018.07.011
54. Zhang Z, Provis JL, Reid A, Wang H (2014) Geopolymer foam concrete: an emerging material for sustainable construction. Constr Build Mater 56:113–127. https://doi.org/10.1016/j.conbuildmat.2014.01.081
55. Somna K, Jaturapitakkul C, Kajitvichyanukul P, Chindaprasirt P (2011) NaOH-activated ground fly ash geopolymer cured at ambient temperature. Fuel 90(6):2118–2124. https://doi.org/10.1016/j.fuel.2011.01.018
56. Jitchaiyaphum K, Sinsiri T, Chindaprasirt P (2011) Cellular lightweight concrete containing pozzolan materials. Procedia Eng 14(2):241–246. https://doi.org/10.1016/j.proeng.2011.07.145
57. Abdullah MMAB, Jamaludin L, Hussin K, Bnhussain M, Ghazali CMR, Ahmad MI (2012) Fly ash porous material using geopolymerization process for high temperature exposure. Int J Mol Sci 13(12):4388–4395. https://doi.org/10.3390/ijms13044388
58. Görhan G, Kürklü G (2014) The influence of the NaOH solution on the properties of the fly ash-based geopolymer mortar cured at different temperatures. Compos Part B Eng 58:371–377. https://doi.org/10.1016/j.compositesb.2013.10.082
59. Bashar II, Alengaram UJ, Jumaat MZ, Islam A (2014) The effect of variation of molarity of alkali activator and fine aggregate content on the compressive strength of the fly ash: palm oil fuel ash based geopolymer mortar. Adv Mater Sci Eng 2014:1–13. https://doi.org/10.1155/2014/245473
60. Nyale S, Babajide O, Birch G, Boke N (2015) The influence of NaOH and NaOCl on the characteristics of fly ash-based foamed geopolymer. In: Proceedings of the 14th international conference on environmental science and technology (CEST2015)

61. Bakkali H, Ammari M, Frar I (2016) NaOH alkali-activated class F fly ash: NaOH molarity, curing conditions and mass ratio effect. J Mater Environ Sci 7(2):397–401
62. Abdullah MAB, Kamarudin H, Norazian MN, Ruzaidi CM, Zarina Y (2013) Microstructure studies on the effect of the alkaline activators ratio in preparation of fly ash-based geopolymer. In: International conference on chemistry and chemical process (IPCBEE), pp 13–17
63. Risdanareni P, Hilmi A, Susanto PB (2017) The effect of foaming agent doses on lightweight geopolymer concrete metakaolin based. In: The 3rd ISM international statistical conference, p 020057
64. Liu MYJ, Alengaram UJ, Jumaat MZ, Mo KH (2014) Evaluation of thermal conductivity, mechanical and transport properties of lightweight aggregate foamed geopolymer concrete. Energy Build 72:238–245. https://doi.org/10.1016/j.enbuild.2013.12.029
65. Zhao Y, Ye J, Lu X, Liu M, Lin Y, Gong W, Ning G (2010) Preparation of sintered foam materials by alkali-activated coal fly ash. J Hazard Mater 174:108–112. https://doi.org/10.1016/j.jhazmat.2009.09.023
66. Masi G, Rickard WDA, Vickers L, Bignozzi MC, Van Riessen A (2014) A comparison between different foaming methods for the synthesis of light weight geopolymers. Ceram Int 40:13891–13902. https://doi.org/10.1016/j.ceramint.2014.05.108
67. Cilla MS, Colombo P, Morelli MR (2014) Geopolymer foams by gelcasting. Ceram Int 40(4):5723–5730. https://doi.org/10.1016/j.ceramint.2013.11.011
68. American Society for Testing and Materials (ASTM) (2012) Standard specification for concrete building brick. The USA. ASTM C55–12, West Conshohocken
69. American Society for Testing and Materials (ASTM) (2011) Standard specification for non-load bearing concrete masonry units 1. The USA. ASTM C129–11, West Conshohocken

Development of Ash-Based and Slag-Based Pressed Geopolymer

Heah Cheng Yong, Ong Shee Ween, Liew Yun Ming, Mohd Mustafa Al Bakri Abdullah, and Ooi Wan En

Abstract Utilizing natural resources as building materials are not suitable for sustainable development as it will lead to the depletion of natural resources in the future. Hence, the use of waste materials is taken into consideration in the context of the construction industry. In terms of converting waste materials to useful products, industrial wastes of various types such as fly ash, bottom ash and blast furnace slag can be used in geopolymer technology. Considering the demand for producing high mechanical properties of geopolymer, recent innovation has led to the development of pressed geopolymer. It combines the benefit of low alkali activator content and pressure compaction proved that pressing method is featured in a simple process, time-saving, cost-saving as well as facile in large scale production. To bring a better understanding of the pressing method, two types of pressing methods are clarified. Furthermore, few types of waste that can be used as starting materials for geopolymer production are suggested in this chapter. Besides, this chapter also discussed the factors that influence the properties of pressed geopolymer.

Keywords Pressing · Geopolymer · Geopolymerization · Parameters

---

H. C. Yong (✉) · O. S. Ween · L. Y. Ming · M. M. A. B. Abdullah · O. W. En
Center of Excellence Geopolymer and Green Technology, School of Materials Engineering, Universiti Malaysia Perlis, 01000 Kangar, Perlis, Malaysia
e-mail: cyheah@unimap.edu.my

O. S. Ween
e-mail: ongsheeween@outlook.com

L. Y. Ming
e-mail: ymliew@unimap.edu.my

M. M. A. B. Abdullah
e-mail: mustafa_albakri@unimap.edu.my

O. W. En
e-mail: wanen2ooi@gmail.com

A. Abdul Kadir et al. (eds.), *Sustainable Waste Utilization in Bricks, Concrete, and Cementitious Materials*, Lecture Notes in Civil Engineering 129,
https://doi.org/10.1007/978-981-33-4918-6_4

1 Introduction

The conventional method of geopolymer production, which is a casting method, has some limitations on strength development attributed to the presence of porosity. It is well known that the porosity of geopolymer results predominantly during the drying process. Throughout this process, water evaporates leaving pores within the big cavities of the polymeric framework [1]. These phenomena contribute to the high capillary pressure in water-filled pores and consequently increases the possibility of drying shrinkage and thermal crack [2]. Besides, air bubbles entrapped in geopolymer paste during polycondensation and dissolution reactions lead to capillary pores formation. The presence of large and inhomogeneous pores acts as stress concentration points that undermine the mechanical performance of resultant specimens. Hence, finding the way to mitigate the destructive effect of pores is the key consideration in manufacturing high strength geopolymers.

Although alkali activator is valuable in dissolving aluminosilicates sources, the use of a great amount of alkali activator is user-hostile due to the large storage requirement. Apart from that, the large quantities of corrosive and viscous alkaline rise its handling difficulties which limit the commercial application of geopolymer. Unless the liquid content of geopolymer is well-controlled, otherwise geopolymer produced from the casting method has a high tendency toward efflorescence associated with the movement of alkali and water towards geopolymer surface during the curing process [3, 4].

To this approach, researchers found that the combination of low alkali activator content and pressure compaction of fresh geopolymer paste might be the applicable solution to the aforementioned problem [5]. Pressing is a powder compaction method that has been widely applied in the production of ceramic and metal (known as powder metallurgy) instead of geopolymer fabrication [6–8]. By using this method, powder is placed in the die cavity, sandwiched by a pair of punch and compacted by uniaxial pressure. The pressing process consists of three basic steps as shown in Fig. 1.

1. Die filling—Dry mix of aluminosilicate materials and alkali activator is poured into the die. Since pressing method is utilized, the workability of the dry mix is not in consideration.
2. Compaction—Pressure is applied in conjunction with trapped air removal. This stage can be conducted at room temperature and/or increased temperature which

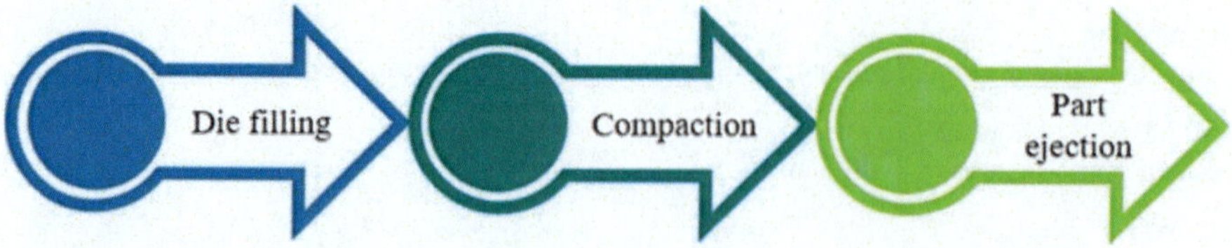

Fig. 1 Flow of pressing process

is known as cold pressing and hot pressing, respectively. Further explanation of these two pressing methods will be discussed in Sect. 2.

3. Part ejection—Green body formed and removed from the die cavity. The performance of geopolymer can be further improved through the curing and sintering process [9].

It can be seen that the pressing process is featured by simple procedures, which high strength geopolymer formed within a tremendously short time. Thus, the present method is facile and suitable for large scale production. As only a minimal amount of alkali activator is required for optimum progress of geopolymerization, this forming method permits efficient production of geopolymer with a high degree of densification with low drying shrinkage as well as thermal cracking. Furthermore, the production cost of pressing geopolymer can be reduced using a lower amount of alkali activator compared to the conventional casting method [10, 11].

2 Types of Pressing Method

In terms of forming methods, pressing is considered a new process that is still rarely used in producing geopolymer. Hence, to date, there are only two types of pressing methods that have been utilized in geopolymer production which are cold pressing and hot-pressing methods. The use of these two pressing methods is summarized in Table 1.

Table 1 Summary of pressed geopolymer prepared by varied aluminosilicate materials and alkali activator

Type of pressing	Aluminosilicate materials	Alkali activator	References
Cold pressing	Fly ash	NaOH, Na_2SiO_3	[12]
	Fly ash, metakaolin	NaOH	[13]
	Blast furnace slag, metakaolin	NaOH, Na_2SiO_3	[14]
	Calcined kaolin processing waste	NaOH, Na_2SiO_3	[10]
	Metakaolin	NaOH	[5]
Hot pressing	Fly ash	NaOH, Na_2SiO_3	[15]
	Fly ash	NaOH, Na_2SiO_3	[16]
	Volcanic ash	NaOH, Na_2SiO_3	[17]
Combination of hot and cold pressing	Volcanic ash	NaOH, Na_2SiO_3	[18]
	Fly ash, waste glass	NaOH, Na_2SiO_3	[11]

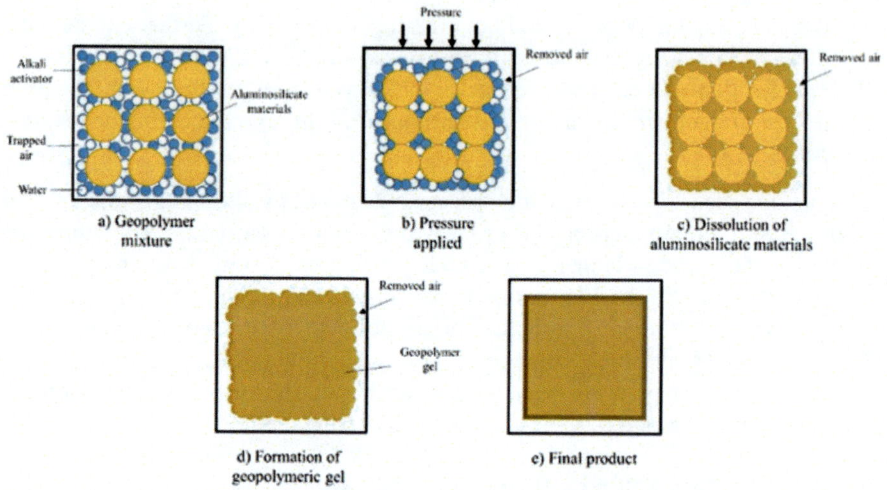

Fig. 2 Schematic diagram of the mechanism of geopolymer formed via cold pressing method [10]

2.1 Cold Pressing

Figure 2 illustrates the schematic mechanism of geopolymer produced from cold pressing method. Basically, there are three steps for pressing method as shown in Fig. 1, including die filling, compaction, and parts ejection. For the production of geopolymer, aluminosilicate materials are needed to mix with alkali activator forming dry mix. The dry mix is then poured in the die cavity, between two punches. Uniaxial pressure is applied to remove the air that is trapped in the mix forming a highly dense and compact matrix. The particles rearranged themselves leading to a well-packed structure. The highly compacted structure inducing the reaction of geopolymerization at the adjacent surface of aluminosilicate materials and alkali activator. Dissolution occurs at the surface of aluminosilicate particles resulting in the formation of Si^{4+} and Al^{3+} ions. These ions combined to form monomers, while monomers polymerized to form oligomers. Lastly, polymerization of oligomers occurred leading to the formation of a tri-dimensional geopolymeric network [10].

2.2 Hot Pressing

Hot pressing is a forming method in which the compaction process of aluminosilicates powder occurs through mechanical densification under high temperature. The process and mechanism of hot pressing are very similar to cold pressing described in Sect. 2.1. The difference is that during hot pressing, the dry mix is subjected to high temperature and pressure simultaneously as demonstrated in Fig. 3. Similar to cold pressing, the applied pressure facilitates in removing a large volume of trapped air. While the

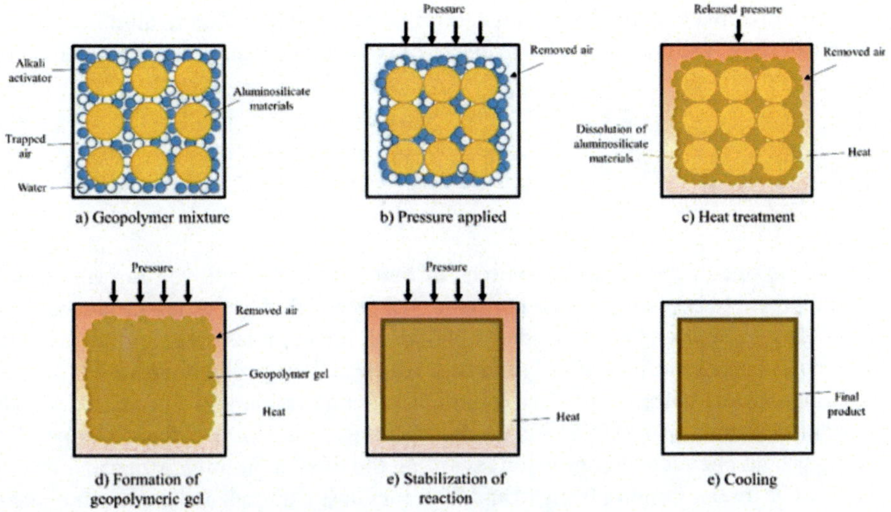

Fig. 3 Schematic diagram of the mechanism of geopolymer formed via hot pressing method [15]

subjected heat (ranging between 100 and 350 °C) helps to accelerate the dissolution of aluminosilicate materials and the formation of geopolymeric gel. During this heating period, free water evaporates from the geopolymer system further improve the specimen's compactness [15]. Besides pore size and volume reduction, Ranjbar et al. [18] stated that this process was able to modify the continuous pore network to small pores.

3 Utilization of Waste Materials in Geopolymer

Instead of using natural resources as building materials, the construction industry is encouraged to seek other alternatives materials. This is due to extracting natural resources as raw materials supply is not suitable for sustainable development [19]. The increasing scarcity of natural resources in the future is taking into considerations in the context of the construction industry [20]. On the other hand, handling a significant amount of industrial waste is a great challenge as disposing waste to sanitary landfills is forbidden due to the potential leaching of contaminants. Rather than building large dump yard to store those industrial waste, utilizing the waste in the building industry seems to be a more economically viable long-term solution [21, 22]. In short, recycling industrial waste not only can reduce the consumption of natural resources, it also resolves the disposal problem of wastes. In terms of converting waste materials to useful products, industrial wastes of various types such as fly ash, bottom ash, and blast furnace slag can be used in geopolymer technology to produce high-performance binder [23]. Any waste material that composes of silica and alumina that can be dissolved in alkali activator will suffice as aluminosilicates raw materials. The chemical compositions of those waste materials are listed in Table 2.

3.1 Fly Ash

Fly ash is one of the common sources of raw materials employed in synthesizing geopolymer. It is a by-product consisting of finely divided ashes produced from electricity generation by the combustion of coal. It comprises particles that are often vitreous, nearly spherical (Fig. 4) and reactive enough to be alkali-activated [31–33]. It composes of majority of silica, alumina, calcium oxide and ferrous oxide. The current standard, ASTM C-618 is used in defining fly ash in two classes, depending on their bulk chemical composition regarding the total of alumina, silica and ferrous oxide. The total is ranging from 50 and 70% for Class C fly ash and greater or equal to 70% for Class F fly ash. Class C fly ash has a higher content of calcium oxide makes it beneficial to be used for synthesizing geopolymer with high strength properties [34].

The reactivity of fly ash in alkaline solution depends on its particle size distribution, glass content and glass composition as well as reactive content of silica [35]. The size of fly ash particles is very crucial in alkali activation of geopolymer as the use of small particle size fly ash is advantageous in producing high strength geopolymer. Furthermore, fly ash with a tiny particle represents high workability and strength at the early duration of the firing process [34, 36]. This statement can be supported by Fernández-Jiménez et al. [37] who stated that the mechanical strength of the final product of geopolymerization increased remarkably due to the removal of particle fraction sized higher than 45 μm.

3.2 Bottom Ash

Figure 5 depicted the SEM micrograph of bottom ash. Similar to fly ash, bottom ash is also the residual waste produced from coal-based electrical power plants. The difference is that fly ash consists of fine particles which dispelled easily from the incinerator and accumulate in electrostatic precipitators, while bottom ash falls at the bottom part of the incinerator [26]. Despite similarities in chemical composition between fly ash and bottom ash, but the difference in strength achievement is evident. The study was done by Chindaprasirt et al. [39] showed that bottom ash geopolymer has lower strength compared to fly ash geopolymer corresponding to the larger particle size of bottom ash which reduces its reactivity and consequently leads to a lower degree of geopolymerization reaction. However, Kim and Lee [40] reported that bottom ash due to its irregular and angular morphology is beneficial

Table 2 Chemical composition of wastes obtained from a different origin

Waste	SiO_2	Al_2O_3	TiO_2	Fe_2O_3	CaO	SO_3	MgO	K_2O	MnO	Na_2O	P_2O_5	FeO	LOI	References
Fly ash	37.02	16.80	–	6.61	29.21	3.38	3.24	1.25	–	0.49	–	–	2.00	[24]
Fly ash	58.40	30.40	2.75	8.44	1.30	–	1.53	1.98	–	1.0	–	–	2.45	[25]
Bottom ash	26.17	15.79	0.31	14.21	28.51	1.50	2.98	1.43	0.12	1.05	0.25	–	7.68	[26]
Volcanic ash	44.19	14.06	2.74	13.22	10.38	–	9.73	1.53	0.18	3.69	0.61	–	0.62	[27]
Volcanic ash	47.74	15.36	2.80	12.88	8.25	–	6.45	1.11	0.17	3.62	0.48	–	0.66	[14]
BFS	36.67	10.31	0.57	0.50	38.82	2.17	1.70	1.03	4.04	0.48	0.04	–	0.12	[28]
BFS	32.77	17.19	0.73	0.27	40.48	–	6.59	0.29	0.32	0.31	–	0.19	0.86	[29]
Kaolin waste	52.68	33.57	0.12	0.93	–	–	–	5.72	–	0.08	–	–	6.75	[30]
Calcined kaolin waste	56.56	33.85	0.16	2.16	0.05	–	0.05	5.15	0.14	0.23	–	–	1.65	[10]

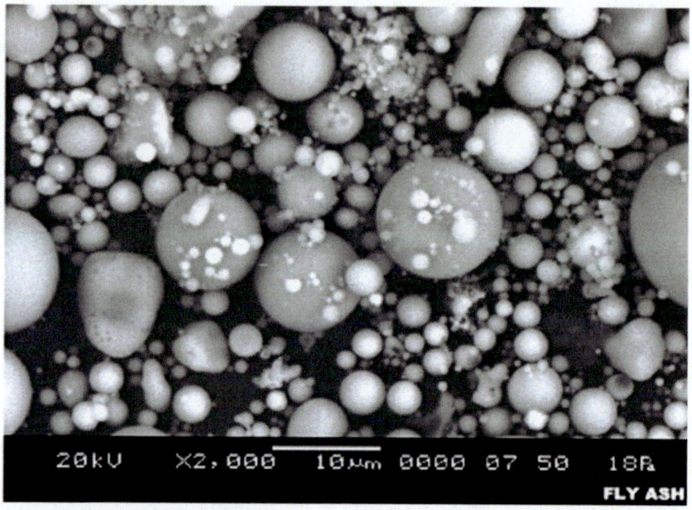

Fig. 4 SEM micrograph of fly ash [38]

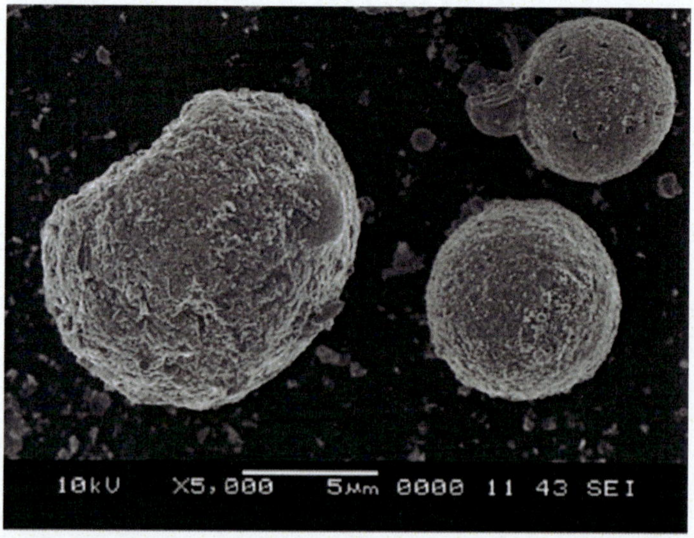

Fig. 5 SEM micrograph of bottom ash [41]

to be utilized as fine and coarse aggregates in synthesizing concrete with desired mechanical properties.

3.3 Volcanic Ash

Volcanic ash is natural waste material deposited at the surface during volcanic eruptions. Unlike another natural resource that requires open-pit mining, volcanic ash is readily accessible which limits the adverse effects on the environment [27]. Utilization of volcanic ash as an aluminosilicates source of geopolymer is of great advantage in economic impacts for countries with huge deposits [42]. From the SEM micrograph of volcanic ash (Fig. 6), it is seen that volcanic ash has complex shapes with a wide range of sizes. The particle size of ash varies depending on the volcano source, but it has been claimed that smaller particle size volcanic ash can be obtained with increasing the distance from the volcano [43].

Due to the low amorphous phase content and low geopolymeric reactivity of this raw materials, previous study indicated that volcanic ash-based geopolymer possessed long setting time, high shrinkage and poor mechanical performance at ambient temperature [44]. However, by implementing hot pressing technique, Ranjbar et al. [18] proved that a greater degree of strength development can be achieved (160 MPa). At pressing pressure and pressing temperature of 98.6 MPa and 350 °C, the authors confirmed that the dissolution of the reactive portion of volcanic ash occurred at a higher rate and thus enhanced the formation of binder gel. This contributed to the generation of condense structure which is very essential in strength achievement of resultant specimens.

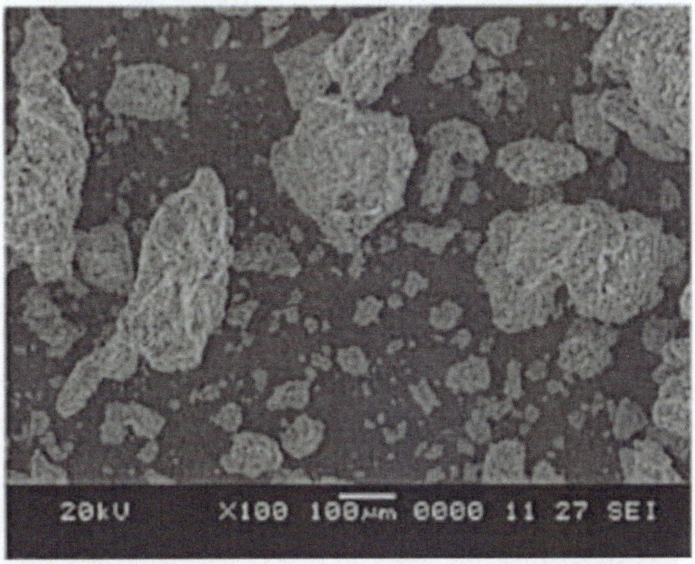

Fig. 6 SEM micrograph of volcanic ash [45]

3.4 Blast Furnace Slag

Blast furnace slag (BFS) is a coproduct that is widely adopted as complementary cementitious materials, which is discharged from iron production [46]. It composed primarily of calcium oxide, silica, alumina and magnesium oxide. BFS is obtained from quenching molten slag in water, where the molten slag is cooled rapidly and undergo solidification forming glassy sand like product. As represented in the SEM micrograph of BFS (Fig. 7), BFS is non-homogeneous with sub-rounded and sub-angular shaped morphology. It consists of bulky particles with sharp edges, while the surface has fibrous morphology detrital grains.

Due to the latent hydraulic reactivity of BFS, it can be activated to produce binder with desirable mechanical properties for the application of the construction industry [47]. Some factors that can affect the activity of BFS include chemical composition, particle distribution and glass content. BFS can be adopted in geopolymerization due to its ability in providing a significant contribution to strength development of geopolymer. This statement was attested by Zawrah et al. [48] who found out that the incorporation of BFS into waste fired clay bricks based geopolymer increases the CaO content in geopolymer matrix and reduce the porosity of specimens, which in turns enhance the mechanical performance of resultant geopolymer.

Kaolin mining and processing cause the formation of immense volume of waste material. By-products are generated during the first and second processing steps. Majority of waste is derived during the first processing step, which produced approximately 70% of the total waste. During the second processing step, mud-like sludge is generated. In order to obtain a pure and finer particle of kaolin, wet sieving is a

Fig. 7 SEM micrograph of blast furnace slag [49]

necessity [30]. Similar to kaolin, kaolin processing waste is also rich in silica and alumina which make it qualify to be utilized in geopolymerization. Low-temperature thermal treatment is required in order to activate kaolin processing waste as starting materials of geopolymer. Compared to inactivated kaolin processing waste, waste that has been calcined has a higher ability to dissolve in alkali activator solution. It thus can be used as aluminosilicate raw materials for the preparation of geopolymer [10].

4 Factors that Affect the Properties of Pressed Geopolymer

According to previous researches, the mechanical and microstructural properties of pressed geopolymers are affected by many factors, such as geopolymer constituents (solid-to-liquid ratio, NaOH concentration, sodium silicate-to-sodium hydroxide ratio), pressing condition (pressing force, hot pressing temperature and hot pressing duration) and curing time [10, 15]. The details about those factors will be clarified in the following sections.

4.1 Geopolymer Constituents

Optimizing the compositional mixing design of geopolymer dry mix plays a vital role in order to confirm the generation of high-quality geopolymer. Hence, in this section, we will focus on the influences of solid-to liquid ratio, NaOH concentration and sodium silicate-to-sodium hydroxide ratio on properties of geopolymer synthesized via pressing method.

4.1.1 Solid-To-Liquid Ratio

The solid-to-liquid (*S*/*L*) ratio discussed here refers to the ratio between solid aluminosilicates raw materials and liquid alkali activator. The alkali activator content should be tailored properly in order to ensure geopolymerization occurred in a favorable manner, at the same time, without compromising the mechanical performance of pressed geopolymer. Ranjbar et al. [15] carried out a study on the effect of *S*/*L* ratio on the mechanical properties of hot-pressed geopolymer. In this study, industrial waste (fly ash) is used as aluminosilicates materials, whereas alkali activator is prepared by mixing sodium hydroxide and sodium silicate solution. From the compressive strength analysis, it was recognized that the strength development is limited when *S*/*L* is exceeded 0.35. This is thought to be indicative of the presence of a significant amount of inert fly ash in geopolymer matrix (Fig. 8), corresponding to the insufficient alkali activator participates in geopolymerization reaction which results in the formation of less dense morphology. As a result, the strength development

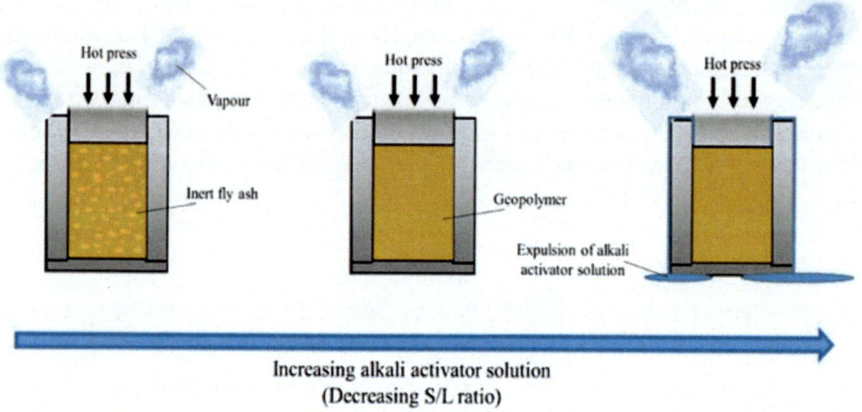

Fig. 8 Schematic diagram of the mechanism of hot pressing using different alkali activator content [15]

of geopolymer diminished. When *S*/*L* is below 0.35, the strength of geopolymer is reduced. This is because water evaporates out from geopolymer matrix throughout the hot pressing process, yielding in volume reduction of geopolymer and hence lowering down the applied pressure. In order to maintain the initial pressing condition, pressure is further applied to the specimen in the hot condition. This prompt to the removal of liquid phase which consists of reactive components of geopolymerization, consequently lower down the strength of hot-pressed geopolymer. It can be interpolated at *S*/*L* of 0.35, the hot-pressed geopolymer should have the optimum amount of alkali activator which facilitates the geopolymerization reaction and thus imparting maximum strength (134 MPa) of geopolymer sample.

A similar trend was well-agreed by Posi et al. [50] who produced lightweight fly ash-OPC geopolymer concrete via pressure compaction of 0.85 MPa. 0–15 wt% of fly ash is used to replace OPC, whereas sodium hydroxide and sodium silicate are served as alkali activator solution. With the 10 wt% of OPC replacement level, the 28 days compressive strength varied between 7.5 and 15 MPa at *S*/*L* of 0.56 to 0.83. The compressive strength increased up to *S*/*L* of 0.63 reaching 15 MPa. However, beyond 0.63 of *S*/*L* ratio, the strength decreased gradually to 12.5 MPa. The key factor that results in the strength reduction at $S/L < 0.63$ is mainly because the excessive liquid content forms a relatively high workability mix which is unsuitable for pressing method. As the *S*/*L* higher than 0.63, the dry mix is difficult to be compacted, attributed to the low liquid volume, this behavior leads to a less dense geopolymer microstructure which is unfavorable for strength development of pressed geopolymer. Therefore, it can be inferred that *S*/*L* ratio is of utmost importance in determining the pressing characteristic of a mixture, this, in turn, will influence the compressive strength of corresponding specimens.

Based on our work, pressed fly ash geopolymer achieved excellent strength of 78.54 MPa after 28 days with *S*/*L* ratio of 5.5. Compared to the hot pressing method aforementioned, it was believed that geopolymer fabricated via cold pressing method required relatively lower alkali activator content in order to achieve high-quality geopolymer.

4.1.2 NaOH Concentration

Regardless of the type of pressing method, NaOH concentration used for pressed geopolymer is generally higher than that of conventional casted geopolymer. The examples of NaOH molarity used in various geopolymer forming methods is given in Table 3.

Ranjbar et al. [15] investigated the effect of NaOH concentration on the mechanical properties of fly ash-based geopolymer synthesized by hot pressing method. The highest strength of 134 MPa was achieved as 16 M of NaOH is incorporated in the geopolymer system. In other words, the pressed geopolymer prepared by 16 M NaOH gained approximately 5% growth of compressive strength compared to that of 8 M. This is consistent with Wongsa et al. [12] who reported that the increase of NaOH concentration (5–15 M) raised the strength of pressed fly ash geopolymer concrete. Likewise, the study carried out by Prasanphan et al. [10] proved that the increment of NaOH concentration from 4 to 10 M improved the microstructure density and homogeneity of pressed geopolymer. These conditions might be an indication of the role of NaOH concentration in enhancing the dissolution abilities of Si^{4+} and Al^{3+} ions from precursors.

In support of the findings of the present study, Wang et al. [53] found out that the dissolution of aluminosilicate raw materials was directly related to the concentration of NaOH solution used in geopolymer preparation. Higher NaOH concentration facilitates the dissolution of raw materials and produces more monomers which are assisted in forming geopolymer framework. Another explanation was made by Posi et al. [50] who highlighted that NaOH concentrations of 10 and 15 M were more suitable in improving the properties of pressed geopolymer due to the extend in setting time of geopolymer mixes. Based on our study, fly ash pressed geopolymer

Table 3 Summary of NaOH molarity used in geopolymer fabrication

Forming method	Aluminosilicate material	NaOH molarity	References
Casting	Fly ash	12	[51]
	Fly ash	8	[52]
	Metakaolin	8	[3]
Hot pressing	Fly ash	16	[18]
	Fly ash	10	[16]
Cold pressing	Fly ash, metakaolin	14	[53]
	Fly ash	15	[12]
	Lunar regolith	10	[54]

with maximum compressive strength of 109.3 MPa was achieved with NaOH 14 M solution. As mentioned before, the high Na^+ and OH^- speeded up the dissolution process of aluminosilicate materials. This condition would be a hindrance to the leaching of Ca^{2+} ions, and thus the reactions were controlled by geopolymerization instead of hydration reactions of calcium [55]. It was known that the geopolymerization process took a longer time than the hydration process. This prolonged the setting time of geopolymer matrix which is beneficial for pressing methods.

4.1.3 Sodium Silicate to Sodium Hydroxide Ratio

Posi et al. [50] reported that the increment of sodium silicate to sodium hydroxide (SS/NaOH) ratio generally increased the density of pressed fly ash-OPC based geopolymer concrete. This was an expected result as sodium silicate solution was denser than that of sodium hydroxide solution. Hence, increasing SS/NaOH tended to increase the density of the alkali activator and subsequently increased the density of resultant concrete. As illustrated in Fig. 9, with the addition of 10% of OPC, the compressive strength of pressed geopolymer concrete maximized to 14.5 MPa when SS/NaOH equal to 1.0 and decreased with subsequent increase in SS/NaOH. The strength improvement suggested the existence of readily available silicate species which help in accelerating the reaction of geopolymerization, but surplus sodium silicate content reduced the workability of mixes and thus impeded the pressure compaction process.

From the result of Morsy et al. [56], it was evident that fly ash-based geopolymer mortar experienced a strength improvement from SS/NaOH of 0.5–1.0. As the ratio went beyond the optimum, strength dropped. The authors pointed out that sodium content in the mixture increased with SS/NaOH. Ascribed to the fact that sodium is

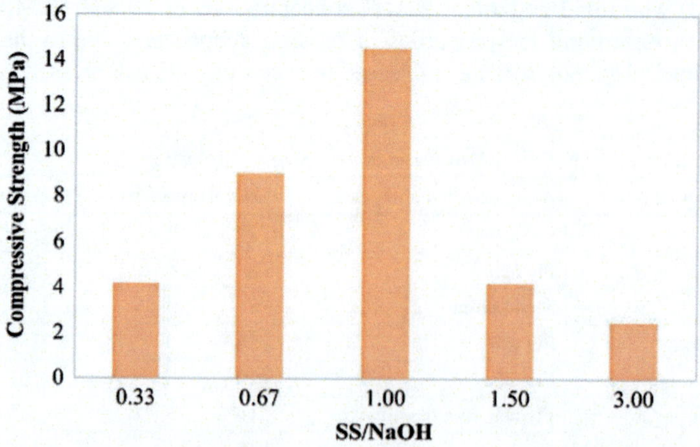

Fig. 9 Compressive strength of 28-day geopolymer concrete with varied SS/NaOH ratio [50]

essential in balancing charge ions, an increase in SS/NaOH promotes the strength growth of geopolymer. Despite that, the excessive amount of sodium silicate obstructs the water from evaporating out of the matrix and hinders the structure formation.

4.2 Pressing Condition

It was well agreed that pressing force, hot pressing temperature and duration have direct consequences on controlling the engineering properties of pressed geopolymer. Among these factors, the pressing force gives the most significant impact on the compressive strength of geopolymer system [15]. To be specific, among these pressing parameters, cold pressing is only affected by pressing force while hot pressing is influenced by pressing force, pressing temperature and pressing duration.

4.2.1 Pressing Force

The study on the influence of pressing force on hot-pressed geopolymer was carried out by Ranjbar et al. [15]. In this study, fly ash-based geopolymer was synthesized by the alkali activation of sodium hydroxide and sodium silicate. When the pressing pressure went up to 27.6 MPa, the geopolymer samples imparted the highest strength reaching 84 MPa. This was undoubtedly the consequence of subjected pressure which aids in removing air bubbles from the geopolymer matrix, thus enhancing structure compactness. Geopolymer with a well compact structure tends to reduce the formation of axial splitting micro-cracks when the compressive load is applied. That is to say, pore size reduction is vital in confirming high-quality engineering properties of geopolymer. On the contrary, further increase in pressing pressure did not show a significant contribution to strength development of geopolymer. This was mainly because almost the entire free pores fraction was removed when 27.6 MPa was applied, the remnant pore fractions were those trapped in hollow spheres that are difficult to be removed.

By utilizing cold pressing method, Khater and Ezzat [14] produced engineering stones based geopolymer using ground granulated blast furnace slag (GGBFS) and metakaolin. Pressing force ranging between 30 and 70 MPa was applied to the geopolymer mixes. When pressing force increased from 30 to 50 MPa, the compressive strength of resultant geopolymer grew from 20.90 to 35.80 MPa. Subsequent increase in pressing force led to strength diminished to 13.33 MPa. It was supposed that structure modification and rearrangement occurred when pressure of compaction increased, this condition increased the structure density and compactness. As a result, the mechanical strength of the pressed geopolymer was improved. Nevertheless, when the pressing force went beyond the optimum level, the matrix was overloaded and provoked the generation of excess micro-cracks. As a consequence, it reflects on the engineering performance of geopolymer. Similar to the finding, Wang et al. [53] noticed that pressed geopolymer gained strength when pressing force increased

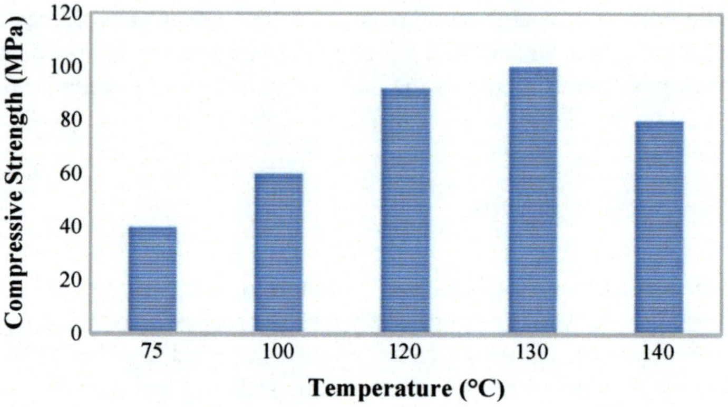

Fig. 10 Compressive strength of geopolymer with different hot-pressed temperatures for 30 min at 200 MPa [16]

to optimum but reduced as force further increased. It was believed that excessive pressing force resulted in the overflow of slurry and reduced the alkali activator content which adversely affects the geopolymerization reaction.

4.2.2 Hot Pressing Temperature

Different from previously discussed geopolymer formation, Takeda et al. [16] produced hot pressed geopolymer by mixing fly ash, sodium hydroxide and sodium silicate solution forming geopolymer slurry. The slurry was heated to form dry starting materials, it was then sieved to form powder. The powder was poured in stainless steel mold and exposed to 200 MPa of pressure compaction while simultaneously heated to temperature between 75 and 140 °C for 30 min. The compressive result (Fig. 10) showed that the pressed geopolymer strengthened to a maximum of 133 MPa as the temperature reached 130 °C, but the strength worsened with subsequent increase in temperature. Based on the study done by Ranjbar et al. [17], it was recognized that temperature contributed to fast dissolution and hardening process. This was corresponding to the rate of evaporation of water and gelation of particles. High rate of water evaporation can be achieved at high temperatures. Once the voids are empty from water, the process of closing up the voids that are trapped between particles can be accelerated. Otherwise, the water occupied voids will hamper the matrix compaction process.

4.2.3 Hot Pressing Duration

During the hot pressing process, water is discharged from geopolymer system, hence reducing the volume of matrix. This leads to the occurrence of continuous pressure loss. Thereby, maintaining the hot-pressing pressure at sufficient duration is a necessity. By applying a constant pressing temperature and force of 130 °C and 200 MPa, respectively, Takeda et al. [16] conducted a study on the effect of hot pressing duration on the engineering properties of fly ash-based geopolymer. By accommodating pressing duration of 15 min, the compressive strength of pressed sample was only 25 MPa. It was observable that pressing the geopolymer mix for 60 min effectively improved the strength to 149 MPa attributed to the adequate pressing duration in densifying geopolymer matrix. This was in agreement with apparent density analysis as geopolymer specimens experienced a marked increase in density when pressing duration extended to 60 min. The measured compressive strength, however, was in contradiction concluded by Ranjbar et al. [15]. The author recognized strength dropped when hot pressing duration prolonged from 20 to 30 min, particularly fly ash-based geopolymer with *S/L* ratio of 2.5. This mainly associated with the existence of small pockets of crystalline phase which led to the decline in strength.

4.3 Curing Time

Curing time is one of the crucial aspects as it impacts the geopolymerization reaction and subsequently mechanical properties of geopolymer. Based on the result (Fig. 11) obtained by Prasanphan et al. [10], calcined kaolin processing waste-based geopolymer experienced a significant strength gain from curing time of 1 day to 28 days. The 28-day strength achievement of pressed geopolymer was about 52.35% higher than that of 1-day strength, associated with the reaction between Si^{4+} and Al^{3+} in the presence of alkali ions. Furthermore, it was believed that the pore filling effect increased over time and hence favored engineering properties of pressed geopolymer.

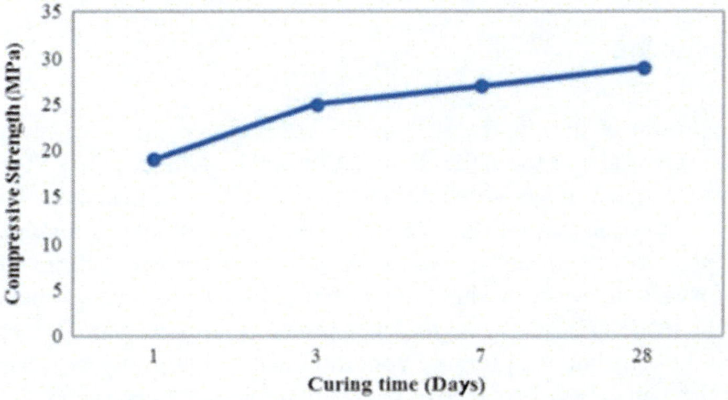

Fig. 11 Compressive strength of calcined kaolin processing waste-based pressed geopolymer with different curing time [10]

Similar to the finding, Tennakoon et al. [57] postulated that there was a great improvement in the formation of reaction products with extended curing time. This is vital in increasing the degree of crosslinking of geopolymer matrix. Evidenced by the FTIR analysis acquired by Catauro et al., [58], the Si–O–Al bonds shifted to a lower wavenumber after 28 days, resulting from the reorganization of aluminosilicate framework during geopolymerization. Whilst, the strength gain after 28 days was relatively small with limited increasing rate [59]. It can be concluded that the extended curing time helps to refine the microstructure and pores of geopolymer, but these effects are insignificant after 28 days.

5 Future Trends

In the foreseeable future, there is a growing emphasis on recycling and reusing the industrial by-product, especially in construction and building sectors which utilize the waste products as supplementary materials. The demand for developing sustainable construction will be the future trend, aiming to reduce environmental impact as well as resource scarcity. Hence, the quality of construction materials is one of the key issues as high strength and high durability construction material tend to improve its sustainability. In this chapter, it has been shown that high strength pressed geopolymer can be synthesized using industrial waste materials. The utilization of pressing methods on geopolymer creates an opportunity to improve mechanical properties of geopolymer at ambient temperature. However, up to now, the use of pressing methods in the field of geopolymer is still very uncommon. Indeed, it has been widely applied in the production of ceramic. Thus, more studies should be carried out to clarify several aspects regarding the waste-based pressed geopolymer.

6 Conclusion

To the approach of developing sustainable construction materials, geopolymer is attracting increased interest in the construction and building industry. A recent innovation, pressed geopolymer combines the benefit of low alkali activator content and pressure compaction proved that pressing method is featured in a simple process, time-saving, cost-saving as well as facile in large scale production. Besides, the demand for utilizing waste products in the production of high-strength building materials can be made. Yet, the strength development of pressed geopolymer is closely related to its constituents, pressing condition and curing time. In order to ensure the good mechanical performance of the resultant geopolymer, such factors must be taken into consideration. Due to the foreseeable advantages of pressed geopolymer, future research is required to fill the gaps of knowledge in this field.

References

1. Sindhunata J, van Deventer SJ, Lukey GC, Xu H (2006) Effect of curing temperature and silicate concentration on fly-ash-based geopolymerization. Ind Eng Chem Res 45(10):3559–3568. https://doi.org/10.1021/ie051251p
2. Collins F, Sanjayan JG (2000) Effect of pore size distribution on drying shrinkage of alkali-activated slag concrete. Cem Concr Res 30(9):1401–1406. https://doi.org/10.1016/S0008-8846(00)00327-6
3. Liew YM, Heah CY, Li L, Jaya NA, Abdullah MMAB, Tan SJ, Hussin K (2017) Formation of one-part-mixing geopolymers and geopolymer ceramics from geopolymer powder. Constr Build Mater 156:9–18. https://doi.org/10.1016/j.conbuildmat.2017.08.110
4. Nematollahi B, Sanjayan J, Shaikh FUA (2015) Synthesis of heat and ambient cured one-part geopolymer mixes with different grades of sodium silicate. Ceram Int 41(4):5696–5704. https://doi.org/10.1016/j.ceramint.2014.12.154
5. Živica V, Balkovic S, Drabik M (2011) Properties of metakaolin geopolymer hardened paste prepared by high-pressure compaction. Constr Build Mater 25(5):2206–2213. https://doi.org/10.1016/j.conbuildmat.2010.11.004
6. Andreola F, Barbieri L, Queiroz Soares B, Karamanov A, Schabbach LM, Bernardin AM, Pich CT (2019) Toxicological analysis of ceramic building materials—tiles and glasses—obtained from post-treated bottom ashes. Waste Manag 98:50–57. https://doi.org/10.1016/j.wasman.2019.08.008
7. Ke S, Wang Y, Pan Z, Ning C, Zheng S (2016) Recycling of polished tile waste as a main raw material in porcelain tiles. J Clean Prod 115:238–244. https://doi.org/10.1016/j.jclepro.2015.12.064
8. Wang H, Zhu M, Sun Y, Ji R, Liu L, Wang X (2017) Synthesis of a ceramic tile base based on high-alumina fly ash. Constr Build Mater 155:930–938. https://doi.org/10.1016/j.conbuildmat.2017.07.049
9. Carter CB, Norton MG (2007) Ceramic materials: science and engineering. Springer, New York
10. Prasanphan S, Wannagon A, Kobayashi T, Jiemsirilers S (2019) Reaction mechanisms of calcined kaolin processing waste-based geopolymers in the presence of low alkali activator solution. Constr Build Mater 221:409–420. https://doi.org/10.1016/j.conbuildmat.2019.06.116
11. Song H, Wei L, Ji Y, Cao L, Cheng F (2018) Heavy metal fixing and heat resistance abilities of coal fly ash-waste glass based geopolymers by hydrothermal hot pressing. Adv Powder Technol 29(6):1487–1492. https://doi.org/10.1016/j.apt.2018.03.013
12. Wongsa A, Siriwattanakarn A, Nuaklong P, Sata V, Sukontasukkul P, Chindaprasirt P (2019) Use of recycled aggregates in pressed fly ash geopolymer concrete. Environ Prog Sustain Energy 39. https://doi.org/10.1002/ep.13327
13. Wang S, Ma X, He L, Zhang Z, Li L, Li Y (2019) High strength inorganic-organic polymer composites (IOPC) manufactured by mold pressing of geopolymers. Constr Build Mater 198:501–511. https://doi.org/10.1016/j.conbuildmat.2018.11.281
14. Khater HM, Ezzat M (2018) Preparation and characterization of engineered stones based geopolymer composites. J Build Eng 20:493–500. https://doi.org/10.1016/j.jobe.2018.08.004
15. Ranjbar N, Mehrali M, Maheri MR, Mehrali M (2017) Hot-pressed geopolymer. Cem Concr Res 100:14–22. https://doi.org/10.1016/j.cemconres.2017.05.010
16. Takeda H, Hashimoto S, Matsui H, Honda S, Iwamoto Y (2014) Rapid fabrication of highly dense geopolymers using a warm press method and their ability to absorb neutron irradiation. Constr Build Mater 50:82–86. https://doi.org/10.1016/j.conbuildmat.2013.09.014
17. Ranjbar N, Kashefi A, Maheri MR (2018) Hot-pressed geopolymer: dual effects of heat and curing time. Cem Concr Compos 86:1–8. https://doi.org/10.1016/j.cemconcomp.2017.11.004
18. Ranjbar N, Kashefi A, Ye G, Mehrali M (2020) Effects of heat and pressure on hot-pressed geopolymer. Constr Build Mater 231:117106. https://doi.org/10.1016/j.conbuildmat.2019.117106

19. Ahmari S, Ren X, Toufigh V, Zhang L (2012) Production of geopolymeric binder from blended waste concrete powder and fly ash. Constr Build Mater 35:718–729. https://doi.org/10.1016/j.conbuildmat.2012.04.044
20. Zhang Z, Provis JL, Reid A, Wang H (2014) Geopolymer foam concrete: an emerging material for sustainable construction. Constr Build Mater 56:113–127. https://doi.org/10.1016/j.conbuildmat.2014.01.081
21. Kim Y, Kim K, Jeong GY (2017) Study of detailed geochemistry of hazardous elements in weathered coal ashes. Fuel 193:343–350. https://doi.org/10.1016/j.fuel.2016.12.080
22. Wan S, Zhou X, Zhou M, Han Y, Chen Y, Geng J, Wang T, Xu S, Qiu Z, Hou H (2018) Hydration characteristics and modeling of ternary system of municipal solid wastes incineration fly ash-blast furnace slag-cement. Constr Build Mater 180:154–166. https://doi.org/10.1016/j.conbuildmat.2018.05.277
23. van Deventer JSJ, Provis JL, Duxson P, Lukey GC (2007) Reaction mechanisms in the geopolymeric conversion of inorganic waste to useful products. J Hazard Mater 139(3):506–513. https://doi.org/10.1016/j.jhazmat.2006.02.044
24. Chousidis N, Ioannou I, Rakanta E, Koutsodontis C, Batis G (2016) Effect of fly ash chemical composition on the reinforcement corrosion, thermal diffusion and strength of blended cement concretes. Constr Build Mater 126:86–97. https://doi.org/10.1016/j.conbuildmat.2016.09.024
25. Kanth UR, Rao PS, Krishna MG (2019) Mechanical behaviour of fly ash/SiC particles reinforced Al-Zn alloy-based metal matrix composites fabricated by stir casting method. J Mater Res Technol 8(1):737–744. https://doi.org/10.1016/j.jmrt.2018.06.003
26. Hanjitsuwan S, Phoongernkham T, Damrongwiriyanupap N (2017) Comparative study using portland cement and calcium carbide residue as a promoter in bottom ash geopolymer mortar. Constr Build Mater 133:128–134. https://doi.org/10.1016/j.conbuildmat.2016.12.046
27. Lemougna PN, MacKenzie KJD, Melo UFC (2011) Synthesis and thermal properties of inorganic polymers (geopolymers) for structural and refractory applications from volcanic ash. Ceram Int 37(8):3011–3018. https://doi.org/10.1016/j.ceramint.2011.05.002
28. Burciaga-Díaz O, Betancourt-Castillo I (2018) Characterization of novel blast-furnace slag cement pastes and mortars activated with a reactive mixture of MgO-NaOH. Cem Concr Res 105:54–63. https://doi.org/10.1016/j.cemconres.2018.01.002
29. Shen X, Zhao H, Xingrong W, Cao F, Wang P, Li L (2019) Effect of steel slag on crystallization and bending strength of glass ceramics based on blast furnace slag. IOP Conf Ser Earth Environ Sci 281:12033. https://doi.org/10.1088/1755-1315/281/1/012033
30. Menezes R, Brasileiro M, Santana L, Neves G, Lira H, Ferreira H (2008) Utilization of kaolin processing waste for the production of porous ceramic bodies. Waste Manag Res 26(4):362–368. https://doi.org/10.1177/0734242X07076947
31. Ayachit AC, Nikam PB, Pise SN, Shah AD, Vinayak H (2016) Mix design of fly-ash based geopolymer concrete. Int J Sci Res Publ 6(2):2250–3153
32. Provis JL, Bernal SA (2014) Geopolymers and related alkali-activated materials. Annu Rev Mater Res 44(1):299–327. https://doi.org/10.1146/annurev-matsci-070813-113515
33. Vickers L, Van Riessen A, Rickard WD (2015) Fire-resistant geopolymers: role of fibres and fillers to enhance thermal properties. Springer, Berlin
34. Laskar S, Mozumder RA, Roy B (2015) Behaviour of geopolymer concrete under static and cyclic loads. In: Advances in structural engineering: materials. Springer, Berlin, pp 1643–1653
35. Tchadjie L, Ekolu S (2018) Enhancing the reactivity of aluminosilicate materials toward geopolymer synthesis. J Mater Sci 53. https://doi.org/10.1007/s10853-017-1907-7
36. Kobayashi T (2016) Applied environmental materials science for sustainability. IGI Global
37. Fernández-Jiménez A, Palomo A, López-Hombrados C (2006) Engineering properties of alkali-activated fly ash. ACI Mater J 103(2):106–112
38. Abdullah MMAB, Kamarudin H, Binhussain M, Nizar K, Yahya Z, Razak R (2012) Ash-based geopolymer lightweight concrete using foaming agent. Int J Mol Sci 13:7186–7198. https://doi.org/10.3390/ijms13067186
39. Chindaprasirt P, Jaturapitakkul C, Chalee W, Rattanasak U (2009) Comparative study on the characteristics of fly ash and bottom ash geopolymers. Waste Manag 29(2):539–543. https://doi.org/10.1016/j.wasman.2008.06.023

40. Kim HK, Lee HK (2011) Use of power plant bottom ash as fine and coarse aggregates in high-strength concrete. Constr Build Mater 25(2):1115–1122. https://doi.org/10.1016/j.conbuildmat.2010.06.065
41. Deraman LM, Abdullah MMAB, Ming LY, Hussin K (2017) Density and morphology studies on bottom ash and fly ash geopolymer brick. AIP Conf Proc 1835:020047. https://doi.org/10.1063/1.4981869
42. Kouamo HT, Mbey JA, Elimbi A, Kenne Diffo BB, Njopwouo D (2013) Synthesis of volcanic ash-based geopolymer mortars by fusion method: effects of adding metakaolin to fused volcanic ash. Ceram Int 39(2):1613–1621. https://doi.org/10.1016/j.ceramint.2012.08.003
43. Djobo JNY, Elimbi A, Tchakoute Kouamo H, Kumar S (2016) Mechanical activation of volcanic ash for geopolymer synthesis: effect on reaction kinetics, gel characteristics, physical and mechanical properties. RSC Adv 6(45):39106–39117. https://doi.org/10.1039/C6RA03667H
44. Tchakoute HK, Elimbi A, Yanne E, Djangang CN (2013) Utilization of volcanic ashes for the production of geopolymers cured at ambient temperature. Cem Concr Compos 38:75–81. https://doi.org/10.1016/j.cemconcomp.2013.03.010
45. Rafiza A, Razak R, Abdullah MMAB, Abdullah B, Kamarudin H, Nizar K, Sandu I, Hardjito D, Yahya Z (2013) Study on radioactivity components, water quality and microstructure characteristic of volcano ash as geopolymer artificial aggregate. Rev Chim Bucharest Orig Ed 64:593–598
46. Huang K, Fan X, Gan M, Ji Z (2019) Use of municipal solid waste incinerator (MSWI) fly ash in alkali activated slag cement. In: Magnesium Technology. Springer Int Publ, Berlin, pp 401–410
47. Liu J, Yu Q, Zuo Z, Yang F, Duan W, Qin Q (2017) Blast furnace slag obtained from dry granulation method as a component in slag cement. Constr Build Mater 131:381–387. https://doi.org/10.1016/j.conbuildmat.2016.11.040
48. Zawrah MF, Gado RA, Feltin N, Ducourtieux S, Devoille L (2016) Recycling and utilization assessment of waste fired clay bricks (Grog) with granulated blast-furnace slag for geopolymer production. Process Saf Environ Prot 103:237–251. https://doi.org/10.1016/j.psep.2016.08.001
49. Aziz IH, Abdullah MMAB, Heah CY, Liew YM (2019) Behaviour changes of ground granulated blast furnace slag geopolymers at high temperature. Adv Cem Res 32(10):465–475. https://doi.org/10.1680/jadcr.18.00162
50. Posi P, Thongjapo P, Thamultree N, Boontee P, Kasemsiri P, Chindaprasirt P (2016) Pressed lightweight fly ash-OPC geopolymer concrete containing recycled lightweight concrete aggregate. Constr Build Mater 127:450–456. https://doi.org/10.1016/j.conbuildmat.2016.09.105
51. Zhang Z, Provis JL, Ma X, Reid A, Wang H (2018) Efflorescence and subflorescence induced microstructural and mechanical evolution in fly ash-based geopolymers. Cem Concr Compos 92:165–177. https://doi.org/10.1016/j.cemconcomp.2018.06.010
52. Alomayri T (2019) Experimental study of the microstructural and mechanical properties of geopolymer paste with nano material (Al2O3). J Build Eng 25:100788. https://doi.org/10.1016/j.jobe.2019.100788
53. Wang H, Li H, Yan F (2005) Synthesis and mechanical properties of metakaolinite-based geopolymer. Colloids Surf Physicochem Eng Asp 268(1):1–6. https://doi.org/10.1016/j.colsurfa.2005.01.016
54. Davis G, Montes C, Eklund S (2017) Preparation of lunar regolith based geopolymer cement under heat and vacuum. Adv Sp Res 59(7):1872–1885. https://doi.org/10.1016/j.asr.2017.01.024
55. Malkawi AB, Nuruddin MF, Fauzi A, Almattarneh H, Mohammed BS (2016) Effects of alkaline solution on properties of the HCFA geopolymer mortars. Procedia Eng 148:710–717. https://doi.org/10.1016/j.proeng.2016.06.581
56. Morsy MS, Alsayed SH, Al-Salloum Y, Almusallam T (2014) Effect of sodium silicate to sodium hydroxide ratios on strength and microstructure of fly ash geopolymer binder. Arab J Sci Eng 39(6):4333–4339. https://doi.org/10.1007/s13369-014-1093-8
57. Tennakoon C, Shayan A, Sanjayan JG, Xu A (2017) Chloride ingress and steel corrosion in geopolymer concrete based on long term tests. Mater Des 116:287–299. https://doi.org/10.1016/j.matdes.2016.12.030

58. Catauro M, Papale F, Lamanna G, Bollino F (2015) Hybrid materials synthesis and investigation of the polymer influence on microstructure and mechanical behavior. Mater Res 18(4):698–705. https://doi.org/10.1590/1516-1439.342814
59. Yadollahi MM, Benli A, Demirboğa R (2015) The effects of silica modulus and aging on compressive strength of pumice-based geopolymer composites. Constr Build Mater 94:767–774. https://doi.org/10.1016/j.conbuildmat.2015.07.052

Development of Fly Ash Concrete Using Glass Bubble for Thermal Insulation Building Application

Noor Fifinatasha Shahedan, Mohd Mustafa Al Bakri Abdullah, Norsuria Mahmed, Andri Kusbiantoro, and Aeslina Abdul Kadir

Abstract Due to the successful sustainable construction materials of precast concrete for green materials, developing quality and durability of geopolymer concrete, including improved comfort capabilities that offer good thermal insulation properties, has been focused. Thus, to achieve the highest possible thermal insulation performance, a new insulation material, which is glass bubble as a solution with low thermal conductivity value, develops with geopolymer concrete. It was shown that the replacement of glass bubbles could only be replaced at 10%, which is most applicable for satisfying the minimum strength requirement of the load-bearing structure. Thermal properties are demonstrated by the development of geopolymer concrete with thermal insulation material having low thermal conductivity, low heat dissipation, and high specific heat, which is relevant to reduce building thermal load. The decrease in thermal conductivity and thermal diffusivity is due to the increase of void ratio from glass bubbles along with thermal conductivity of the glass bubble itself helps to improve thermal properties. The presence of a glass bubble internal

N. F. Shahedan · N. Mahmed
Faculty of Chemistry Engineering Technology, Universiti Malaysia Perlis, 02600 Arau, Perlis, Jejawi, Malaysia
e-mail: fifinatasha@unimap.edu.my

N. Mahmed
e-mail: norsuria@unimap.edu.my

M. M. Al Bakri Abdullah (✉)
Geopolymer & Green Technology, Centre of Excellence (CEGeoGTech), Universiti Malaysia Perlis, 01000 Kangar, Perlis, Malaysia
e-mail: mustafa_albakri@unimap.edu.my

A. Kusbiantoro
Faculty of Engineering Technology, Universiti Tun Hussein Onn Malaysia, 86400 Parit Raja, Batu Pahat, Johor, Malaysia
e-mail: andri@uthm.edu.my

A. A. Kadir
Faculty of Civil Engineering and Built Environment, Universiti Tun Hussein Onn Malaysia, 86400 Parit Raja, Batu Pahat, Johor, Malaysia
e-mail: aeslina@uthm.edu.my

A. Abdul Kadir et al. (eds.), *Sustainable Waste Utilization in Bricks, Concrete, and Cementitious Materials*, Lecture Notes in Civil Engineering 129,
https://doi.org/10.1007/978-981-33-4918-6_5

gas core contributes to the high specific heat, which results in a longer time for the heat to be stored before the heat leaks out of the building.

Keywords Glass bubble · Geopolymer concrete · Fly ash · Thermal conductivity · Thermal diffusivity · Specific heat

1 Introduction

Glass bubble is a new material that has been utilized for most applications, for example, in polyproline composite, epoxy-matrix composite and cement composite as filler, additive, aggregate and cement replacement [1–8]. The unique properties of the spherical hollow shape of a glass bubble, high crush strength, low density, easy to work with, chemically stable, excellent water and oil resistance, and excellent thermal insulation properties have made glass bubble very useful in a wide range of industrial sectors including construction industry [8, 9]. With these advantages, the glass bubble has barely been used as insulation material in regular concrete.

Due to the rapid economic growth in Malaysia with current housing demand, building construction has shifted its focus to transform the conventional method to prefabrication method using industrialized building system (IBS). The most common IBS component used in Malaysia is precast concrete. Environmentally, precast concrete search for green materials has increased of central concern to minimize the use of ordinary Portland cement (OPC) and emissions of carbon dioxide (CO_2) not to risk the needs of future generations.

Recycling and reusing waste materials have become an increasingly important research area in recent years [10]. Geopolymer concrete is a new construction material with great potential to replace conventional concrete mostly produced by OPC [11]. The binder materials used for geopolymer concrete, such as, fly ash are mostly industrial waste or by-products containing high silica content and alumina, which acted as precursors for geopolymerization. Geopolymer concrete provides comparable properties in strength and workability compared with normal concrete and has excellent potential to be used in the precast industry [12].

Therefore, concerning the demands of energy efficiency of the building, concepts like passive houses, zero-emission building and thermal insulation of the building can play important roles. It is crucial to develop new beneficial insulation materials for the building's concrete. Several unique properties of glass bubble are low density (125 kg/m^3), low thermal conductivity (0.044 W/mK), which benefit geopolymer concrete. In order to achieve the highest possible thermal insulation performance in the geopolymer concrete system itself without adding another layer to the concrete, new insulation materials such as the glass bubble have become a new development to the building industry, including in the precast concrete industry.

2 Characteristics of Raw Materials Fly Ash and Glass Bubble

Fundamentally, the characteristics of raw materials fly ash and glass bubble are strongly influenced by the performance of the concrete and thermal insulation properties of the concrete itself. The basic characterization usually characterizes the raw materials through several analyses testing using Particle Size Analysis (PSA), X-Ray Fluorescence (XRF), Scanning Electron Microscope (SEM) and Fourier Transform Infrared Spectroscopy (FTIR) for size distribution, chemical composition analysis, morphology analysis, and functional group analysis, respectively.

2.1 Particle Size Analysis

The distribution of the particle size of fly ash raw materials is shown in Fig. 1. According to the results, almost 90% of particles distributed in the range of 4–70 μm and 10% of particles distributed in the range of 80–550 μm. The distribution appeared to be bimodal. The first mode is located at 60 μm, whereas the second smallest mode at around 400 μm. The mean particle size (X50) of fly ash was 46.22 μm.

Figure 2 shows the distribution of particle size of glass bubble raw materials. The results illustrated almost 92.5% of particles distributed in the range of 4–80 μm and 7.5% of particles distributed in the range of 80–200 μm. The mode is located at 80 μm while the mean particle size (X50) of glass bubble was 66.68 μm.

Particle size analysis of source materials is important physical properties impacting the reactivity and affects the geopolymerization rate, thus influence the strength of the product in the synthesis of geopolymer, which also has agreed by Chindaprasirt et al. [13]. Smaller particle size of the materials increased the available

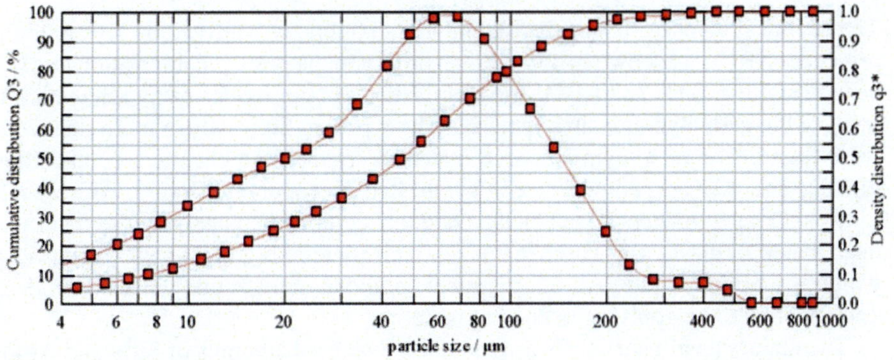

Fig. 1 Particle size analysis of fly ash

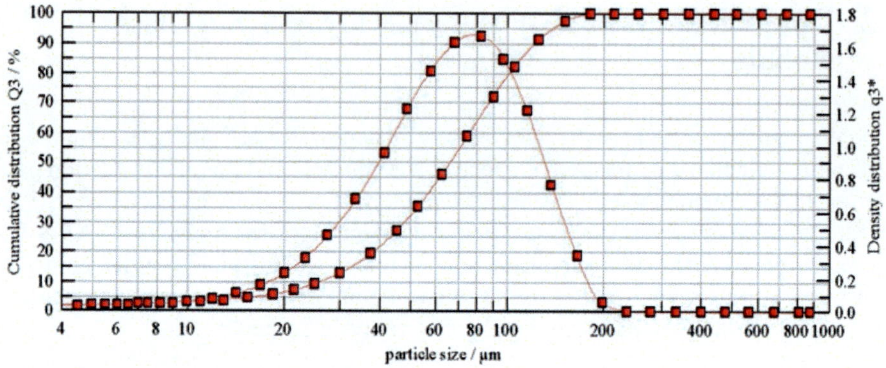

Fig. 2 Particle size analysis of glass bubble

surface area, thus improving particles' reactivity and promoting faster dissolution and formation of geopolymer gel during geopolymerization.

In terms of insulation properties, particle size has great importance in improving the insulation properties of the concrete itself, which is related to the size of the insulation materials that consist of a greater amount of air by glass bubble into geopolymer concrete leads to delay or low the thermal conductivity properties. The particle size in geopolymer concrete has also influenced the heat transfer or convection in the thermal insulation system where there is no convection if the particle size is less than 4 mm. Hence, the mean size of the glass bubble (66.68 μm) can be assumed to have no convection heat, which this condition made the heat transfer difficult because of the heat absorbed by the glass bubble as illustrated in Fig. 3.

2.2 Chemical Composition Analysis

The chemical composition of fly ash and glass bubble, as analyzed by X-Ray Fluorescence (XRF), are shown in Table 1. According to the present composition that has been examined, SiO_2 and Al_2O_3 showed the most major oxides of fly ash, which is most of the geopolymer source materials show rich in SiO_2, where more than 70% were found. The geopolymer source materials, rich in the chemical composition of SiO_2 and Al_2O_3 are important in synthesizing geopolymerization, which links to strength performance. Based on the chemical composition analysis, thermal insulated materials glass bubble used in this research contains high SiO_2, ClO_2 and Na_2O, which is more than 70%. All the chemical composition of geopolymer materials is good enough to be applying in coating materials.

The major constituents of fly ash include substantial amounts of SiO_2 and Al_2O_3 more than 80% ($SiO_2 + Al_2O_3 = 84\%$), where the percentage of CaO was less than 10%. As referred to ASTM C618, this fly ash is classified as Class F fly ash or low calcium fly ash. In contrast, the high content of SiO_2, Na_2O and ClO_2 is shown

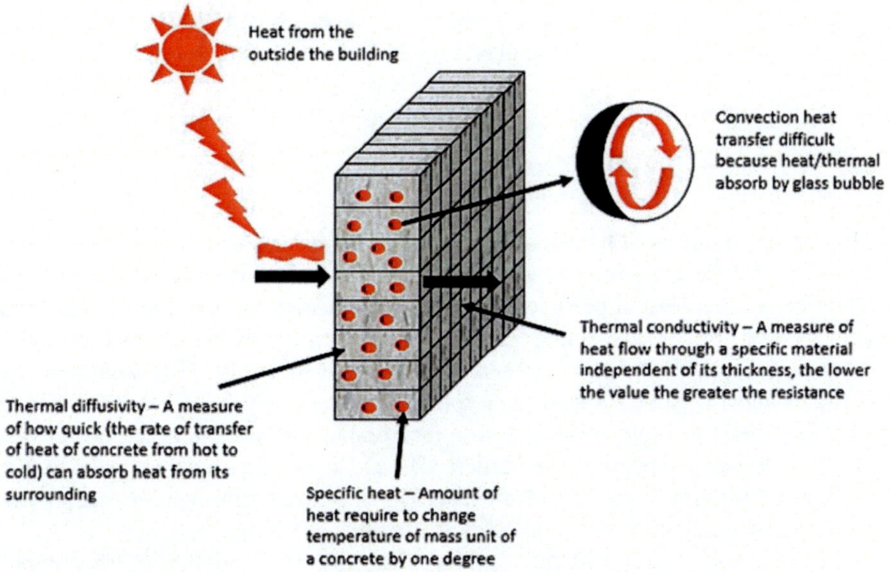

Fig. 3 The illustration of convection heat and the function of thermal insulation properties

Table 1 Chemical composition of raw materials fly ash and glass bubble

Chemical composition	Fly ash (%)	Glass bubble (%)
SiO_2	55.9	68.5
Al_2O_3	28.1	–
CaO	3.84	0.19
Fe_2O_3	6.97	0.53
Na_2O	–	9.90
ZrO_2	0.14	0.09
TiO_2	2.21	0.15
K_2O	1.55	0.92
V_2O_5	0.09	–
ClO_2	–	15.6
SO_3	–	2.66
RuO_2	–	0.77

in glass bubble by chemical composition analysis. The glass bubble or soda-lime glass results of approximately 75% SiO_2 and Na_2O which are also similar findings obtained by Yun et al. [7].

As the glass bubble is expected to remain in its original spherical shape, the chemical composition does not influence much on the concrete matrix. However, the main chemical composition of the glass bubble is Si, which also the main backbone of the geopolymerization process. Thus, some properties will be affected because of the reaction between Si from the glass bubble and the geopolymer binder.

2.3 Morphology Analysis

Morphology analysis of fly ash in Fig. 4 has illustrated a series of spherical vitreous particles of different sizes like a shell structure. Most of the visually distinguishable fine solid spherical particles have a shell structure formed. Due to the inflatable melting drop, these spherical particles are formed by gases released by the combusting gas which has also been found by Alehyen et al. [14]. Agglomerated particles and irregularly shaped amorphous particles may have been due to inter particle contact or rapid cooling during the combustion process at the power plant [15]. In another perspective, the particle size and shape are important characterizations in contributing to the pozzolanic reactivity of fly ash performance in geopolymer concrete, as mentioned by Abdullah et al. [16].

Figure 5 shows the SEM micrograph of the raw glass bubble where it depicts spherical particles at different sizes and can be described as a spherical hollow shape. The size of the spherical particles is ranged from 16 to 34 μm. The raw material glass bubble typically consists of a larger spherical hollow shape compared to raw material fly ash.

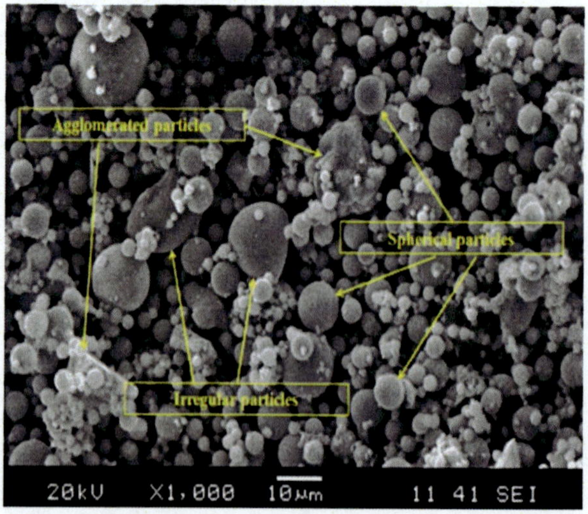

Fig. 4 SEM micrograph analysis of raw material fly ash

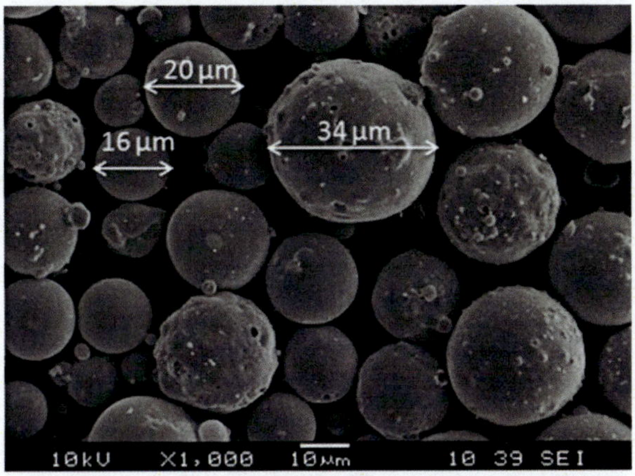

Fig. 5 SEM micrograph and EDX analysis raw material glass bubble

2.4 *Elemental Analysis*

Conventionally, EDX is a powerful technique allowing an elemental analysis associated with electron microscopy based on the generation of characteristic X-rays that reveals the presence of elements present in the samples' surface. EDX analysis of fly ash in Fig. 6a reveals that fly ash contains Al and Si peaks, moderate Ca and Fe peaks, and a minimal Ca peak similar to the result found by Kutchko and Kim [15]. Meanwhile, Fig. 6b shows the glass bubble's predominant elements are Si and Cl with a moderate peak of Na. The EDX spectrum results show that the chemical composition obtained from XRF analysis can be used to distinguish between glass bubbles and fly ash because both have similar spherical particle shapes.

2.5 *Functional Group Identification Analysis*

The transmission mode of Fourier-Transform Infrared Spectroscopy (FTIR) spectra for raw materials fly ash and glass bubble is shown in Figs. 7 and 8. The infrared spectroscopic result is tabulated in Table 2 for molecular vibrations of different bonds present in fly ash and glass bubble with their possible assignment. The infrared spectroscopic result for fly ash shows transmission bands at 3415, 1660, 1519, 1025 and 734 cm^{-1} (main peaks are at 1025 and 734 cm^{-1}). The IR spectrum of glass bubble also shows the transmission bands at 3278, 1649, 1448 and 980 cm^{-1} (main peaks are at 1448 and 980 cm^{-1}).

The IR spectrum of fly ash bands at 3415 and 1660 cm^{-1}, while for the glass bubble, it bands at 3278 and 1649 cm^{-1} corresponded to O–H vibration. The peak appears around 1519 cm^{-1} for fly ash and 1448 cm^{-1} for the glass bubble attributed to the weak bond of O–C–O stretching carbonates which vanished after geopolymerization reaction. A peak at 1025 cm^{-1} for fly ash corresponded to Si–O and Al–O

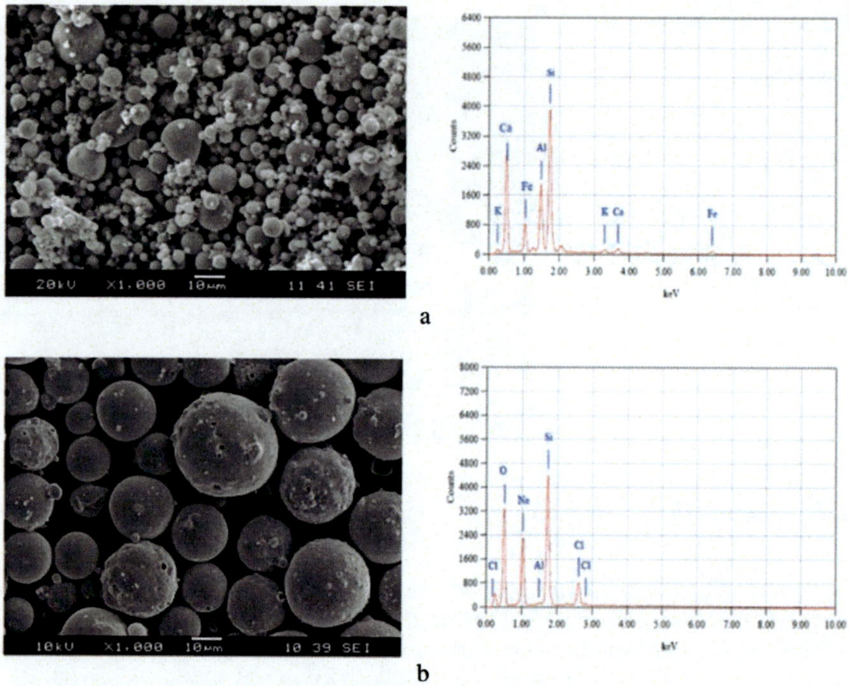

Fig. 6 EDX analysis of **a** raw material fly ash, **b** raw material glass bubble

bonds while band at 980 cm^{-1} refers to the glass bubble. Another band of fly ash at 734 cm^{-1} was attributed to the Al–O–Si in symmetric bending.

The functional group identification of fly ash and glass bubble relates to bending and rocking of bonds in the fly ash and glass bubble used as raw materials of geopolymer synthesis. FTIR has demonstrated a scalable spectrum to allow direct comparison of the spectrum with measured intensity varies due to the radiation variation of the two raw materials.

3 Mechanical Analysis of Geopolymer Concrete Performance with Replacement Volume Fraction of Glass Bubble

This section discusses the measured data in optimizing the proportion of glass bubble in geopolymer concrete that satisfies the minimum strength requirement of loading bearing structure starting with workability, density, compressive strength, water absorption, microstructure, identification functional group, and thermal insulation properties.

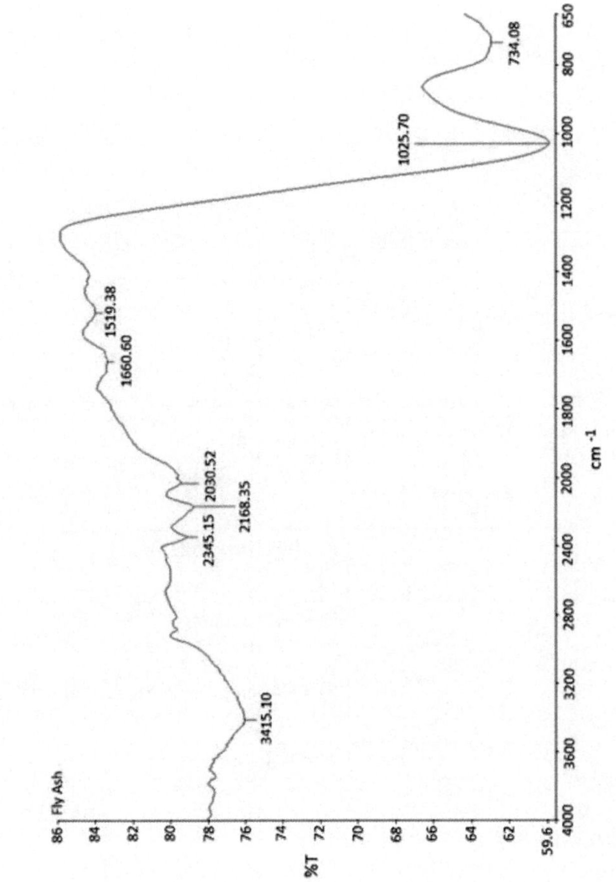

Fig. 7 Spectra of fly ash

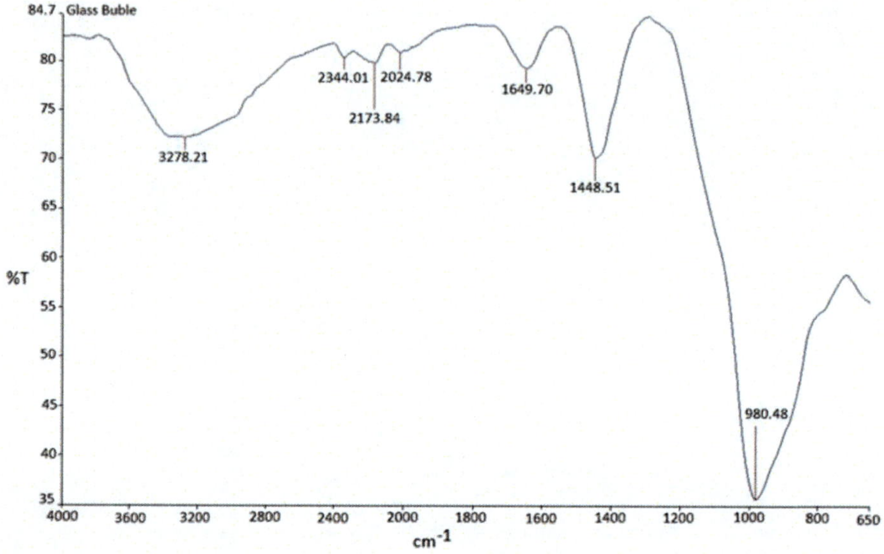

Fig. 8 Spectra of glass bubble

Table 2 Important IR bonds of fly ash and glass bubble with their possible assignment

Wave number (cm^{-1})	Fly ash	Glass bubble	Assignment
3415	✓		– OH (stretching)
3278		✓	
1660	✓		H–O–H (bending)
1649		✓	
1519	✓		O–C–O streching (Carbonates)
1448		✓	
1025	✓		Si–O–Si and Al–O–Si (Asymmetric streching)
980		✓	
734	✓		Al–O–Si (stretching)

3.1 *Workability*

Regarding workability performance, fresh concrete was found to be applicable until 30% of the glass bubble replacement in geopolymer concrete. This is due to the volume increase of solid part compare to liquid part. The workability of geopolymer concrete with glass bubble replacement from 0 to 30% was found to range between 60 and 100 cm and the results showed an increase of almost 16% of geopolymer concrete without glass bubbles (Fig. 9).

By comparing with and without replacement, the volume fraction of the glass bubble significantly affects the workability of the concrete from more slurry to poor

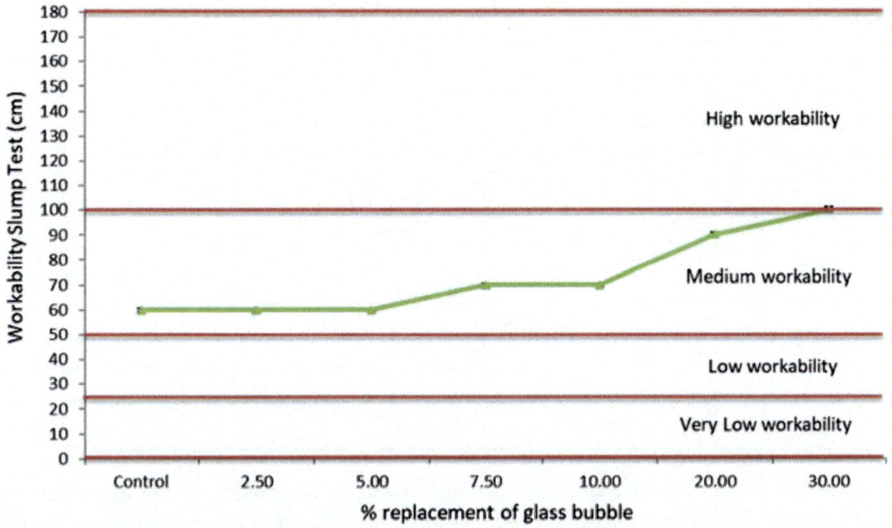

Fig. 9 Measured slumps of geopolymer concrete with replacement volume fraction of glass bubble (0, 2.5, 5.0, 7.5, 10, 20 and 30%)

slurry. This result may contribute to the increased friction between the glass bubble and geopolymer concrete. Increasing the solid part proportion of the glass bubble in the geopolymer concrete will cause an increase in viscosity. Subsequently, it decreased as the flow of the fresh mixture is hindered by high friction between particles. Besides, the spherical shape of fly ash, glass bubble particles, and fineness have beneficial effects on the workability. The shape reduces the friction binder, produces a ball-bearing effect at the point of contact and allows the concrete to move more freely. This justification is also agreed by Ranjbar et al. [17].

3.2 *Density*

The density results are shown in Fig. 10 where the density was decreased from 2247, 2242, 2234, 2231, 2165, 2128 and 2118 kg/m^3 with replacement of 0, 2.5, 5.0, 7.5, 10, 20 and 30% volume fraction of glass bubble. From the results, the lowest density was recorded at 30% replacement while normal concrete density is around 12%. The upper and lower bound are calculated from the calculation of rule mixture by Wang and Davidson [18] where it compares whether measured or experimental result lies between the predicted results or outliers from the predicted results. It also can be used as guidelines on how much the experimental results outliers from the prediction results.

The density of concrete typically depends upon the volume fractions of constituent materials and their volume of voids present in the concrete. The increasing percentage

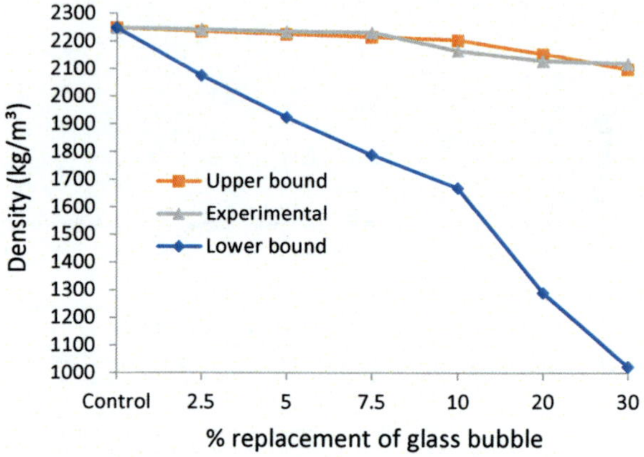

Fig. 10 Measured density (experimental) with upper bound assumption indicated for geopolymer concretes with a replacement volume fraction of glass bubble (0, 2.5, 5.0, 7.5, 10, 20 and 30%)

of glass bubble in the concrete eventually lowered the density of concrete. Air void or bubbles produced by air entrainment from their closed-shell structure was also the reason for decreased density in this study.

The experimental result was recorded far from the lower bound or minimum value of predicted density and nearly lightweight. The density of geopolymer concrete is almost near to lightweight concrete due to the percentage of glass bubble reacting with a geopolymer binder. Density properties also play a significant role in lowering the thermal conductivity of concrete. A decrease in the density of 100 kg/m^3 may decrease the thermal conductivity of 0.04 W/mK [19]. It is found that the variation replacement of the glass bubble might affect the density in concrete, which also affects the thermal insulation behaviour of concrete.

3.3 Compressive Strength

The difference in compressive strength can be calculated from the gap between experimental results and the predicted results (lower bound). The amount of glass bubble incorporated into the geopolymer concrete is limited to 7, 13, 19, 22, 39 and 50%, with the glass bubble's total volume fraction of 2.5, 5.0, 7.5, 10, 20 and 30% is selected, respectively. The replacement on the effect of volume fraction of glass bubble (0, 2.5, 5.0, 7.5, 10, 20 and 30%) on compressive strength with respect to the density of concrete for 28 days is described in Fig. 11.

The result showed that the highest compressive strength was obtained at 71.4 MPa (control geopolymer concrete) and the lowest strength was recorded at 16.4 MPa with 30% of glass bubble replacement. The difference strength between control and with the volume fraction of glass bubble replacement is about 55 MPa. From the result, the optimum replacement of the glass bubble in geopolymer concrete satisfies the minimum strength requirement of a load-bearing structure by ASTM, which is >17 MPa [20].

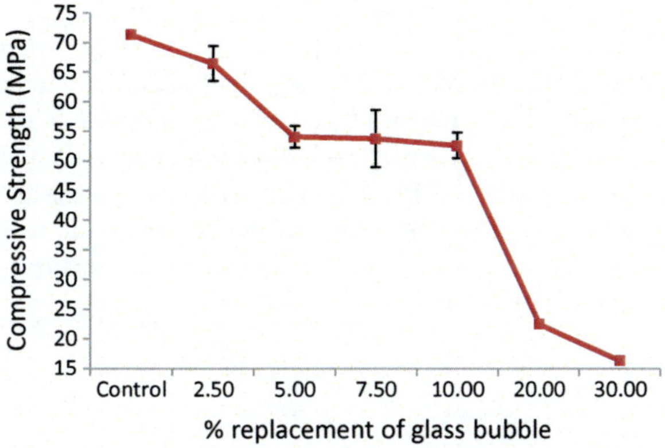

Fig. 11 Compressive strength of geopolymer concrete with replacement volume fraction of glass bubble (0, 2.5, 5.0, 7.5, 10, 20 and 30%)

The measured result shows that the geopolymer concrete matrix that holds the glass bubble will decrease. The glass bubble will get loose from the matrix without changing deformation due to the compression load, causing the compression strength of composite to decrease. Most of the studies directly correlate properties such as density, compressive strength and thermal conductivity as the key factors determining the insulation application [21–24]. Therefore, it is essential to identify the capability of compressive strength in the insulation application so that the development of geopolymer concrete will promote better strength at the same time good in thermal insulation properties.

The analysis of compressive strength with different glass bubble replacement has also been studied based on the effect of aging days (28, 60, and 90 days). The results obtained in this study are important to analyze the properties of glass bubble in geopolymer concrete in long-term effects. The result in Fig. 12 shows the measured result of compressive strength has increased with aging time due to the mechanical properties and long-term durability of geopolymer concrete, similar results as found

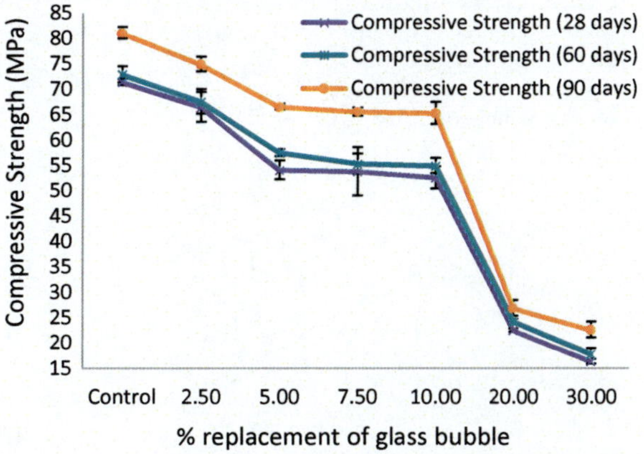

Fig. 12 Compressive strength of geopolymer concrete with replacement volume fraction of glass bubble (0, 2.5, 5.0, 7.5, 10, 20 and 30%) with increment of aging days

by Olivia and Nikraz [25]. In one case by Hardjito et al. [26], geopolymer concrete significantly increased with the age where the behavior and strength of geopolymer concrete are seemed to have a similar standard to existing concrete.

The development of geopolymer concrete is suitable for load-bearing structure with the presence of glass bubble for insulation application since the optimal volume fraction glass bubble in geopolymer concrete is set to 10% (65.25 MPa for 90 days). The strength of geopolymer concrete is even improved with the compressive strength of existing insulation, such as expanded perlite (20.68 MPa) and palm oil shell (27.17 MPa).

3.4 *Water Absorption*

The replacement volume fraction of glass bubble from 0, 2.5, 5.0, 7.5, 10, 20 and 30% into geopolymer concrete has shown the ability of concrete to absorb water. Figure 13 shows the results of water absorption for geopolymer concrete with the replacement volume fraction of glass bubble. The result shows that water absorption of geopolymer concrete is 3.55 to 3.95% for 0 and 30% replacement with an increment of 10% absorption.

In this study, water absorption represents the main mechanism for water and water vapour transport in concrete. Usually, water absorption was performed to evaluate the water absorption characteristics of concrete to increase their life span of the structure. From the result, water absorption was slightly increased with the percentage of air voids from glass bubble in geopolymer concrete but still within the comparable percentage 3–6.5%, similarly observed by Zmijewski and Sokołowski [27]. Luhar and Khandelwal [28] indicate that the water absorption of geopolymer concrete is less compared to control concrete. That means the addition of glass bubble could maintain the adequate performance of the water absorption of geopolymer concrete.

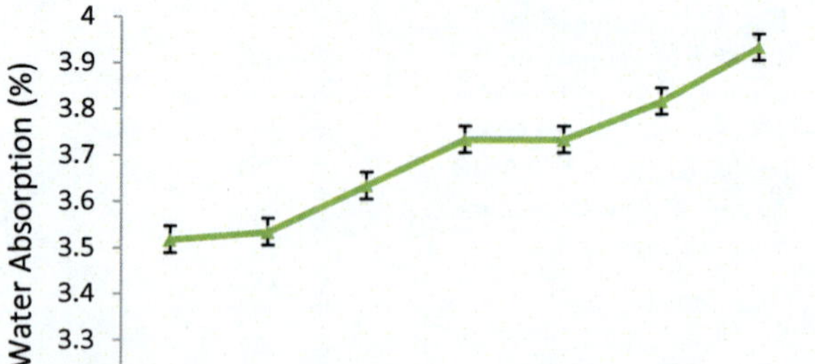

Fig. 13 Comparison of water absorption between geopolymer concrete with replacement volume fraction of glass bubble (0, 2.5, 5.0, 7.5, 10, 20 and 30%)

It is evaluated that increasing glass bubble content would reduce the other constituent such as aggregates, which eventually reduce the capillary voids and decrease water absorption. Besides, a slight increase in water absorption may also contribute to the spherical shapes of fly ash that may block the capillary voids, thus preventing the water from transport or slowing down the water progress.

3.5 Morphology Analysis

The morphology analysis presented the microstructure image of raw materials and geopolymer matrix with the presence of glass bubble by using Scanning Electron Microscope (SEM). The illustration SEM image in Fig. 14 has zoomed out with 1000× magnification at a scale of 10 μm. Figure 14a, b show the particles of fly ash and glass bubble occurs with different size and mostly with closed-cell spherical particles with the appearance of bright crystals [16, 29]. According to Shahedan et al. [30] fly ash contains smaller particles compared to the glass bubble particle. In Fig. 14c, fly ash is well embedded, connected to the matrix and consists of a non-porous microstructure. The figure also demonstrated a homogenous geopolymer matrix with a completely geopolymerization process as it the best ratio of solid/liquid and Na_2SiO_3, as obtained by Hardjito et al. [26], Abdullah et al. [16] and Sumajouw et al. [31]. According to Wang et al. [32], the non-porous microstructure is a result of the extensive dissolution of aluminosilicate species that occurs before polycondensation commences. Thus, by enhancing the dissolution of fly ash particles and polycondensation of aluminosilicate compound, it has created an excellent bonding between these two elements.

Besides, with replacement 30% of glass bubble, Fig. 14d shows the presence of glass bubble was bonded with a geopolymer matrix and still retains the shape of closed spherical particle and finally contributes to the better enhancement of thermal properties of geopolymer concrete. The glass bubble maintained their properties of hollow spherical shape in the geopolymer matrix, which could be observed in Fig. 14e as this property needs due to the existence of interior gas core lead to low thermal conductivity.

The SEM image in Fig. 15a illustrates the glass bubble with the hollow structure properties required to improve thermal insulation properties. The figure also presents a closed pore of spherical shape glass bubble which still maintains its shape about 85% from the total volume fraction in the concrete, comparable with the study by Liu et al. [33]. However, the SEM image in Fig. 15b revealed that there is in termination of the geopolymer matrix into the glass bubble. This can be clarified from the data analysis that some percentage of the glass bubble has reacted with a geopolymer matrix since the main backbone of geopolymerization process is Si, which is also the main component of glass bubble [14]. Figure 15c shows the idea of how glass bubble should work if the glass bubble does not terminate in geopolymer concrete.

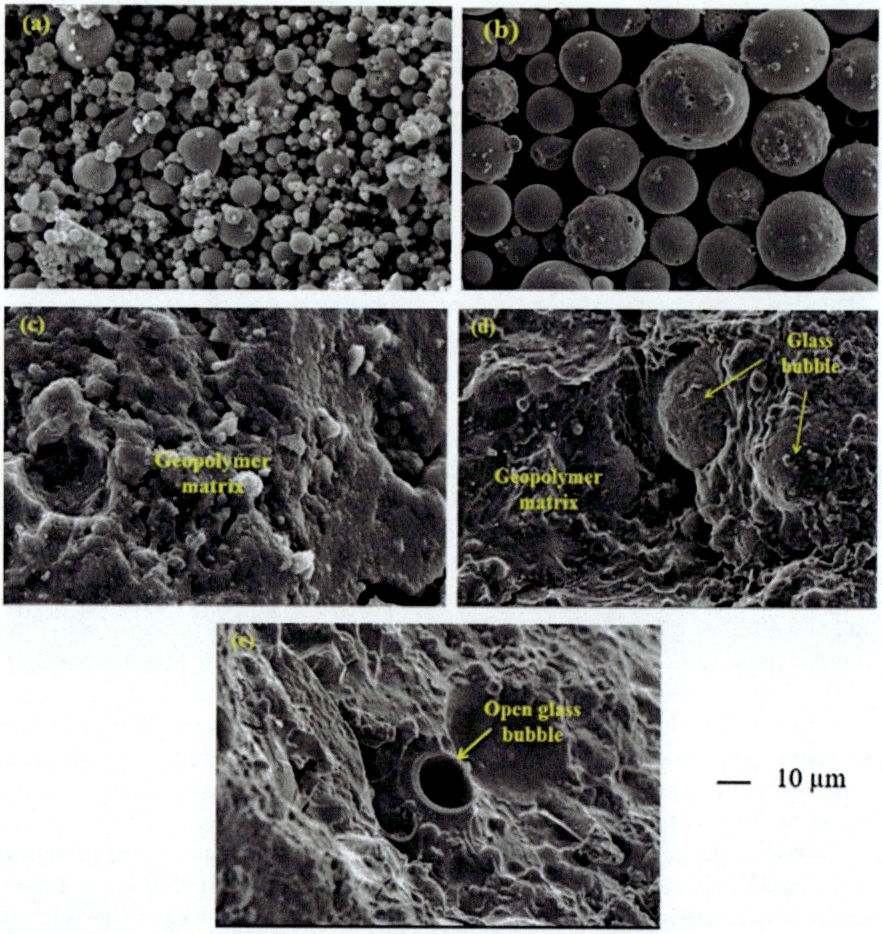

Fig. 14 SEM images of geopolymer concrete samples with glass bubble content: **a** raw material fly ash, **b** raw material glass bubble, **c** geopolymer concrete without glass bubble, **d** geopolymer concrete with 30% of glass bubble and **e** cross-section geopolymer concrete with glass bubble

3.6 Functional Group Identification Analysis

The functional group identification of fly ash geopolymer concrete is carried out to identified the change of bonds and direct comparison of spectra with different measured intensities due to variations in the beam with the replacement volume fraction of glass bubble. The replacement volume fraction of glass bubble on geopolymer concrete is typical remained in the spectra, but the appearance bands have slightly changed due to the presence of glass bubble in the geopolymer concrete matrix. The infrared spectra of fly ash geopolymer concrete with the replacement volume fraction of glass bubble as insulation material in the range of 600–4000 cm^{-1} are

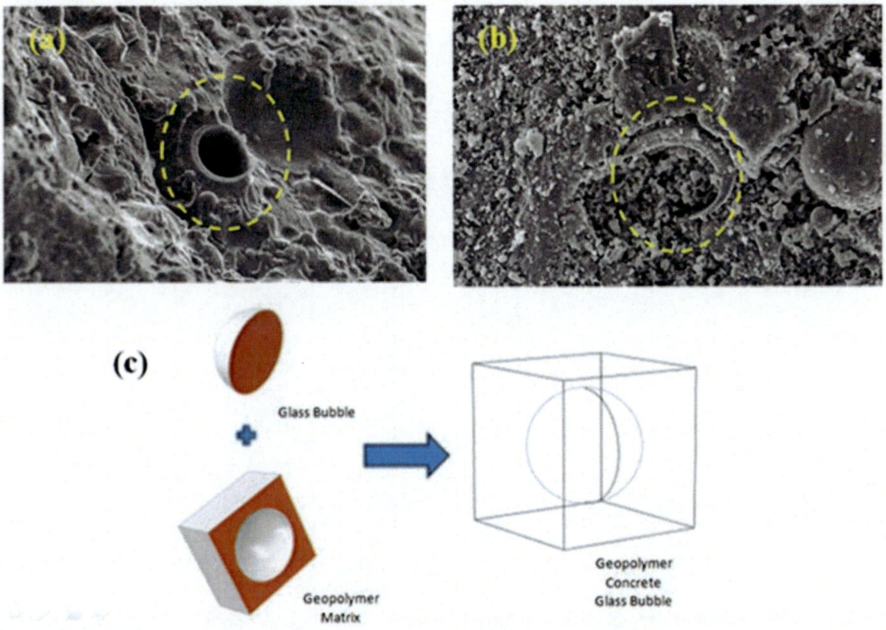

Fig. 15 SEM image of cross section geopolymer concrete **a** with open pore glass bubble, **b** with interminate pore glass bubble and **c** illustration glass bubble should be in geopolymer concrete

presented in Fig. 16. The spectrum of absorption peaks of geopolymer concrete with replacement glass bubble (0, 2.5, 5.0, 7.5, 10, 20 and 30%) contains absorption bands 3800–3600 cm^{-1} and around 1600 cm^{-1} correlated with the O–H vibration.

The strongest vibration centred is nearly at 1000 cm^{-1}, referred to as a major geopolymer fingerprint, which is the asymmetric stretching of Si–O–Si (Al) bridges. The wavenumber in geopolymer concrete with replacement glass bubble (0, 2.5, 5.0, 7.5, 10, 20 and 30%) IR spectra within 600 and 1000 cm^{-1} show a slight difference and shifted wavelength compared to the IR spectra of geopolymer concrete without glass bubble due to the presence of glass bubble in replacement of volume of geopolymer concrete. The details of the IR spectra of geopolymer concrete with replacement glass bubble (0, 2.5, 5.0, 7.5, 10, 20 and 30%) are shown in Table 3.

Due to the reduction in the volume fraction of geopolymer concrete matrix that holds or bind the materials, it could lower the micro-reaction inside the system, which leads to the lower density and strength of the products [34]. The shift in wavelength of the samples with the replacement of glass bubble (0, 2.5, 5.0, 7.5, 10, 20 and 30%) was attributed to the low chemical bond of the geopolymer concrete matrix. It should be noted that the relative intensity of the bands around 1600 and 600 cm^{-1} subjected to 20 and 30% glass bubble replacement eventually turned to disappear, as shown by the reduction of compressive strength of the geopolymer concrete.

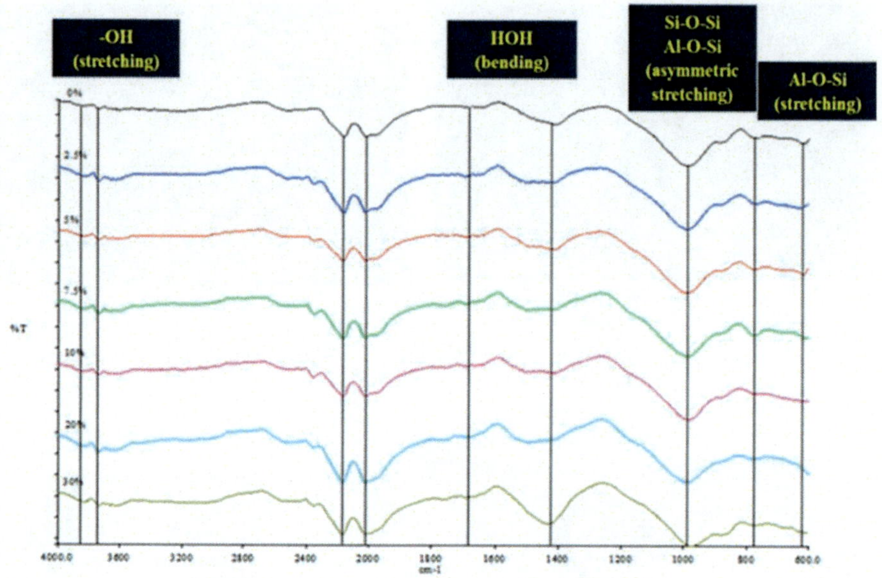

Fig. 16 IR spectra of geopolymer concrete with replacement glass bubble (0, 2.5, 5.0, 7.5, 10, 20 and 30%)

4 Thermal Insulation Analysis of Geopolymer Concrete with Substitution of Glass Bubble

Thermal insulation properties of concrete are relevant to the insulation application, especially when it relates to the structure with desired low thermal conductivity and thermal diffusivity, high specific heat and capability of heat losses or gain through the building envelope. The measured result for thermal properties includes specific heat, thermal conductivity and thermal diffusivity in geopolymer concrete that from different volume fraction of glass bubble will be discussed in detail in this section.

4.1 Thermal Conductivity

The results thermal conductivity of geopolymer concrete with standard deviations indicated by error bars are presented in Fig. 17 for different replacement of insulation materials at room temperature. The control geopolymer concrete (without replacement of glass bubble) recorded thermal conductivity of 1.65 W/mK with a density of 2247 kg/m^3. With increasing glass bubble content, the thermal conductivity of the concrete was increased in their performance. The thermal conductivity decreases to 1.473, 1.406, 1.347, 1.196, 1.337 and 1.312 W/mK for 2.5, 5.0, 7.5, 10.0, 20.0 and 30.0% of glass bubble respectively and maximum decrement is 18.8%. As we can

Table 3 Main IR spectra absorption peaks of geopolymer concrete with replacement glass bubble (0, 2.5, 5.0, 7.5, 10, 20 and 30%)

% replacement of glass bubble	Wave number (cm^{-1})					
	– OH (stretching)	C=C (variable intensity)	HOH (bending)	Si–O-Si Al–O–Si (asymmetric stretching)	Al–O–Si (stretching)	Al–O–Si (symmetric bending)
0	3837, 3602	2167, 2024, 1412	1685	989	769	622
2.5	3837, 3738	2164, 2025, 1505	1690	992	777	626
5.0	3837, 3739	2166, 2025, 1421	1695	991	779	620
7.5	3838, 3739	2165, 2025, 1505	1697	989	780	615
10	3837, 3739	2164, 2022, 1421	1697	987	781	610
20	3820, 3733	2168, 2024, 1496	–	992	778	–
30	3841, 3733	2168, 2022, 1435	–	974	775	–

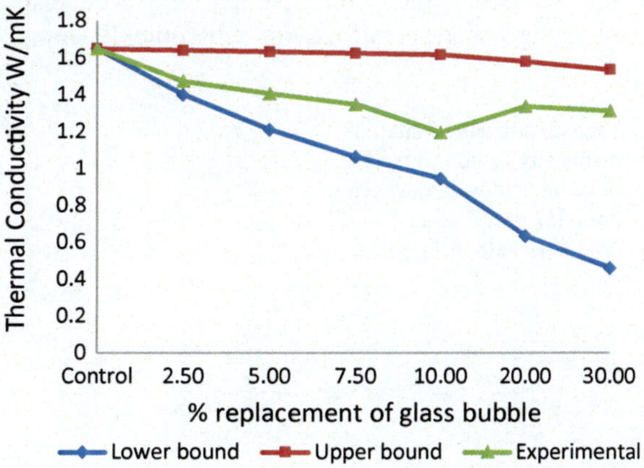

Fig. 17 Measured thermal conductivities (with standard deviations indicated by error bars) for geopolymer concretes with different % replacement of glass bubble

see, the measured result of thermal conductivity lies in between the upper and lower bound as per expected.

The results revealed that the effective thermal conductivity of geopolymer concrete decreases with the increased volume content of glass bubble. Saygili and Baykal [35] have stated that the decrease in thermal conductivity is due to the increase of void ratio that decreased the unit weight of concrete [35]. Besides, the thermal conductivity of the glass bubble itself helps the improvement of thermal conductivity of the geopolymer concrete where thermal conductivity of glass bubble is 0.044 W/mK while thermal conductivity of glass bubble of the solid wall recorded as 1.03 W/mK, as found by Wang et al. [36]. Furthermore, lower themal conductivity is also attributed to the existence of an interior gas core in glass bubble (proved by morphology analysis in Sect. 3.5) since air is the poorest conductor compared to solid and liquid due to its molecular structure which also agreed by Ng and Low [37].

4.2 Thermal Diffusivity

The results of thermal diffusivity of geopolymer concrete with standard deviations are indicated by error bars and presented in Fig. 18. The results show that increasing replacement of the glass bubble would decreased thermal diffusivity from 1.881 to 1.541, 1.310, 1.204, 1.016, 1.008 m^2/s. While the control sample without replacement of glass bubble was recorded at 0.998 m^2/s of thermal diffusivity. Compared to the control sample, thermal diffusivity with a maximum replacement of 30% of glass bubble (thermal diffusivity = 0.998 m^2/s) was seen to decrease 46%.

The results have indicated that diffusivity decreased by increasing glass bubble replacement, which consequently improved insulation behaviour [38]. Concrete with high thermal diffusivity will quickly adjust its temperature to match with

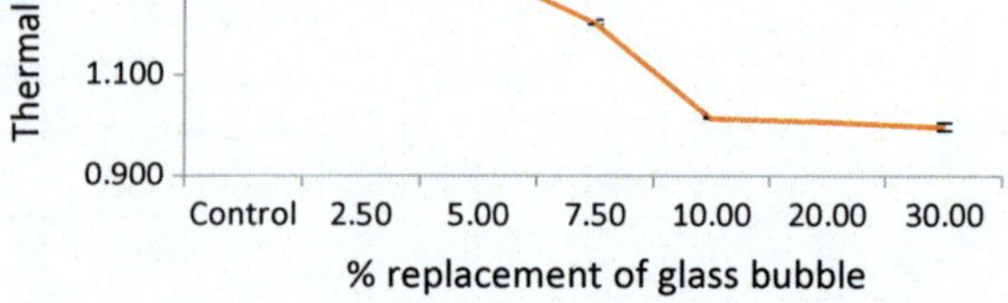

Fig. 18 Measured thermal diffusivity (with standard deviations indicated by error bars) for geopolymer concretes with different % replacement of glass bubble

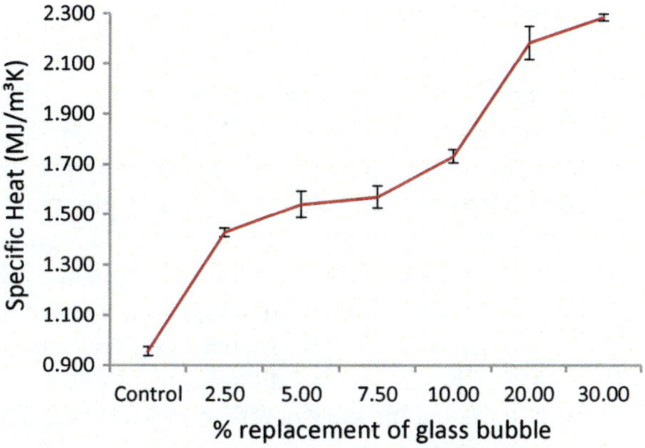

Fig. 19 Measured specific heat (with standard deviations indicated by error bars) for geopolymer concretes with different % replacement of glass bubble

the surroundings. Thus, low thermal diffusivity is preferred to buffer temperature experienced during extremes or hash conditions [39].

The obtained result of thermal diffusivity is also influenced by the thermal insulation properties of the concrete [40, 41]. Thus, increasing replacement volume fraction of glass bubble as constituents in geopolymer concrete has improved the thermal diffusivity which also agreed by Bouguerra et al. [42]. Meanwhile, Bouguerra et al. [42] stated that thermal conductivity and diffusivity are decreased while the heat capacity increases with the increasing volume fraction of thermal insulation material in building materials.

4.3 Specific Heat

The results of specific heat of geopolymer concrete with standard deviations are indicated by error bars and presented in Fig. 19. The measured specific heat of control (without replacement of glass bubble) was 0.955 MJ/m^3K for 2247 kg/m^3 while 30% replacement of glass bubble was 2.284 MJ/m^3K for 2118 kg/m^3 with increment at 58%.

Specific heat of geopolymer concrete has increased with the presence of glass bubble as a thermal insulating material due to the existence of an interior gas core, which can absorb more heat inside before transfer through the geopolymer concrete itself. Fly ash as raw material also has little influence on the measured specific heat [43]. When compared to normal concrete, it only achieved a specific heat at 1.0 MJ/m^3K for 2300 kg/m^3 density. Insulating materials with high specific heat efficiently help in the passive energy storage system, which can benefit both the building owner and the energy supplier. Besides, high specific heat for building materials does not need constant heat supply, as high specific heat it takes a long time for the stored and leaks out heat in buildings [44].

5 Conclusion

The general aim is to develop geopolymer concrete with a glass bubble for thermal insulation application. The focus is to obtain an optimum formulation of replacement volume fraction of the glass bubble in geopolymer concrete that satisfies the minimum strength requirement of the load-bearing structure. From the results and discussion obtained, the measured result has obtained optimum at 10% of volume replacement because it achieved minimum strength at 52.587 MPa (28 days), 54.916 MPa (60 days) and 65.250 MPa (90 days) with measured results of workability at 70 cm, water absorption at 3.73%, thermal conductivity at 1.196 W/mK, specific heat at 1.731 MJ/m^{3}K, and thermal diffusivity at 1.016 mm^2/s. It was shown that the replacement of glass bubble only could be replaced at 10% with most applicable to satisfy the minimum strength requirement of the load-bearing structure.

From the results obtained, the decrease in the thermal conductivity and thermal diffusivity is due to the increase of the void ratio from the glass bubble plus the thermal conductivity of the glass bubble itself, which helps improve the thermal properties of the concrete. The existence of interior gas core of glass bubble contributed to the high specific heat, which leads to more heat to be absorbed and the longer time is taken for the stored the heat before leaking out heat in buildings. For the insulating construction materials, the lower thermal conductivity, thermal diffusivity and higher specific heat promoting the better thermal properties. So that less energy required for the building to cool down and consequently save energy and cost. The optimum proportion obtained in the present work shows the potential for application of load-bearing structure for the precast industry.

References

1. Liang JZ, Li FH (2006) Measurement of thermal conductivity of hollow glass-bead-filled polypropylene composites. Polym Test 25(4):527–531. https://doi.org/10.1016/j.polymertesting.2006.02.007
2. Oreshkin D, Semenov V, Rozovskaya T (2016) Properties of light-weight extruded concrete with hollow glass microspheres. Procedia Eng 153:638–643. https://doi.org/10.1016/j.proeng.2016.08.214
3. Ren S, Liu J, Guo A, Zang W, Geng H, Tao X, Du H (2016) Mechanical properties and thermal conductivity of a temperature resistance hollow glass microspheres/borosilicate glass buoyance material. Mater Sci Eng A 674:604–614. https://doi.org/10.1016/j.msea.2016.08.014
4. Shahidan S, Aminudin E, Noor K, Hannan NIR, Bahari MS (2017) Potential of hollow glass microsphere as cement replacement for lightweight foam concrete on thermal insulation performance. MATEC Web Conf 103(1014):1–9. https://doi.org/10.1051/matecconf/201710301014
5. Swetha C, Kumar R (2011) Quasi-static uni-axial compression behaviour of hollow glass microspheres/epoxy based syntactic foams. Mater Des 32(8):4152–4163. https://doi.org/10.1016/j.matdes.2011.04.058
6. Peres CYU, Munhoz AH, Miranda LF, Neto AC, Zandonadi AR, Peres RM, Diaz FRV (2015) Concrete with addition of hollow glass microspheres. Mater Sci Forum 820:509–514. https://doi.org/10.4028/www.scientific.net/MSF.820.509

7. Yun TS, Jeong YJ, Han TS, Youm KS (2013) Evaluation of thermal conductivity for thermally insulated concretes. Energy Build 61:125–132. https://doi.org/10.1016/j.enbuild.2013.01.043
8. Yung KC, Zhu BL, Yue TM, Xie CS (2009) Preparation and properties of hollow glass microsphere-filled epoxy-matrix composites. Compos Sci Technol 69(2):260–264. https://doi.org/10.1016/j.compscitech.2008.10.014
9. Zhu BL, Ma J, Wang J, Wu J, Peng D (2012) Thermal, dielectric and compressive properties of hollow glass microsphere filled epoxy-matrix composites. J Reinf Plast Compos 31:1311–1326. https://doi.org/10.1177/0731684412452918
10. Zaman AU (2014) Measuring waste management performance using the 'Zero Waste Index': the case of Adelaide, Australia. J Clean Prod 66:407–419. https://doi.org/10.1016/j.jclepro.2013.10.032
11. Zhang Z (2014) The effects of physical and chemical properties of fly ash on the manufacture of geopolymer foam concretes. Dissertation, University of Southern Queenland
12. Aleem A, Arumairaj P (2012) Geopolymer concrete—a review. Int J Eng Sci Emerg Technol 1:118–122. https://doi.org/10.7323/ijeset/v1_i2_14
13. Chindaprasirt P, Jaturapitakkul C, Chalee W, Rattanasak U (2009) Comparative study on the characteristics of fly ash and bottom ash geopolymers. Waste Manag 29(2):539–543. https://doi.org/10.1016/j.wasman.2008.06.023
14. Alehyen S, El Achouri M, Taibi M (2017) Characterization, microstructure and properties of fly ash-based geopolymer. J Mater Environ Sci 8(5):1783–1796
15. Kutchko BG, Kim AG (2006) Fly ash characterization by SEM–EDS. Fuel 85(17):2537–2544. https://doi.org/10.1016/j.fuel.2006.05.016
16. Abdullah MMAB, Kamarudin H, Abdulkareem OA, Ruzaidi C, Razak R, Mohamed Noor N (2012) Optimization of alkaline activator/fly ash ratio on the compressive strength of manufacturing fly ash-based geopolymer. Appl Mech Mater 110–116:734–739. https://doi.org/10.4028/www.scientific.net/AMM.110-116.734
17. Ranjbar N, Behnia A, Alsubari B, Moradi Birgani P, Jumaat MZ (2016) Durability and mechanical properties of self-compacting concrete incorporating palm oil fuel ash. J Clean Prod 112:723–730. https://doi.org/10.1016/j.jclepro.2015.07.033
18. Wang X, Davidson N (2010) The upper and lower bounds of the prediction accuracies of ensemble methods for binary classification. In: International conference on machine learning and applications, pp 373–378
19. Budaiwi I, Abdou A, Al-Homoud M (2002) Variations of thermal conductivity of insulation materials under different operating temperatures: Impact on envelope-induced cooling load. J Archit Eng 8(4):125. https://doi.org/10.1061/(ASCE)1076-0431(2002)8
20. American Society for Testing and Materials (ASTM) (2011) Standard specification for loadbearing concrete masonry. West Conshohocken, The USA ASTM C90–11
21. Abdou A, Budaiwi I (2005) Comparison of thermal conductivity measurements of building insulation materials under various operating temperatures. J Build Phys 29(2):171–184. https://doi.org/10.1177/1744259105056291
22. Abu-Jdayil B, Mourad A-H, Hittini W, Hassan M, Hameedi S (2019) Traditional, state-of-the-art and renewable thermal building insulation materials: an overview. Constr Build Mater 214:709–735. https://doi.org/10.1016/j.conbuildmat.2019.04.102
23. Al-Ajlan SA (2006) Measurements of thermal properties of insulation materials by using transient plane source technique. Appl Therm Eng 26(17):2184–2191. https://doi.org/10.1016/j.applthermaleng.2006.04.006
24. Kontoleon KJ, Theodosiou TG, Tsikaloudaki KG (2013) The influence of concrete density and conductivity on walls' thermal inertia parameters under a variety of masonry and insulation placements. Appl Energy 112:325–337. https://doi.org/10.1016/j.apenergy.2013.06.029
25. Olivia M, Nikraz H (2011) Strength and water penetrability of fly ash geopolymer concrete, 6
26. Hardjito D, Wallah S, Sumajouw D, Rangan B (2004) Factors influencing the compressive strength of fly ash-based geopolymer concrete. Civ Eng Dimens 6
27. Zmijewski K, Sokołowski M (2010) efficiently about the power industry (Part 1). Energ Budynek Energy Build 38(7):12–15. https://doi.org/10.1201/9781420009248

28. Luhar S, Khandelwal U (2015) A Study on water absorption and sorptivity of geopolymer concrete. SSRG Int J Civ Eng
29. Fernández-Jiménez A, Palomo Á, Revuelta D, Torroja E (2009) Alkali activation of industrial by-products to develop new earth-friendly cements. In: 11th International conference on non-conventional materials and technologies (NOCMAT), pp 1–5
30. Shahedan N, Abdullah MMAB, Mahmed N, Kusbiantoro A, Bouaissi A (2018) Characterization of fly ash geopolymer concrete with glass bubble for thermal insulation application. In: 4th international conference on green design and manufacture. AIP conference proceedings, p 20294
31. Sumajouw D, Hardjito D, Wallah S, Rangan B (2007) Fly ash-based geopolymer concrete: study of slender reinforced columns. J Mater Sci 42:3124–3130. https://doi.org/10.1007/s10853-006-0523-8
32. Wang Y, Hu S, He Z (2019) Mechanical and fracture properties of fly ash geopolymer concrete addictive with calcium aluminate cement. Materials (Basel) 12:2982. https://doi.org/10.3390/ma12182982
33. Liu B, Wang H, Qin Q (2018) Modelling and characterization of effective thermal conductivity of single hollow glass microsphere and its powder. Materials (Basel) 11:133. https://doi.org/10.3390/ma11010133
34. Yong H, Kamarudin H, Abdullah MMAB, Binhussain M, Musa L, Nizar K, Ruzaidi C, Ming L (2012) Study on solids-to-liquid and alkaline activator ratios on kaolin-based geopolymers. Constr Build Mater 35. https://doi.org/10.1016/j.conbuildmat.2012.04.102
35. Saygılı A, Baykal G (2011) A new method for improving the thermal insulation properties of fly ash. Energy Build 43(11):3236–3242. https://doi.org/10.1016/j.enbuild.2011.08.024
36. Wang H, Hou F, Zhao X (2015) Simulation of thermal behavior in hollow-glass-microsphere-filled cement composites. Am J Mater Sci 4(1):1–11
37. Ng SC, Low KS (2010) Thermal conductivity of newspaper sandwiched aerated lightweight concrete panel. Energy Build 42(12):2452–2456. https://doi.org/10.1016/j.enbuild.2010.08.026
38. Schiavoni S, D'Alessandro F, Bianchi F, Asdrubali F, (2016) Insulation materials for the building sector: A review and comparative analysis. Renew Sustain Energy Rev 62:988–1011. https://doi.org/10.1016/j.rser.2016.05.045
39. Amos SE (2015) Hollow glass microspheres for plastics, elastomers and adhesives compounds. William Andrew Publishing, Oxford
40. Carman APN (2007) Thermal conductivity and diffusivity of concrete. In: Carman RANAP
41. Nambiar EKR (2008) Models for strength prediction of foam concrete. Mater Struct 41:247–254. https://doi.org/10.1617/s11527-007-9234-0
42. Bouguerra A, Aït-Mokhtar A, Amiri O, Diop MB (2001) Measurement of thermal conductivity, thermal diffusivity and heat capacity of highly porous building materials using transient plane source technique. Int Commun Heat Mass Transf 28(8):1065–1078. https://doi.org/10.1016/S0735-1933(01)00310-4
43. Bentz D, Peltz M, Durán-Herrera A, Valdez P, Juarez C (2011) Thermal properties of high-volume fly ash mortars and concretescxdf.1177/1744259110376613
44. Nagy B, Nehme SG, Szagri D (2015) Thermal properties and modeling of fiber reinforced concretes. Energy Procedia 78:2742–2747. https://doi.org/10.1016/j.egypro.2015.11.616

Characteristic of One-Part Geopolymer as Building Materials

Liew Yun Ming, Ooi Wan En, Heah Cheng Yong, Mohd Mustafa Al Bakri Abdullah, and Ong Shee Ween

Abstract Safety concerns in handling alkali solutions for traditional geopolymers have prompted the development of one-part geopolymer in which solid activators replace activating solution. The characteristics of one-part geopolymer are influenced by the types of material used. Waste materials such as fly ash, blast furnace slag, rice husk ash, red mud, geothermal silica, and albite were employed in one-part geopolymer studies. Solid activators namely sodium hydroxide, sodium silicate, sodium carbonate, sodium oxide, sodium aluminate, calcium hydroxide, potassium carbonate and so on were used to produce dry binders that can initiate geopolymerization when water is added. This chapter clarifies the characteristics of fresh and hardened one-part geopolymer based on materials, water content, additives and thermal treatment involved. The reaction of fabricated one-part geopolymers towards different extreme environments was also explained thoroughly in this chapter.

Keywords One-part geopolymer · Solid activators · Additives · Thermal · Durability

---

L. Y. Ming (✉) · O. W. En · H. C. Yong · M. M. A. B. Abdullah · O. S. Ween
Center of Excellence Geopolymer and Green Technology, School of Materials Engineering, Universiti Malaysia Perlis, 01000 Kangar, Perlis, Malaysia
e-mail: ymliew@unimap.edu.my

O. W. En
e-mail: wanen2ooi@gmail.com

H. C. Yong
e-mail: cyheah@unimap.edu.my

M. M. A. B. Abdullah
e-mail: mustafa_albakri@unimap.edu.my

O. S. Ween
e-mail: ongsheeween@outlook.com

A. Abdul Kadir et al. (eds.), *Sustainable Waste Utilization in Bricks, Concrete, and Cementitious Materials*, Lecture Notes in Civil Engineering 129,
https://doi.org/10.1007/978-981-33-4918-6_6

1 Introduction

The main driving force of inorganic polymer development is the urge to substitute ordinary Portland cement (OPC) as the major building materials due to its negative impact on the environment. Research found out that geopolymer, also known as alkali-activated material, exhibits a wide variety of potential application and possesses similar or better performance than OPC [1–4]. Thereafter, geopolymer has been considered as an alternative to substitute OPC in construction materials. It has been reported that 44–64% of greenhouse gas reduction could be achieved with the replacement of OPC with geopolymer [5]. Most importantly, geopolymer utilized secondary industrial products, which could help to solve waste disposal issues, reduce by-products management cost and minimize waste landfills.

Geopolymer is an inorganic polymer material. Geopolymer is made by aluminosilicates precursor dissolved and condensed in the highly alkalized environment, subsequently an amorphous to semi-crystalline 3-dimensional silico-aluminate material is developed [6]. Aluminosilicates are materials rich in aluminium and silicon that can be obtained from industrials' waste, such as rice husk ash, fly ash, blast furnace slag and red mud. While high pH activators used are usually alkali hydroxide and alkali metal silicate. The production of traditional geopolymer is by the mixing of aluminosilicate materials with a highly alkaline solution, followed by curing at ambient or slightly higher temperature conditions.

Traditional geopolymer (two-part geopolymer) processing involved liquid chemical ingredients that are corrosive and irritant. Hence, there is a great demand for geopolymer to be presented in dry solid form, like OPC, in order to simplify the handling process of geopolymer binders, at the same time, without compromise on the quality. Consequently, research and development have been focusing on the manufacturing of user-friendly geopolymer binders that its geopolymerization reaction can be initiated by just adding water. To date, geopolymer manufactured by waste aluminosilicate materials and dry activators has been proven to have positive potential in construction applications. Solid materials (normally in powder form) are first dry-mixed, then water is added to the dry mix and further mixing before casting and compaction. Geopolymer developed by such a method has been named as "one-part geopolymer". Although the processing manner is different from the traditional geopolymer, the philosophy of geopolymerization reaction is similar. Figure 1 clarifies the differences in process flow between traditional geopolymer and one-part geopolymer.

Same as traditional geopolymer, one-part mixture comprises two primary constituents: aluminosilicate source and dry alkali activators. Major contribution of aluminosilicate materials for geopolymer production is from industrial by-products. Due to variation of sources, waste materials possess different chemical compositions and physical properties, thus prompting one-part geopolymer of different properties. In this approach, water is functioned to dissolve solid alkali activators at the beginning of the reaction to create a high alkalinity environment. Solid aluminosilicate then undergoes dissolution to produce aluminate and silicate species. Consequently,

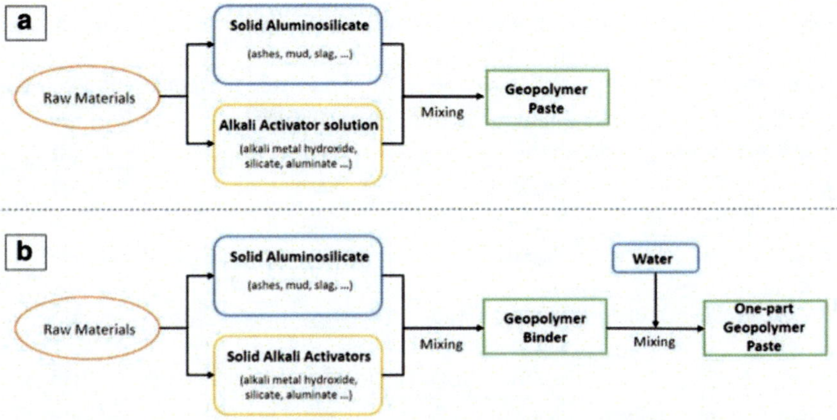

Fig. 1 Processing flow chart of **a** traditional geopolymer; and **b** one-part geopolymer

slurry matter developed. The concentrated geopolymer paste starts to develop into geopolymeric matrix. The system generates oligomers with a large network through condensation phase and releases water that was consumed beforehand. After gelation, the oligomers arrange and connect themselves to form the three-dimensional aluminosilicate network. Resulting gel-like matrix is known as geopolymer [7].

Nevertheless, geopolymeric matrix development varies according to the types of aluminosilicate materials, solid activators, additives, and curing conditions, subsequently give rise to one-part geopolymer with different characteristics. This chapter presents a comprehensive overview of waste materials used for the production of one-part geopolymer mixtures and respective properties.

2 Precursor for One-Part Geopolymer

Various waste materials were used for one-part geopolymer development. Table 1 summarized the literature of raw materials employed for one-part geopolymer studies.

The most common waste materials employed in the manufacturing geopolymer is fly ash. Fly ash is the waste residue obtained from the incineration process of finely ground coal in electric power plants [41]. Fly ash is commonly rich in silicon and aluminium. It can be classified into three (3) classes, which are class N, F and C according to its chemical composition and physical properties. Generally, fly ash used in geopolymer production is of class F which has 70% minimum sum of SiO_2, Al_2O_3 and Fe_2O_3 content [42].

Class F-fly ash as the sole precursor and sodium silicate (Na_2SiO_3) powder as alkali activator was used by Yang et al. [9] to develop one-part geopolymer. However, the resulting properties were not too encouraging. One-part geopolymer

Table 1 Summary of one-part geopolymer binders used in experimental work in accordance to year of study

Geopolymer precursors	Solid activators	Additives, fibers and aggregates	References
Geothermal silica	$NaAlO_2$	N/A	[8]
Fly ash, BFS	Na_2SiO_3	Sand	[9]
Fly ash, BFS	Na_2SiO_3, NaOH	Sand	[10]
BFS	Na_2SiO_3	Expanded clay granules, natural sand	[11]
Fly ash, geothermal silica	NaOH, $NaAlO_2$, amorphous alumina	N/A	[12]
Albite	NaOH, Na_2CO_3	N/A	[13]
Microsilica, silica residue	$NaAlO_2$	N/A	[14]
BFS	Na_2CO_3	Silica fume, slaked lime, dolomite stone and sand, sodium lignosulfonate	[15]
Fly ash, slag, calcium hydroxide	Na_2SiO_3, NaOH	PCE superplasticizer	[16]
Red mud	NaOH	N/A	[17]
Red mud, silica fume	NaOH	Sodium lignosulphonate	[18]
Microsilica, silica residue	$NaAlO_2$	N/A	[19]
Fly ash, slag	Na_2SiO_3	PVA fiber	[20]
Fly ash	NaOH, red mud	N/A	[21]
Fly ash, slag	Na_2SiO_3	Hydrophosphate	[22]
Fly ash	Na_2SiO_3, NaOH	N/A	[23]
RHA	$NaAlO_2$	N/A	[24]
Fly ash, slag	Na_2SiO_3	PVA, PE fiber	[25]
Fly ash, slag	Na_2SiO_3	PVA fibers	[26]
Fly ash, BFS	Na_2SiO_3	N/A	[27]
Fly ash, BFS	Na_2SiO_3	Sand, glass fine	[28]
BFS	Paper sludge, NaOH	N/A	[29]
Silica materials, rice husk ash	$NaAlO_2$	BFS, sand	[30]
Fly ash, slag	Na_2SiO_3, $Ca(OH)_2$, Na_2O,LiOH, K_2CO_3	PCE superplasticizer	[31]
BFS	Na_2SiO_3	Sand, ceramic waste	[32]
Fly ash, BFS	NaOH, KOH, Na_2SiO_3	N/A	[33]

(continued)

Table 1 (continued)

Geopolymer precursors	Solid activators	Additives, fibers and aggregates	References
Fly ash, BFS	Na_2SiO_3	Sodium tetraborate decahydrate-borax, sodium triphosphate, polycarboxylate; sodium gluconate, sodium lignosulphate, calcium lignosulphate	[34]
Fly ash, RHA, municipal solid waste	KOH, Na_2CO_3	Silica sand	[35]
BFS	Na_2SiO_3	SP superplasticizer, sand, fibers (PVA, steel, cellulose, basalt)	[36]
Fly ash, BFS	Na_2SiO_3, Na_2CO_3	N/A	[37]
Fly ash, BFS	Na_2SiO_3	Ceramic waste, natural sand, fibers (PVA, PP, basalt)	[38]
Fly ash, BFS	Na_2SiO_3	Fibers (steel, PVA, basalt)	[39]
Fly ash, BFS	Na_2SiO_3	Sand	[40]
N/A	=Not available		

developed by fly ash without heat curing can only be demoulded after 3 days of casting attributed by slow setting rate. Furthermore, 28 days compressive strength reached 9.45 MPa. No potential outcome obtained from this mix design. They modified the geopolymer binders by adding sodium hydroxide (NaOH) in the system [10]. Regardless of how, the compressive strength dropped to 3.5 MPa, which is worse than the previous record. Nematollahi et al. [16] attempted to provoke the performance of fly ash-based one-part geopolymer by supplying calcium hydroxide ($Ca(OH)_2$), also known as hydrated lime, in the mix. With 11% of $Ca(OH)_2$ to replace fly ash and 0.015 wt% of Na_2SiO_3 powder, one-part geopolymer successfully achieved 14.2 MPa of compressive strength. The result is supported by Askarian et al. [43], where the addition of $Ca(OH)_2$ in fly ash-based one-part geopolymer system reached 18.1 MPa of compressive strength whereas the same mix without $Ca(OH)_2$ recorded only 10.9 MPa.

Besides fly ash, blast furnace slag (BFS) was employed as aluminosilicate precursor. BFS is a molten material that is collected above pig iron at the bottom of the blast furnace during the metallurgy of iron. Production of 1 tonne of iron produces 0.6 tonnes of slag by-product [41]. In Yang et al. [9] study, slag consists of approximately 40% of CaO, much higher than CaO content in fly ash, which is also the main factor that drives different properties in respective products. Performance of slag-based one-part geopolymer mortar is comparable to OPC mortar. Geopolymer activated by solid sodium silicate possesses 28-days strength of 50.6 MPa, when OPC exhibited 28-days strength of 44.6 MPa. This is due to the high composition of Ca^{2+} ions in BFS that yields calcium silicate hydrate (C–S–H) gel during the chemical reaction. That is also the key binding phase in OPC in conjunction with geopolymer matrix.

There are various types of slag available in the industries. Behaviour of one-part geopolymer paste made out of "typical" slag and "gypsum-free" slag was compared. Nematollahi et al. [26] reported higher workability of one-part geopolymer involved with "gypsum-free" slag due to the absence of 2% gypsum compared to "typical" slag. "Gypsum-free" slag supplied into the one-part geopolymer system resulted in lower elastic modulus by 12–29%, lower crack tip toughness by 18–24% and lower fracture toughness by 17–21%, compared to "typical" slag. "Gypsum-free" slag delivers more brittle structure.

Partial replacement of fly ash with BFS has also proven to improve the mechanical strength of fly ash-based one-part geopolymer. In the study of Askarian et al. [31], properties of fly ash-based one-part geopolymer with and without the participation of commercially available slag has been experimented. Their fly ash-based one-part geopolymer recorded 28-days strength of 9.9–31.2 MPa (depending on activators) while achieving 24.3–38 MPa of strength when fly ash was replaced by 50% of slag. This is due to a higher composition of calcium in BFS that promotes crystalline calcium aluminium silicate hydrate (C–A–S–H) binding phase, thus enhancing strength development. While fly ash-based one-part geopolymer mainly consists of sodium aluminosilicate hydrate (N–A–S–H) gel that results in weaker strength. Also, it has been reported by Nematollahi et al. [31] that fly ash and slag mixed one-part geopolymer paste exhibited thixotropic property.

Besides that, the substitution of aluminosilicate source by other waste products has been done by researchers with the aim to further ascertain the potential of one-part geopolymer. In fact, reactive alumina and silica can be supplied by 2 separate bodies. Geothermal silica is the secondary product from the electrical generation process by extracting steam from geothermal fields. The silica waste is rich in silica, comprises alkalis and possesses a relatively large surface area which is beneficial for geopolymerization reaction [44]. Instead of landfilling geothermal wastes, Hajimohammadi et al. [8] applied geothermal silica as silicate source and solid $NaAlO_2$ as alkali and aluminate source to develop one-part mixture. The raw silica was first milled, followed by water lixiviation to remove salts, dried, milled, then finally mixed with solid sodium aluminate. Research reported rapid dissolution of aluminate source and relatively slower silica dissolution rate in the early stages of geopolymerization. The scenario was caused by the dissolved aluminate species attached on a silica surface, subsequently inhibiting geothermal silica dissolution. They suggested that high early silica concentration could enhance geopolymer network formation instead. The influence of activators dissolution rate is crucial in geopolymer study, further explanation about the dissolution rate will be discussed in the next section.

Apart from geothermal silica, other silica residues were also being studied with the aim to replace aluminosilicate sources in one-part mixture. It is recorded that a chemical plant produces 5000–10,000 tonnes of residues per annum and these by-products are mostly landfilled. Hence few research teams utilized silica residues from waste treatment of chlorosilane production, as well as micro-silica (silica fume), a by-product from silicon alloys production, to develop geopolymers [14, 19, 45]. Sturm et al. [19] work involved the mixing of solid $NaAlO_2$ with silica residue and micro-silica (silica fume). Research reported a similar degree of geopolymerization reaction of both silicas. The materials reacted approximately 100% in one-part geopolymer when SiO_2/Al_2O_3 ratio = 2.0. It is interesting to note that the mechanical strength increases consistently even when SiO_2/Al_2O_3 ratio further increases to 6.0. This was attributed to the remaining silica particles that act as micro-aggregates in the structure.

Red mud is an alkaline residue from the production of alumina extraction through Bayer's process. The estimated red mud production is 120 million tonnes per year. As red mud possesses silica and alumina composition, it fulfils the requirement for the production of geopolymer. Al-rich red mud together with NaOH pellets has been pre-treated to produce one-part geopolymer precursor [17, 18]. Based on Ke et al. [17], inclusion of 10% NaOH was able to develop red mud-based one-part geopolymer with compressive strength of 9.8 MPa after 7 days. However, mechanical strength of red mud-based one-part geopolymer demonstrated deterioration after the first week due to carbonation and efflorescence. Efflorescence products (white alkaline deposits) were observed on the surface of cube specimens after 14 days of curing. This could cause depolymerisation of geopolymer, consequently led to the loss of strength. Another reason for the collapsed strength is the low Si/Al molar ratio that causes unstable geopolymerization. To increase SiO_2/Al_2O_3 molar ratio, Ye et al. [18] added 5–30 wt% of silica fumes in their mix. As shown in Fig. 2, the red mud-based one-part geopolymer without silica fume has higher strength after 3 or 7 days, but strength dropped after 28 days, and the inclusion of silica fume successfully promoted stable growth of long-term. The impact obtained was due to the slow dissolution of silica that only raised SiO_2/Al_2O_3 molar ratio at the later stage of geopolymerization.

Rice husk ash (RHA) is obtained from the rice milling industry. It is estimated that about 70 million tonnes of RHA are generated per year [46]. The physical and chemical properties of RHA depend on the temperature and duration of combustion

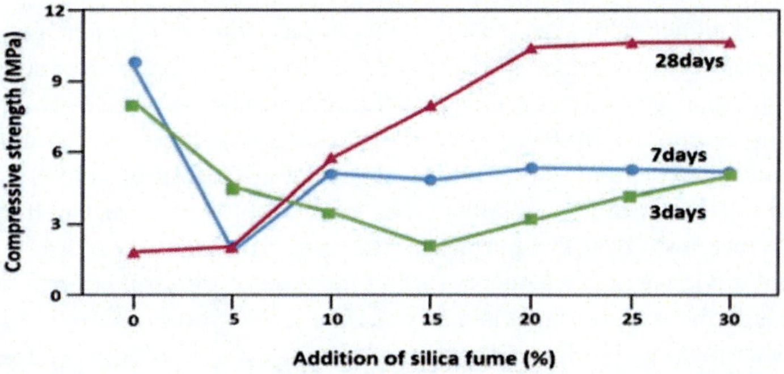

Fig. 2 Compressive strength of red mud-based one-part geopolymer in accordance to silica fume percentage [18]

of rice husk [47]. RHA is rich in silica, thus favourable towards geopolymer production. Hajimohammadi and van Deventer [24] stated that the mixture of RHA with $NaAlO_2$ developed dry precursor with desired Si/Al ratio, which in turn manipulate the characteristics of geopolymer. One-part geopolymer by these materials with Si/Al molar ratio of 2.5 possessed 22 MPa of strength after 3 weeks. Though the result was contradicted to Hajimohammadi and van Deventer [23], declaring that one-part geopolymer made from fly ash has lower compressive strength when Si/Al ratio increases. The report suggested increasing in Si/Al ratio reduced Si contribution in the final geopolymer gel formation, hence reduced in strength. To conclude, different materials behaved differently in accordance with Si/Al ratio.

Albite (sodium feldspar) is the major component in mine tailing, and the storage of ore waste of mine is a major environmental issue. The feldspar obtained by Feng et al. [13] consists of approximately 71% SiO_2 and 17% Al_2O_3, which is suitable to be treated as geopolymer precursor. However, the experimental work showed that albite was not able to set after 7 days of mixing with 7.5 M of NaOH solution, while albite powder calcined at 1000 °C exhibited poor geopolymerization with NaOH solution and only resulted in 2.5 MPa of strength after 28 days of mixing. This is due to the low extent of dissolution and leaching of Si and Al of albite at room temperature. Alkali-thermal treatment of albite with dry activators and heat curing of geopolymer specimens has been applied to initiate the effective chemical reaction. The approach will be further clarified in the following section.

With the attempt to reduce carbon footprint and landfilling issues globally, various types of waste materials that consist of silica and/or alumina content have been utilized by researchers to fabricate one-part geopolymer. It is believed that more experimental works will be carried out in this aspect to explore more suitable waste materials, develop an ideal mix design for construction application in order to bring them into service.

3 Alkali Activators for One-Part Geopolymer

Different alkali activators were used in the investigation of one-part geopolymer to suit a wide variety of aluminosilicate materials, as well as to discover potential properties of one-part geopolymer, thus promoting the viability of one-part geopolymer in building materials. Material such as solid sodium hydroxide, sodium silicate, sodium aluminate, calcium hydroxide, magnesium oxide, lithium hydroxide, potassium carbonate and others (Table 1), has been employed as one-part geopolymer binders in recent studies. The purpose of the on-going discovery of a new alkali activator is to provide necessary ions or further increase of alkalinity of reaction mixture which in turn disrupts aluminosilicate bonds, facilitates the dissolution of geopolymer precursors, thus improve the performance of one-part geopolymer systems.

When Na_2SiO_3 alone or with NaOH was utilized as the activators of fly ash one-part geopolymer mortar, compressive strength increased with increasing activators content, due to increasing alkali quality coefficient, QA (Eq. 1) [9, 11]. QA is an indicator that depends on the weight of sources of oxides in source materials involved in a system, allowing the understanding of the relationship between the binders' properties and the products' performances. Subsequent conclusion of one-part geopolymer was reported as such: increasing in QA resulted in reduced initial flow, increased rate of flow loss, improved compressive strength and increased rupture modulus.

$$Q_A = \frac{\frac{Na_2O}{(SiO_2)^2} \cdot Al_2O_3 \cdot CaO}{\text{total weight of binder}} \quad (1)$$

Besides the weight of oxides sources, dissolution of alkali activators also plays an important role in developing long-term performances. A study conducted by Hajimohammadi et al. [12] investigated alumina release rate of solid activators during geopolymerization, towards the formation of high strength and durable one-part geopolymers. Research found out that $NaAlO_2$ revealed a faster alumina release rate, promoted geopolymerization and rapid strength development at the early stage of reaction, consequently strength development became slow after the first week. Diversely, amorphous alumina with slower alumina release rate produced weak one-part geopolymer at the early stage but achieved similar strength after the third week. Alumina species are used in the formation of geopolymer matrix, where matrix network forms faster when aluminium is available. However, the silica particles were sorbed by large amount of dissolved alumina, obstructed the dissolution of silica, led to silicon deficit, then impeded more gel formation and further strength development.

Addition of calcium hydroxide ($Ca(OH)_2$) as an activator in fly ash-based one-part geopolymer system reached 18.1 MPa of compressive strength while activated by Na_2SiO_3 alone only achieved 10.9 MPa [43]. $Ca(OH)_2$ can increase in alkalinity of the system then promote the dissolution of ions from fly ash. Therefore, when the amount of $Ca(OH)_2$ increased from 9 to 12 wt%, the 28-days strength was raised by 19%. However, the workability of one-part geopolymer paste mix with Na_2SiO_3 and $Ca(OH)_2$ is about 40% worse than that with Na_2SiO_3 alone, even the water/binder ratio was increased by 90%. To enhance the workability of one-part mixes while securing the mechanical properties, 6 wt% of lithium hydroxide (LiHO) was introduced into the system. Other than beneficial towards workability, the compressive strength of fly ash-based geopolymer with Na_2SiO_3, $Ca(OH)_2$ and LiHO also increased. The combination demonstrated higher degree of silicon substitution by aluminium attributed by increased alkalinity, thus delivering a larger amount of hydration product and also a more compact microstructure.

Solid alkaline carbonates were tested in one-part geopolymer mixes. Solid potassium carbonate (KCO_3) has been proven to deliver positive effect in fabrication of OPC-geopolymer cement [43]. Hence, Askarian et al. [43] employed KCO_3 in fly ash-based one-part geopolymer study. However, addition of 6 and 10 wt% of the powder activator reduced the paste's workability and strength of the mixes of all ages. The alkali carbonate with lower pH values inhibited the release of Si and Al from fly ash, followed by the reduction of reaction rate, thus led to formation of porous structure. Similar outcome obtained when Ma et al. [37] replaced Na_2SiO_3 solid alkali activator with sodium carbonate (Na_2CO_3) in slag-based one-part geopolymer system partially. Participation of Na_2CO_3 also reduced the pH, led to moderate geopolymerization and prolonged setting time of one-part mix. They clarified that lesser $[SiO_4]^{4-}$ ions present for geopolymerization reaction when Na_2CO_3 was involved, prompted a less effective setting. The specimens also revealed a large proportion of unfavorable pores (pore size larger than 20 mm), thus strength decreased from approximately 75–65 MPa after the application of Na_2CO_3. Indeed, from our research work, the increasing content of Na_2CO_3 led to deterioration of the mechanical strength too. As the Na_2CO_3/Na_2SiO_3 ratio increased from 0.5 to 1.5, the 28 days compressive strength of fly ash-based one-part geopolymer slid from 83.77 to 22.31 MPa.

Waste material itself can also be treated as a solid alkali activator in one-part geopolymer systems. Instead of usual sodium hydroxide pellets or sodium metasilicate powder, Choo et al. [21] replaced synthetic alkali activators with red mud. The analysis reflected pH value of red mud range from 10 to 12.5, due to NaOH residue in red mud during Bayer's process. Since an alkaline environment is required for dissolution of silica and alumina ions in geopolymer precursor, the addition of red mud promotes alkaline environment, subsequently the geopolymerization reaction [48]. The dissolution and polycondensation process was enhanced, consequently promoted strength development. Resulting compressive strength of one-part geopolymer activated by NaOH and red mud are comparable. The author concluded 3–5 wt% of NaOH is equivalent to 40–60 wt% of red mud content in one-part geopolymer binder.

Besides red mud, paper sludge consists of chemical compositions that permit them to be used as a dry activator. Waste paper mill sludge is the by-product obtained during paper production. Adesanya et al. [29] utilized paper sludge that consists of 95.1 wt% $CaCO_3$ as the activator of slag-based one-part geopolymer. Nevertheless, paper sludge has to undergo pre-treatment to become a dry activator. Paper sludge was first mixed with 6.3 M of NaOH, dried and then milled. The chemical reactions during pre-treatment are shown in Eqs. (2) and (3).

$$2NaOH + CaCO_3 \rightarrow Na_2CO_3 + Ca(OH)_2 \tag{2}$$

$$Na_2CO_3 + CaCO_3 + 2H_2O \rightarrow Na_2Ca(CO_3)_2 \cdot 2H_2O \tag{3}$$

The pre-treatment generated Calcite ($CaCO_3$), Pirssonite ($Na_2Ca(CO_3)_2 \cdot 2H_2O$), Portlandite $Ca(OH)_2$ and Natrite (Na_2CO_3). Setting time retarded due to the increasing amount of calcite and pirssonite that lowered the pH of paste, causing slow dissolution of Ca^{2+} from BFS. This is also the reason that one-part geopolymer experienced low early mechanical strength when paper sludge content is high. However, high strength developed after 7 days when calcium from BFS is released to react with silicate that forms the C–A–S–H phase. Paper sludge also acted as fillers in geopolymer structures that reduce pore size and water absorption (less than 5%). Lastly, high drying shrinkage was accredited by the water absorbed on the organic matter of the paper sludge after the addition of water. Regardless of how, no visible cracks were observed by the researcher. In short, the experimental work ascertained that the increase of paper sludge content prolonged the setting time, reduced water absorption values, increased compressive strength and drying shrinkage of one-part geopolymer.

4 Water Content in One-Part Geopolymer

Water is required for the transportation of ions among materials, initiating chemical reactions. In the effort to develop one-part geopolymer, water is a key factor that attributes to paste workability, materials dissolution, geopolymeric gel formation and the resulting structure properties. This parameter is important, whereby inorganic polymer formation through the "just add water" method. However, different types of source materials require different amount of water to achieve appropriate workability of the fresh mix.

Hajimohammadi and van Deventer [23] investigated the characteristics of fly ash-based one-part geopolymer based on the water content. Upon increasing water content, the alkalinity of the system was decreased, leading Al-rich gel formation at the early stage of geopolymerization. The condition contributed negative impact towards the microstructure and mechanical strength. Microstructure of specimen with H_2O/Na_2O molar ratio 14 exhibited less pore structure compared to microstructure of specimen with H_2O/Na_2O molar ratio of 12.

From the investigation work of Oderji et al. [33], increasing w/b ratio from 0.28 to 0.32 in fly ash and BFS mixed one-part geopolymer caused increased workability and setting time, as well as decreased compressive strength. The decrease in compressive strength is due to the reactivity of the system. Binder with higher w/b ratio resulted in higher amount of unreacted particles while binder with lower w/b ratio presented more homogeneous microstructure and large portion of gel formation. The different explanation given by different researchers could be due to the different materials used, as well as the limited range of water content that had been worked on.

When aggregates are involved, properties of aggregates are factors that influence the water demand for geopolymerization. Yang et al. [11] found out that air-dried lightweight aggregates consumed 20% of water introduced, thus reducing the water consumed by hydration reaction at the early stage of one-part geopolymer production. As a result, higher strength growth would occur at the beginning due to alkalinity factor, but unfavourable for long-term strength development. From the experimental work, a maximum of 0.5 w/b ratio was set to allow the development of slag-based one-part geopolymer concrete with practical purpose.

From our work, the grout spread diameter of fly ash-based one-part geopolymer activated by the mix of Na_2SiO_3 and $NaAlO_2$ was 126 mm larger than those activated solely by Na_2SiO_3 even the w/b ratio was higher by 0.15. Besides, with w/b ratio of 0.25, the fluidity of paste activated by mixture of Na_2SiO_3 and Na_2CO_3 was always better than those activated solely by Na_2SiO_3. The result also did not illustrate that the systems with higher water content resulted in lower mechanical properties. In fact, the water requirement for optimum geopolymerization differs from one system to another. Nevertheless, all research work came to an understanding that the increasing water content for workability enhancement brings positive results but is not always the best approach as the negative impact towards mechanical strength might be significant. To conclude, water requirement for optimum properties strongly depends on the raw materials.

5 Additives Addition in One-Part Geopolymer

Additives serve to modify the characteristics of building materials to suit varying applications. This section lists out the types of admixtures, fibers and aggregates added into one-part geopolymer systems, and their impact towards the one-part geopolymers.

One-part geopolymer cement containing BFS and 3.5–4.5 wt% of Na_2CO_3 delivered 28 days compressive strength of 29–42 MPa. Though, the samples could not be demoulded after a day of casting due to the low setting rate. This is because Na_2CO_3 yields lower pH value, reducing the activation ability of BFS at early stage [49]. As such, the combination required accelerating admixture to enhance initial setting and early strength development. According to Kovtun et al. [15], 6 wt% silica fume and slaked lime (calcium hydroxide) associated with geopolymer mix to act as an accelerating admixture were able to shorten the hardening of one-part geopolymer at early age. Accelerated early strength by this approach is contributed by C–S–H formed and increased alkalinity environment by the rapid reaction and dissolution of silica fume and slaked lime.

Either water or plasticizer can be added in a geopolymer mix to achieve proper workability level and setting ability for one-part geopolymer paste. Increase of water content in geopolymer is not encouraged due to impair strength development caused by increasing porosity [49]. So, plasticizers such as sodium lignosulfate or polycarboxylate ether (PCE) superplasticizer were usually employed (Table 1). A project carried out by Kovtun et al. [15] included 5 wt% sodium lignosulfonate powder in slag-based one-part geopolymer provided positive outcome. This plasticizer enhanced the workability of paste, prolonged setting time and increased strength performance. Plasticizer can reduce water demand of the binders, as well as slow down the geopolymerization at early stage, thus developing better hydration products.

Polyvinyl alcohol (PVA) fibers and ultra-high molecular weight polyethylene (UHMWPE) fibers were incorporated in dry precursor to develop one-part strain hardening geopolymer composite (SHGC), with the aim to evaluate its micromechanics properties [25, 26]. Result shows that participation of 2% PVA fibers enhanced the samples' compressive strength while PE fibers caused reduction in sample's compressive strength, this suggested PE fibers instilled more damage effect (air entrapping effect) in the composite compared with PVA fibers. They studied the fiber-matrix interface properties, declaring that the fiber and matrix interaction depended on the types of fibers involved. Elastic modulus of PVA fibers is 3 times less than that of PE fibers. Fibers with lower elastic modulus are more easily damaged in the composite, leading to a jamming effect during fiber slippage when load is applied. Hence, the resulting ultimate tensile strain of PVA-SHGC was 48.7 MPa, while 44.3 MPa for PE-SHGC. The SHGCs in those studies displayed multiple racking patterns. The crack patterns corresponded to high tensile strain capacity of the composites.

Previous results regarding the improved compressive strength with the addition of PVA fibers were supported by Abdollahnejad et al. [38], where they reported 20% of increment in compressive strength when 1% of PVA fibers is involved in BFS-based one-part geopolymer. Besides, the participation of steel fibers also showed 25% of strength increment while cellulose and basalt fibers led to 20 and 50% of strength reduction, respectively. Nonetheless, the incorporation of fibers increased all the specimens' porosity and water absorption by immersion. Steel and PVA fibers resulted in more interconnected pores and tortuous samples while cellulose and basalt fibers produced one-part geopolymer composite with larger pores. These pore structures of one-part geopolymer composites are affected by the fibers interaction with the matrix, as well as the mechanical properties. The response of composites towards loading is different. Some fibers were debonded from the matrix, while some were abraded due to the strong connection with geopolymer.

Influences of lightweight aggregates (expanded clay granules) in one-part geopolymer mortar and concrete were investigated by Yang et al. [11]. Experimental results were attributed to the interaction of aggregates with one-part geopolymer matrix. As fine lightweight aggregates possess larger surface area for reaction, initial workability was enhanced, but also shortened the concrete setting time. Besides having lower density and porous structure, particle distribution of lightweight aggregates was analyzed. Analysis results showed discontinuous grading (gap-grading). Lightweight aggregates have higher porosity and the discontinuous grading delivered porous zone at the aggregates-matrix interface, increased internal voids, subsequently rapid continuous crack pattern along the interfaces were formed when load was applied. These factors reduced compressive strength and splitting tensile strength of one-part geopolymer mortar and concrete, as well as deteriorated modulus of rupture and elasticity of concrete.

Ceramic waste can also be treated as recycled aggregate instead of using natural aggregates in building materials, especially in the inorganic polymer as it possesses aluminium and silica that can facilitate gel formation. Application of ceramic aggregates in fly ash and slag mixed binder fabricated by Abdollahnejad et al. [38] successfully attain 100 MPa of compressive strength after 28 days of curing. Whereas the use of natural sand in the same mix only exhibited compressive strength of around 70 MPa. The ceramic aggregates have smaller grain sizes, as well as the ability to absorb water available in the mixture, increase pH value and reduce water-to-binder ratio. These are better characteristics that contributed to one-part geopolymer mix compared to the participation of natural sand.

6 Thermal Treatment of One-Part Geopolymer

Due to the concern of slow reaction kinetics cause of slow dissolution rate of solid alkali activators, effort has been done by researchers to accelerate the leaching ability of alkali content as well as chemical reaction in one-part mixture. With the urge to produce binders that contain high amounts of alkali that can be easily leachable when water is added, alkali-thermal activation of raw materials has been executed. While heat curing of specimens was done to accelerate geopolymerization reaction.

In the experimental work of Ke et al. [17], Al-rich red mud underwent alkali-thermal preactivation together with NaOH pallets in a muffle furnace of 800 °C for an hour. Table 2 shows few FTIR absorption peaks of red mud and alkali-thermal activated precursor added with NaOH. Bands at 3622, 3525, 3430 cm^{-1} and bands between 1500 and 1300 cm^{-1} corresponded to the hydroxyl and carbonate groups were reduced after thermal treatment at 800 °C. This is caused by the absorption of CO_2 and H_2O from the atmosphere before thermal treatment, as red mud binders possess hygroscopic and alkaline properties. Next, the bands in between 900 to 1200 cm^{-1} (attributed by stretching vibrations of Si–O–T bonds) were shifted to lower wavenumbers after alkali-thermal treatment indicates that the binders are more prepared for geopolymerization reaction. These transformations in infrared spectra after alkali-thermal treatment favours the chemical reaction compared to raw red mud.

Aforementioned that one-part geopolymer by albite with NaOH or Na_2CO_3 does not provide desired outcome, hence Feng et al. [13] provided pretreatment before and after the mixing process. The alkali-thermal activation of albite with solid NaOH or Na_2CO_3 at 1000 °C, generated matte formed product, then was pulverized and mixed with water to make geopolymer paste. Substantial heat was released when water was added to the alkali-albite glassy powder, promote rapid dissolution and hardening. 28 days strength of NaOH and Na_2CO_3 alkali-thermal treated albite were 44.2 MPa and 42.6 MPa respectively. The approach was a success due to albite structure breaking down, formed amorphous, depolymerized geopolymer precursor during alkali-thermal activation.

Silica-based one-part geopolymer activated by $NaAlO_2$ was exposed to increasing temperature to determine the effect of heat treatment on the mechanical and structural characteristics [19]. Since the XRD analysis showed amorphous gel phase and large portion of crystalline phases formed in the product, the author named the

Table 2 Major FTIR bands of unreacted red mud to alkali-thermal activated red mud [17]

Assignment	Wavenumbers (cm^{-1})	
	Unreacted red mud	Red mud after thermal treatment
O–H	3622, 3525, 3430	3430
Carbonate group	1504, 1387	1504, 1434
Si–O–T	1116, 1035, 998	1110, 993

one-part mix as geopolymer-zeolite composite. Strength testing displayed improvement when specimens were treated from 100 to 200 °C, with the highest strength of 28 MPa. However, the geopolymer-zeolite composite experienced reduction in strength, from 200 to 700 °C. Dilatometry was used to determine the volume shrinkage of geopolymer exposed up to 900 °C. A linear shrinkage was observed up to 700 °C with maximum of 2–3% of shrinkage measured. The minor volume shrinkage was attributed to the presence of "zeolitic" water instead of "structural" water in the specimens. The behaviour is extremely beneficial for high-temperature applications. Whereas a sudden increase of shrinkage occurs when temperature raised above 700 °C that was caused by viscous sintering or densification of geopolymer.

Please be noted that not all geopolymers experience improvement in properties after thermal treatment. As for fly ash and slag mixed one-part geopolymer fabricated by Nematollahi et al. [26], heat curing does not result in significant effect on the elastic modulus. Moreover, heat curing reduced the compressive strength and fracture toughness of specimens, compared to heat-cured specimens at 60 °C for 1 day. Their unloaded sample exhibited saturated multiple crack behaviour. The crack was contributed by the coarse microstructure when cured at ambient temperature, leading to fracture toughness of 0.316 MPa $m^{1/2}$, while heat-cured specimens recorded 0.287 MPa $m^{1/2}$. The small crack width is a big leap in the field of geopolymer development as brittle geopolymer usually show larger crack width. Thermal treatment for this system delivered brittle characteristics, also reduced crack tip toughness by 9–18% compared to ambient cured specimens.

Further investigation was done to determine the effect of thermal treatment towards geopolymer composite, therefore PVA fibers were added in the fly ash and slag-mixed one-part geopolymer. The higher fracture toughness of geopolymer matrix cured at ambient temperature delivered better composite performance, compared to specimens that underwent heat curing, the samples delivered matrix with lower fracture toughness. Fracture surface roughness of composite dependent on fracture toughness of matrix. As fracture surface roughness increases along with fracture toughness, the fiber-matrix interface area is raised, so does the fiber-matrix interface friction. Larger friction required to fully debond fibers from the surrounding matrix produced one-part geopolymer with higher ultimate tensile strength [26].

In short, thermal treatment is not the only solution for property enhancement. Some one-part geopolymer delivered positive outcomes with thermal pre-treatment or heat curing while some received negative impact. Furthermore, thermal treatment of materials requires higher workforce and cost, as well as release of more greenhouse gases, compared to ambient fabrication of one-part geopolymer. Therefore, thermal treatment is suggested to be conducted only when necessary.

7 Durability Performances of One-Part Geopolymer

To date, studies regarding the durability performance of one-part geopolymer are limited. The impact of building materials on damaging environments is essential for users to understand their suitability of materials in practice. Effect of acid attack, sulphate attack, chloride attack, freeze/thaw, high temperature, carbonation and efflorescence on one-part geopolymer should be made clear. This section goes into the characteristics of one-part geopolymer exposed to harmful conditions.

Despite other potential properties as the alternative to OPC, the efflorescence effect is a concerning matter in one-part geopolymer. Alkali activators that react with atmospheric CO_2 would generate a white carbonate salt known as efflorescence. Since efflorescence is not a precipitate, the carbonate salt does not form a protective layer for the materials, in fact, alkaline pore solution will migrate to the surface continuously, deteriorating the protection for reinforcements. In the study of Dong et al. [50], the efflorescence effect of solid Na_2SiO_3 activated one-part geopolymer is influenced by the fineness of metasilicate particles and Na_2O content in the activator. Lower amount of efflorescence obtained from one-part geopolymer fabricated with coarser metasilicate. That was due to lower surface area of coarse metasilicate granules for reaction at mixing stage, left air voids in the structure, and led to ingress of moisture and CO_2 from the atmosphere, as shown in Fig. 3. In addition, the amount of efflorescence increased with increasing Na_2O_3. The efflorescence formation increased significantly once the metasilicate content is higher than $Na_2O = 6\%$, so did the deterioration of compressive strength. Also, the carbonate content increased along with curing age.

Sturm et al. [30] executed a durability study on one-part geopolymer against sulphuric attack. Rice husk ash (RHA)-based one-part geopolymer that subjected to sulphuric acid environment (pH = 1) for 70 days showed higher compressive strength (77.9 MPa) compared to those specimens that immersed in water (32.0 MPa)

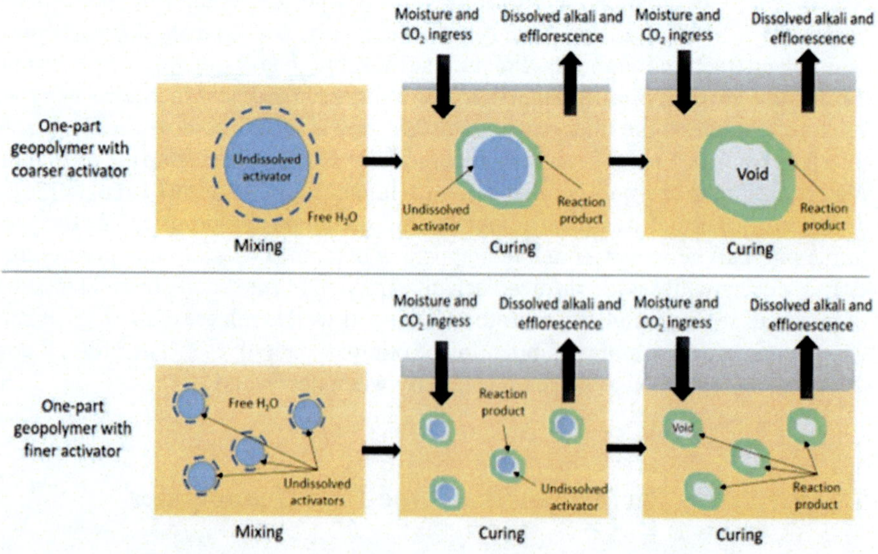

Fig. 3 Schematic diagram of efflorescence progress in one-part geopolymer [50]

for the same duration. RHA that has high silica content released silica into acid solution during immersion. As silica has low solubility in an acidic environment, silica gel precipitates at the mortar/acid interface, forming a protective layer that hinders further disintegration of materials from specimens. Besides that, unreacted silica that presents in the structure acts as filler to the matrix, further sustaining the strength. The explanation was in line with the SEM image whereby the sample underwent acid attack exhibited 'debris-like' surface, which was the precipitated silica gel. The experimental work also pointed out that presence of CaO content in the one-part binders was unfavourable towards acid resistance. Since leached calcium could react with sulphate ions, generate gypsum and expansion, followed by the crack formation in the structure.

Sulphate and acid resistance of fly ash and slag mixed one-part geopolymer mortar has been carried out by Zhou et al. [40]. The experimental work concluded that a larger portion of fly ash improved that durability performance due to more homogeneous microstructure when fly ash is involved. Sulphate attack resulted in a slight decrease of compressive strength after 60 days of immersion, then increased gradually until 120 days of immersion as shown in Fig. 4a. The deterioration of performance was not severe because of the low dissolution ability of the stable aluminosilicate network of geopolymer when immersed in sodium sulphate solution. Moreover, as time goes, sodium sulphate induced secondary geopolymerization reaction, especially in fly ash containing one-part geopolymer (G75F25, G50F5 and G25F75), whereby the 120 days compressive strength is even higher than strength before sulphate exposure. On the other hand, compressive strength showed significant deterioration when

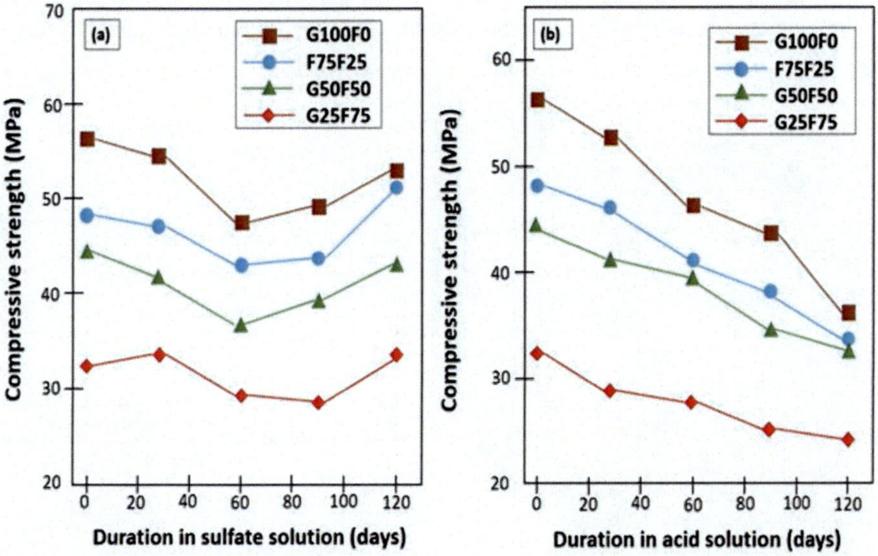

Fig. 4 Compressive strength of one-part geopolymer mortar after immersed in **a** sulphate solution; **b** acid solution [40]

mortar samples experienced acid attack (pH = 2) (Fig. 4b). This is due to the dissolution of hydroxide ions from the systems that lowered the pH values. Also, leaching of sodium hydroxide when samples were exposed to nitric acid, inhibited the formation of N–A–S–H gel, leading to decreased mechanical properties.

Presence of additives modified the durability performance of one-part geopolymer. Reinforced one-part geopolymer with fibers fabricated by Abdollahnejad et al. [36] has proven to exhibit excellent acid resistance. PVA, steel, basalt and cellulose fibers incorporated in slag-based one-part geopolymer experienced mass loss of < 2% after exposure with 3% sulphuric acid (pH = 1) for 9 days. The durability performance of one-polymer composite also depends on the resistance of fibers against an acidic environment.

Freeze/thaw resistance of building materials is essential to seasonal places. Freeze/thaw resistance of one-part geopolymer relies on its pore structure. Geopolymer with higher porosity has a high water absorption ability, which may lead to a larger volume of ice formation inside the pores. Formation of ice causes stress on internal matrix walls while melting of ice releases the stress against the walls. There would be a change of pore structure and generation of connected pores depending on the elastic strain of materials. Such cycles might eventually lead to cracks formation around the pores. The cycle is illustrated in Fig. 5. Anyhow, the fatigue loading cycle worsens the strength of slag-based one-part geopolymer. The deterioration can be controlled by the participation of additives [36, 38]. The addition of PVA fiber limited the loss of strength to lower than 1% in slag-based one-part geopolymer mortar, and around 10% in fly ash and slag mixed one-part geopolymer mortar.

In short, durability performance of one-part geopolymer is not widely studied. More experimental works should be done to widen the knowledge in this discipline.

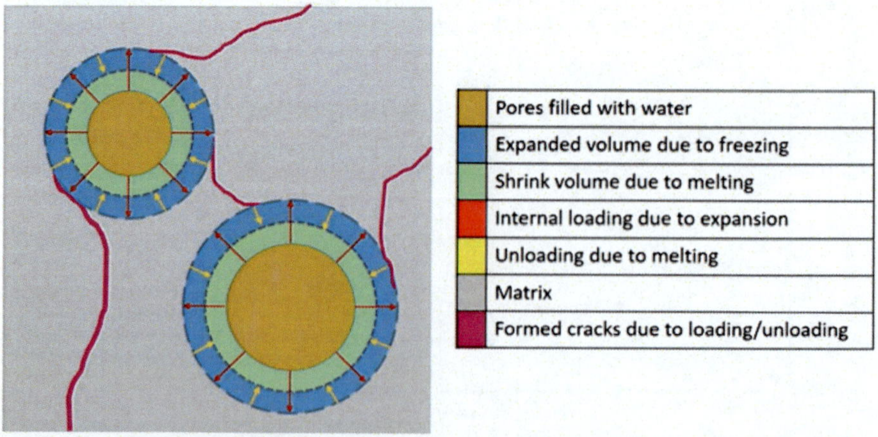

Fig. 5 Schematic diagram of the impact of freeze/thaw cycle towards pores of one-part geopolymer [36]

8 Future Trends

Unlike OPC and traditional geopolymer, one-part geopolymer does not necessarily require high-temperature processing and transportation of highly alkaline solutions. In addition, one-part geopolymer can obtain raw material directly from industrial waste and is worker-friendly. As future development orients towards a more sustainable and environmentally friendly concept, shift from OPC to one-part geopolymer in practice can be foreseen. It is not difficult to understand the chemistry of geopolymer as the philosophy was consolidated a few decades ago. However, due to the fact that one-part geopolymer is a more recent approach in this field, manipulation of type and dosage of raw materials would be massively studied in the forthcoming years. Further research work regarding the cost and environmental impact of one-part geopolymer will be conducted in succession. Besides that, more durability tests should be carried out to promote the application of one-part geopolymer in various conditions, such as freeze–thaw, carbonation, wet-dry, acid, sulphate, chloride and sea water attack. Until then, it is advisable to go through the guide for the development of OPC and traditional geopolymer, then apply those strategies to advance this new material.

9 Conclusion

Employment of recycling materials in one-part geopolymer investigations delivered promising results. Most of the materials used for the research of one-part geopolymer are industrial by-products, such as fly ash, blast-furnace-slag, rice husk ash, red mud, geothermal silica and albite. Depending on the local availability, as long as materials consist of aluminium and/or silica, they could be developed into one-part geopolymer. Such an effort is to reduce environmental issues all around the globe. There are also various types of solid activators applied, subject to the types of precursor used. Solid activators have to be easily soluble in water to create an alkaline environment, only then able to promote geopolymerization. Geopolymerization reaction is also highly influenced by water content. The increased amount of water usually delivers positive fresh properties but negative mechanical properties. Additives such as admixture, aggregates and fibers were used to alter the properties of one-part geopolymer. Furthermore, thermal treatment enhanced activator dissolution, delivered one-part geopolymer with improved strength. However, this approach does not always deliver a positive outcome. More concern should be taken into consideration when thermal treatment is involved.

References

1. Bakharev T (2005) Resistance of geopolymer materials to acid attack. Cem Concr Res 35(4):658–670. https://doi.org/10.1016/j.cemconres.2004.06.005

2. Luukkonen T, Abdollahnejad Z, Yliniemi J, Kinnunen P, Illikainen M (2018) One-part alkali-activated materials: a review. Cem Concr Res 103:21–34. https://doi.org/10.1016/j.cemconres.2017.10.001
3. Okoye FN, Prakash S, Singh NB (2017) Durability of fly ash based geopolymer concrete in the presence of silica fume. J Clean Prod 149:1062–1067. https://doi.org/10.1016/j.jclepro.2017.02.176
4. Zhuang XY, Chen L, Komarneni S, Zhou CH, Tong DS, Yang HM, Yu WH, Wang H (2016) Fly ash-based geopolymer: clean production, properties and applications. J Clean Prod 125:253–267. https://doi.org/10.1016/j.jclepro.2016.03.019
5. Mclellan B, Williams R, Lay J, van Riessen A, Corder G (2011) Costs and carbon emissions for geopolymer pastes in comparison to ordinary Portland cement. J Clean Prod 19:1080–1090. https://doi.org/10.1016/j.jclepro.2011.02.010
6. Davidovits J (2008) Geopolymer chemistry and applications. Geopolymer Institute, France
7. Duxson P, Fernández-Jiménez A, Provis J, Lukey G, Palomo A, Van Deventer J (2007) Geopolymer technology: the current state of the art. J Mater Sci 42:2917–2933. https://doi.org/10.1007/s10853-006-0637-z
8. Hajimohammadi A, Provis J, Van Deventer J (2008) One-part geopolymer mixes from geothermal silica and sodium aluminate. Ind Eng Chem Res 47(23):9396–9405. https://doi.org/10.1021/ie8006825
9. Yang KH, Song JK, Ashour AF, Lee ET (2008) Properties of cementless mortars activated by sodium silicate. Constr Build Mater 22(9):1981–1989. https://doi.org/10.1016/j.conbuildmat.2007.07.003
10. Yang KH, Song J (2009) Workability loss and compressive strength development of cementless mortars activated by combination of sodium silicate and sodium hydroxide. J Mater Civ Eng 21(3):119–127. https://doi.org/10.1061/(ASCE)0899-1561(2009)21:3(119)
11. Yang KH, Song J, Lee JS (2010) Properties of alkali-activated mortar and concrete using lightweight aggregates. Mater Struct Constr 43(3):403–416. https://doi.org/10.1617/s11527-009-9499-6
12. Hajimohammadi A, Provis J, Van Deventer J (2010) Effect of alumina release rate on the mechanism of geopolymer gel formation. Chem Mater Chem Mater 22(18):5199–5208. https://doi.org/10.1021/cm101151n
13. Feng D, Provis J, Deventer J, Scherer G (2012) Thermal activation of albite for the synthesis of one-part mix geopolymers. J Am Ceram Soc 95. https://doi.org/10.1111/j.1551-2916.2011.04925.x
14. Sturm P, Greiser S, Gluth G, Jaeger C, Brouwers HJH (2015) Degree of reaction and phase content of silica-based one-part geopolymers investigated using chemical and NMR spectroscopic methods. J Mater Sci 50(20):6768–6778. https://doi.org/10.1007/s10853-015-9232-5
15. Kovtun M, Kearsley E, Shekhovtsova J (2015) Chemical acceleration of a neutral granulated blast-furnace slag activated by sodium carbonate. Cem Conc Res 72:1–9. https://doi.org/10.1016/j.cemconres.2015.02.014
16. Nematollahi B, Sanjayan J, Shaikh FUA (2015) Synthesis of heat and ambient cured one-part geopolymer mixes with different grades of sodium silicate. Ceram Int 41(4):5696–5704. https://doi.org/10.1016/j.ceramint.2014.12.154
17. Ke X, Bernal S, Ye N, Provis J, Yang J (2015) One-part geopolymers based on thermally treated red mud/NaOH blends. J Am Ceram Soc 98(1):5–11. https://doi.org/10.1111/jace.13231
18. Ye N, Yang J, Liang S, Hu Y, Hu J, Xiao B, Huang Q (2016) Synthesis and strength optimization of one-part geopolymer based on red mud. Constr Build Mater 111:317–325. https://doi.org/10.1016/j.conbuildmat.2016.02.099
19. Sturm P, Gluth GJG, Simon S, Brouwers HJH, Kühne HC (2016) The effect of heat treatment on the mechanical and structural properties of one-part geopolymer-zeolite composites. Thermochim Acta 635:41–58. https://doi.org/10.1016/j.tca.2016.04.015
20. Nematollahi B, Sanjayan J (2016) Ambient temperature cured one-part engineered geopolymer composite: a sustainable alternative to engineered cementitious composite. In: Paper presented

at the 9th Rilem International Symposium on Fiber Reinforced Concrete (BEFIB 2016), Vancouver Canada
21. Choo H, Lim S, Lee W, Lee C (2016) Compressive strength of one-part alkali activated fly ash using red mud as alkali supplier. Constr Build Mater 125:21–28. https://doi.org/10.1016/j.conbuildmat.2016.08.015
22. Wang K, Du L, Lv X, He Y, Cui X (2017) Preparation of drying powder inorganic polymer cement based on alkali-activated slag technology. Powder Technol 312:204–209. https://doi.org/10.1016/j.powtec.2017.02.036
23. Hajimohammadi A, van Deventer J (2017) Characterization of one-part geopolymer binders made from fly ash. Waste Biomass Valor 8(1):225–233. https://doi.org/10.1007/s12649-016-9582-5
24. Hajimohammadi A, Van Deventer J (2017) Solid reactant-based geopolymers from rice hull ash and sodium aluminate. Waste Biomass Valor 8(6):2131–2140. https://doi.org/10.1007/s12649-016-9735-6
25. Nematollahi B, Sanjayan J, Qiu J, Yang EH (2017) High ductile behavior of a polyethylene fiber-reinforced one-part geopolymer composite: a micromechanics-based investigation. Arch Civ Mech Eng 17(3):555–563. https://doi.org/10.1016/j.acme.2016.12.005
26. Nematollahi B, Sanjayan J, Qiu J, Yang EH (2017) Micromechanics-based investigation of a sustainable ambient temperature cured one-part strain hardening geopolymer composite. Constr Build Mater 131:552–563. https://doi.org/10.1016/j.conbuildmat.2016.11.117
27. Ma C, Long G, Shi Y, Xie Y (2018) Preparation of cleaner one-part geopolymer by investigating different types of commercial sodium metasilicate in China. J Clean Prod 201:636–647. https://doi.org/10.1016/j.jclepro.2018.08.060
28. Hajimohammadi A, Ngo T, Kashani A (2018) Sustainable one-part geopolymer foams with glass fines versus sand as aggregates. Constr Build Mater 171:223–231. https://doi.org/10.1016/j.conbuildmat.2018.03.120
29. Adesanya E, Ohenoja K, Luukkonen T, Kinnunen P, Illikainen M (2018) One-part geopolymer cement from slag and pretreated paper sludge. J Clean Prod 185:168–175. https://doi.org/10.1016/j.jclepro.2018.03.007
30. Sturm P, Gluth GJG, Jäger C, Brouwers HJH, Kühne HC (2018) Sulfuric acid resistance of one-part alkali-activated mortars. Cem Concr Res 109:54–63. https://doi.org/10.1016/j.cemconres.2018.04.009
31. Askarian M, Tao Z, Samali B, Adam G, Shuaibu R (2019) Mix composition and characterization of one-part geopolymers with different activators. Constr Build Mater 225:526–537. https://doi.org/10.1016/j.conbuildmat.2019.07.083
32. Abdollahnejad Z, Luukkonen T, Mastali M, Giosue C, Favoni O, Ruello ML, Kinnunen P, Illikainen M (2019) Microstructural analysis and strength development of one-part alkali-activated slag/ceramic binders under different curing regimes. Waste Biomass Valor 11(6):3081–3096. https://doi.org/10.1007/s12649-019-00626-9
33. Oderji SY, Chen B, Ahmad MR, Shah SFA (2019) Fresh and hardened properties of one-part fly ash-based geopolymer binders cured at room temperature: effect of slag and alkali activators. J Clean Prod 225:1–10. https://doi.org/10.1016/j.jclepro.2019.03.290
34. Oderji SY, Chen B, Shakya C, Ahmad MR, Shah SFA (2019) Influence of superplasticizers and retarders on the workability and strength of one-part alkali-activated fly ash/slag binders cured at room temperature. Constr Build Mater 229:116891. https://doi.org/10.1016/j.conbuildmat.2019.116891
35. Almalkawi AT, Balchandra A, Soroushian P (2019) Potential of using industrial wastes for production of geopolymer binder as green construction materials. Constr Build Mater 220:516–524. https://doi.org/10.1016/j.conbuildmat.2019.06.054
36. Abdollahnejad Z, Mastali M, Falah M, Shaad KM, Luukkonen T, Illikainen M (2020) Durability of the reinforced one-part alkali-activated slag mortars with different fibers. Waste Biomass Valor 1–15. https://doi.org/10.1007/s12649-020-00958-x
37. Ma C, Zhao B, Guo S, Long G, Xie Y (2019) Properties and characterization of green one-part geopolymer activated by composite activators. J Clean Prod 220:188–199. https://doi.org/10.1016/j.jclepro.2019.02.159

38. Abdollahnejad Z, Mastali M, Woof B, Illikainen M (2020) High strength fiber reinforced one-part alkali activated slag/fly ash binders with ceramic aggregates: microscopic analysis, mechanical properties, drying shrinkage, and freeze-thaw resistance. Constr Build Mater 241:118129. https://doi.org/10.1016/j.conbuildmat.2020.118129
39. Shah SFA, Chen B, Oderji SY, Aminul Haque M, Ahmad MR (2020) Comparative study on the effect of fiber type and content on the performance of one-part alkali-activated mortar. Constr Build Mater 243:118221. https://doi.org/10.1016/j.conbuildmat.2020.118221
40. Zhou S, Ma C, Long G, Xie Y (2020) A novel non-Portland cementitious material: mechanical properties, durability and characterization. Constr Build Mater 238:117671. https://doi.org/10.1016/j.conbuildmat.2019.117671
41. Davidovits J (1994) Global warming impact on the cement and aggregates industries. World Resour Rev 6:263–278
42. American Society of Testing and Materials (2008) Standard specification for coal fly ash and raw or calcined natural pozzolan for use in concrete (ASTM C618)
43. Askarian M, Tao Z, Adam G, Samali B (2018) Mechanical properties of ambient cured one-part hybrid OPC-geopolymer concrete. Constr Build Mater 186:330–337. https://doi.org/10.1016/j.conbuildmat.2018.07.160
44. Escalante JI, Mendoza G, Mancha H, López J, Vargas G (1999) Pozzolanic properties of a geothermal silica waste material. Cem Concr Res 29(4):623–625. https://doi.org/10.1016/S0008-8846(98)00238-5
45. Gluth GJG, Lehmann C, Rübner K, Kühne H-C (2013) Geopolymerization of a silica residue from waste treatment of chlorosilane production. Mater Struct 46(8):1291–1298. https://doi.org/10.1617/s11527-012-9972-5
46. Rice Husk Ash (2008) About rice husk ash. Retrieved 21 Feb 2020, from https://www.ricehuskash.com/product.html
47. Bakar BH, Putra JR, Hamidi A (2011) Malaysian rice husk ash—improving the durability and corrosion resistance of concrete: pre-review. In: Asia Pacific structural engineering conference (APSEC). Concrete Research Letters
48. Davidovits J (1991) Geopolymers: inorganic polymeric new materials. J Therm Anal Calorim 37:1633–1656. https://doi.org/10.1007/BF01912193
49. Wang SD, Scrivener KL, Pratt PL (1994) Factors affecting the strength of alkali-activated slag. Cem Concr Res 24(6):1033–1043. https://doi.org/10.1016/0008-8846(94)90026-4
50. Dong M, Elchalakani M, Karrech A (2020) Development of high strength one-part geopolymer mortar using sodium metasilicate. Constr Build Mater 236:117611. https://doi.org/10.1016/j.conbuildmat.2019.117611

Performance on Physical and Mechanical Properties of Fired Clay Brick Incorporated with Palm Kernel Shell for Lightweight Building Materials

Noor Amira Sarani, Aeslina Abdul Kadir, Nur Fatin Nabila Hissham, Mohd Ikhmal Haqeem Hassan, and Nurul Nabila Huda Hashar

Abstract This paper presents the potential of palm oil mill waste incorporated into fire clay bricks for lightweight building materials. Palm kernel shell (PKS) from oil palm processing was used as a clay replacement for brick making with a percentage of 1, 5, and 10% by volume. The firing temperature was set to 1050 °C at 1 °C/min heating rates. The effect of incorporation PKS on the bricks was assessed by physical and mechanical properties analysis. The results showed that bricks with 5% of PKS are significant to reach equilibrium between positive (decrease of weight and increase of porosity) and negative (increase of water absorption and decrease of mechanical resistance) effects. In conclusion, optimizing PKS into fired clay brick is beneficial in reducing the disposal challenges and promoting lightweight building materials that readily meet the standard requirements.

Keywords Fired clay brick · Recycling palm oil mill waste · Building material · Lightweight materials

1 Introduction

In recent years, the growth of the agricultural sector has risen in Malaysia and across the globe to meet the world's growing population. The importance of the agricultural

N. A. Sarani · A. A. Kadir (✉) · N. F. N. Hissham · M. I. H. Hassan · N. N. H. Hashar
Faculty of Civil Engineering and Built Environment, Universiti Tun Hussein Onn Malaysia, 86400 Parit Raja, Batu Pahat, Johor, Malaysia
e-mail: aeslina@uthm.edu.my

N. F. N. Hissham
e-mail: nabilahissham95@gmail.com

M. I. H. Hassan
e-mail: mohd.ikhmal.haqeem@gmail.com

N. N. H. Hashar
e-mail: nurulnabilahudahashar@gmail.com

A. Abdul Kadir et al. (eds.), *Sustainable Waste Utilization in Bricks, Concrete, and Cementitious Materials*, Lecture Notes in Civil Engineering 129,
https://doi.org/10.1007/978-981-33-4918-6_7

sector is undeniable as it supports the economy and development of our country. However, activities during the processing of agricultural products generate a significant amount of agricultural waste wherein the accumulation of this waste can be very severe and damaging towards the environment [1–3]. Agricultural waste usually can be obtained from raw material processing of agricultural products such as wheat, cereals, seeds, spices, fruits, vegetables, poultry, meat, and dairy products.

Malaysia is one of the leading countries in the agricultural sector, especially in the oil palm plantation sector, contributing to Malaysia's gross national income. According to the recent statistics by Malaysian Palm Oil Board (MPOB), the oil palm planted area in Malaysia reached approximately 5.85 million hectares in 2018, yielding almost 17.16 tonnes per hectare of fresh fruit bunches in the same year. Approximately 0.98 tonnes per hectare of palm kernel shell, 2.42 tonnes per hectare of mesocarp fiber, 3.9 tonnes per hectare of empty fruit bunch, and 0.89 tonnes per hectare of palm oil fuel ash are produced annually [4]. Pahang recorded the largest cultivation areas in Peninsular Malaysia, with a total area of 12.9%, while Sarawak accumulated a total planting area of 26.9% [5]. According to Awalludin et al. [2], palm oil made up only 10% of the total after extraction while the remaining 90% of palm oil mill waste is discarded as solid waste. Unfortunately, the waste is disposed of in staggering amounts without commercial returns. This very matter can lead to environmental problems and trigger a negative impact on human health.

Palm kernel shell (PKS) or endocarp is a fibrous shell fraction roughly 5–8 mm in diameter to about half a shell derived from crude palm oil processing, which enclosed the nut of palm kernel fruit after the crushing process [6]. Over the past decade, most research has emphasized the use of PKS for a source of fuel in wood cogeneration plants, thermal power stations, biomass power stations, and industrial plants as an alternative energy [7–9]. However, PKS is not fully utilized because the silicates from the shell form a scale in the boiler due to excess feed in the furnace [10]. Besides, PKS is complicated to degrade naturally and potentially emit excessive smoke during combustion [11]. Even though the combustion value is substantial, the process of burning PKS releases significant volatiles and particulates, which pose pre-ignition and pollution concerns. Hence, there is a need to find a suitable solution to dispose of the waste to reduce the impact towards the environment.

Therefore, many researchers and developers have shifted their interest in recycling agricultural waste in the construction sector, particularly as a replacement for building materials. Agricultural waste can be in the form of ashes, solid wastes, and sludges of varying elemental compositions. The example of agricultural waste such as sugarcane bagasse ash (SBA), rice husk ash (RHA), coffee ground ash, washed olive pomace ash, sawdust, sugarcane bagasse, rice husk, paddy crop, pomace waste and spent shea waste [12–19]. Building material especially fired clay brick, has gained attention as an alternative method to incorporate with agricultural waste to fulfil the demand of clay bricks due to urbanization [20, 21]. Fired clay brick is preferable due to the simple processing technique, the flexibility of its composition with any materials incorporated and significantly less maintenance.

Fig. 1 Grounded PKS

2 Materials and Methods

2.1 Raw Materials and Their Preparation

The clay soil was collected at Yong Peng Batu Bata Company, Johor. Upon delivery, the soil was oven-dried at 105 °C for 24 h to remove water content in the soil. It should be noted that water content cannot be easily eliminated in the natural environment. Therefore, drying the soil at a temperature above 105 °C is indispensable. The oven-dried soil was then stored in closed containers to maintain the consistency of the moisture content. Before conducting the test, the dried clay soil was grounded, crushed, and sieved to yield homogeneous particles that are able to pass through a 500 μm sieve.

Meanwhile, PKS, as shown in Fig. 1, was obtained from a palm oil mill located in Kluang, Johor. Upon arrival, the raw materials were oven-dried at 105 °C for 24 h to remove water content. Drying at 105 °C is sufficient to eliminate free form water. Nevertheless, it will not cause a significant loss of organic matter at the same time. Due to the nature of PKS, either crushing or grinding, was necessary.

2.2 Chemical and Geotechnical Properties

Raw materials also went through X-Ray Fluorescence Spectrometry (XRF) test in order to analyze the chemical composition while X-Ray Diffraction (XRD) was operated to access their mineralogical phase. XRF will give the detailed chemical composition of the sample. However, it is incapable of indicating the phases present in the sample. Therefore, X-Ray Diffraction (XRD) was further conducted to identify the shortcomings of the XRF test.

XRD was conducted with a powder method to achieve a representative of the whole sample. The sample was prepared by grinding and passing the sample through a 63 μm mesh sieve. The sample was then placed in a sample holder and flattened using a glass slide. While flattening, any excess powder should be cleaned to prevent clusters forming on the sample holder. The XRD was carried out using D8, Bruker with Cu Kα radiation in the range of 10–90°, scan step angle of 0.02°.

Geotechnical tests, including the Atterberg Limit test, specific gravity test, standard proctor test, and loss of ignition, were carried out to determine the general characteristics of raw materials used in this study. The Atterberg limit and specific gravity were determined according to BS 1377-2 [22], while a standard proctor test was performed to determine the optimum moisture content (OMC) of control bricks and clay-sludge bricks under BS 1377-4 [23]. The loss of ignition was determined by igniting raw materials at 1000 °C in a laboratory furnace for 4 h following BS 1377-3 [24].

2.3 *Methods for the Preparation of Brick Samples*

In this study, four different percentages of waste were incorporated into fired clay bricks with 0, 1, 5, and 10%. These percentages were chosen based on the experience of industrial manufacturers and previous researchers. Several studies reported positive results meeting or overtaking required percentages of waste used from 1 to 50% of waste [25–27]. However, waste content in this study was limited up to 10% because any further addition would be decreasing their physical and mechanical properties. The mixture proportions of raw materials for brick manufacturing are listed in Table 1.

Control brick was prepared by gradually adding predetermined water (calculated from optimum moisture content (OMC)) with clay soil until a homogeneous mixture was obtained. Meanwhile, PKS bricks were prepared by replacing clay soil with 1, 5 and 10% of PKS. During this preparation, a mechanical mixer with a 10 L capacity was used during the mixing procedure. Afterward, it was pressed into moulds with the following size; 215 mm × 102.5 mm × 65 mm and compressed at 2000 psi. High pressure is required to convert the loose clay mixture into dense and very compact bricks. The bricks were then kept at room temperature for 24 h to reduce the moisture in the bricks by 2–4%. The bricks were later placed in an oven at 105 °C for another

Table 1 Mixture identification of PKS brick

Mixture (%)	Clay (g)	PKS (g)	Water (mL)
CB-0	2800	0	476
PKSB1	2780	20	493
PKSB5	2710	93	518
PKSB10	2600	200	532

Fig. 2 Selected brick samples were positioned vertically to ensure even firing in a furnace

24 h. As mentioned by Muñoz et al. [28], slow drying should be done carefully, as it will affect the physical and mechanical properties of bricks.

The dried bricks were positioned vertically in a furnace and fired at 1050 °C at heating rates of 1 °C/min, as shown in Fig. 2. The temperature was selected based on the current industrial practice where a temperature of 1050 °C is sufficient to meet the optimum vitrification point [29–31]. A 2-h soaking time was applied at the finishing temperature to ensure that the whole brick attains uniformity. At the final stage, the bricks were left to undergo a slow cooling process to ensure that the liquid completely solidifies and bonds other particles together. The slow cooling rate is essential to avoid excessive thermal stress in the bricks while minimizing bricks' firing shrinkage. The manufactured bricks were stored in a secure place before further testing was performed.

2.4 Properties of Manufactured Brick

The initial test was performed by measuring the dimension of each brick before and after firing. Later, the brick samples were tested in terms of physical and mechanical properties. The physical and mechanical properties consisted of firing shrinkage, dry density, initial rate of absorption, water absorption, compressive strength.

The firing shrinkage of bricks is defined as the decrease in length after the firing process. The longitudinal shrinkage of the samples was directly measured to the nearest millimetre. The linear shrinkage was calculated in percentage to the nearest 0.5%. Ten clay brick samples for each mixture were tested for drying and firing shrinkage, and their values were averaged to represent the total linear shrinkage for the mixture. This shrinkage can be determined by using Eq. 1.

$$F_s = \left(\frac{L_s}{L}\right) \times 100 \tag{1}$$

where F_s is the percentage of shrinkage (%), L_s is the difference between actual length and dry length (m), and L is the actual length (m).

Meanwhile, dry density of manufactured bricks was calculated directly by dividing the mass of brick over the volume of bricks. Each brick was weighed as an ambient mass. The volume of brick was calculated by measuring the overall dimension (length, width, and height) of bricks. The volume was then used to calculate the dry density, ρd using Eq. 7.2.

$$\rho_d = \left(\frac{m_0}{V}\right) \times 1{,}000{,}000 \tag{2}$$

where ρ_d is the dry density of brick (kg/m^3), m_o is the ambient mass (g), and V is the volume of brick (m^3).

For the initial rate of absorption, ten samples of bricks were dried until a constant mass was achieved in a ventilated drying oven at 110 ± 8 °C and weighed after they were cooled to room temperature. The weight of the samples was determined to the accuracy of 0.001 g. Two bars (each not more than 6 mm wide) were placed to support the bed face of the test samples, which happened to be positioned parallel to each other and 3 mm beneath the water surface. The water level in the tank was maintained at 3 ± 1 mm above the bars. The samples were then placed on the bars for a period of 60 ± 1 s, measured from the time when the specimens come into contact with water. Upon removal from the water tank, the samples were immediately wiped dry and then reweighed within a minute. The initial rate of absorption due to the gross area of immersion (IRA), in kg/m^2 min, was calculated using Eq. 3.

$$\text{IRA} = \frac{1000(m_2 - m_1)}{A} \tag{3}$$

where m_1 is the weight of dry brick (g), m_2 is the weight of wet brick (g), and A is the net area of the contact surface of the brick with water (mm^2).

Furthermore, water absorption was conducted in accordance with BS EN 772:7 [32]. Brick samples were prepared and dried at 110 ± 8 °C in a drying oven and weighed at intervals of not less than 4 h until a constant weight was achieved (change in mass not greater than 0.1%). Measurements of the weight were taken after the samples were cooled to room temperature. The 5 h boiling test procedures were later followed in compliance with the standards. The water absorption of bricks can be computed using Eq. 4.

$$A = \left(\frac{m_2 - m_1}{m_1}\right) \times 100 \quad (4)$$

where m_1 is the dry mass of brick (g), m_2 is the wet mass of brick (g).

The compressive strength of brick in dry specimen condition was measured using a compression machine, Model NL 4000 X/006-A 002, with a maximum capacity of 2000 kN. Ten bricks were selected randomly from each batch. The bricks were oven-dried at 105 °C for at least 24 h and set to cool at room temperature for another 4 h until a constant mass was attained. The sample is positioned in the centre of the thrust of the testing machine to ensure uniform loading over the total area of the bricks. The specimen was initially applied with a convenient rate of loading and the rate was later adjusted after achieving half of the expected maximum load. Once the loading reaches failure, the strength of each specimen is calculated by dividing the maximum load achieved by its loaded area.

3 Results and Discussion

3.1 Properties of Raw Materials

The chemical composition of clay soil is presented in Table 2. The clay soil in this study consists mainly of silicon dioxide (SiO_2), aluminium oxide (Al_2O_3), and iron oxide (Fe_2O_3). As a primary constituent in soil, the abundant amount of silica (SiO_2) gives strength to clay brick after firing [33]. The studied clay also presents a practical limit of alumina (Al_2O_3), which is below 30%. It indicates a high quality of bricks [33]. In the meantime, high amounts of iron oxide in clay soil influences the reddish colour of brick in this study and it was also observed previously by Mueller et al. [33] and ILO [34]. The presence of iron oxide was also claimed to contribute to energy saving during the firing process by lowering the bricks vitrification temperature [33].

Meanwhile, minor contents in clay soil are sodium oxide (Na_2O) and calcium oxide (CaO). Due to the low composition of CaO but rich in silica-alumina content, clay soil in this study is categorized as non-calcareous clay, as studied by El Ouahabi et al. [35]. Consequently, non-calcareous clay has the potential to form mullite subjected to high temperature, nevertheless, under conditions that the CaO content should be low. The alkali oxides such as MgO, K_2O, ZnO and Na_2O can also affect brick properties. As can be noted, alkali oxides in clay soil are very minimal, hence prevent the brick from the excessive firing shrinkage. Barbieri et al. [36], in their

Table 2 Chemical composition of raw materials

Oxide content	Raw clay soil (%)	PKS (%)
SiO_2	55.7	47.1
Al_2O_3	24.4	15.19
Na_2O	0.3	0.94
K_2O	2.24	0.25
Fe_2O_3	4.46	12.2
CaO	0.25	7.59
MgO	1.2	0.93
PbO	n.a	1.01
P_2O_5	n.a	0.68
ZnO	n.a	0.99
TiO_2	0.94	n.a
MnO	0.04	1.16
LOI	1.92	9.23

study, also mentioned that the alkali oxides could be regarded as fluxing agents to lower the initial melting point to reduce energy consumption during firing brick.

The loss of ignition (LOI) of clay soil was recorded as 1.9%, which is expected to be low with organic matter and clay minerals, as precisely discussed by Monazam et al. [37] and Monteiro and Vieira [38]. The LOI of clay soil is also depending on the site location as well as type of soil used for making bricks. As a result, the lower LOI value of clay soil in this study supported the fact that the clay used is non-calcareous.

The chemical composition of the palm kernel shell (PKS) is tabulated in Table 2. From the table, PKS was found high in silicon dioxide (SiO_2), aluminium oxide (Al_2O_3) and iron oxide (Fe_2O_3). Similar to clay soil, the presence of iron oxide in PKS potentially provides the reddish colour of brick when incorporated into a clay brick. Additionally, minor components such as potassium oxide (K_2O), phosphorus oxide (P_2O_5) and magnesium oxide (MgO) were detected in PKS. The presence of these alkali oxides was presumably from the impurities during the palm oil process, hence increased the potential of these elements to attach on the samples. It was observed that the chemical composition obtained in this study resembled the result from Huang et al. [39] and Nizamuddin et al. [40].

Meanwhile, loss of ignition result indicated that PKS sample contained a high amount of organic matter with 9.23% because PKS is lignocellulosic material. Arsenovic et al. [41] emphasized that the incorporation of high LOI value into fired clay bricks could be advantageous in terms of energy-saving and promoting porosity after firing, which consequently improving the thermal conductivity of bricks. Regarding fiber content, Table 3 shows PKS is mainly composed of cellulose (24.9%), hemicellulose (19.8%) and lignin (35.8%). The results of fiber content obtained in this study are closed to previous studies, where the content of lignocelluloses is also around 23–28% (cellulose), 21–24% (hemicellulose) and 44–60% (lignin) [39, 40].

XRD pattern of clay soil is presented in Fig. 3. The crystalline phases identi-

Table 3 Properties of PKS

Chemical component	PKS (wt%)
Cellulose	24.9
Hemicellulose	19.8
Lignin	35.8

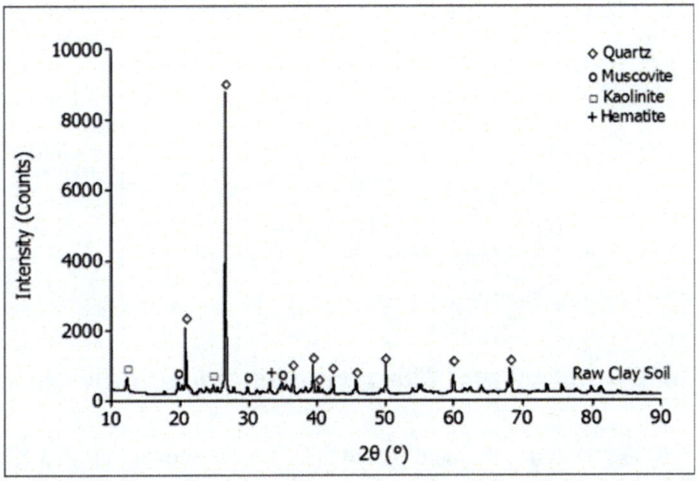

Fig. 3 XRD of clay soil

fied quartz (SiO_2) as a primary crystalline structure present in clay soil at scanning degrees of 20.8 and 26.6° 2θ. As well-agreed by Akinshipe and Kornelius [42] and Ingham [43], quartz is the common mineral found in soil. Meanwhile, minor minerals present in clay soil were muscovite [$KAl_2(AlSi_3O_{10})(F,OH)_2$], kaolinite [$Al_2(SiO_2O_5)(OH)_5$] and hematite ($Fe_2O_3$) at 19.8, 12.3 and 33.1° 2θ, respectively. Muscovite and kaolinite are the original minerals that exist in clay soil. In the meantime, the presence of hematite resulted from the crystallization of iron in clay soil, is similar to those observed in the previous studies [38]. As a final product, the presence of hematite in clay soil influenced the red colour of bricks. Therefore, the crystalline spectrum observed in clay soil is verifying the results of XRF in Table 2.

The mineralogical composition of PKS is displayed in Fig. 4. From the figure, the following mineralogical phases present in PKS were quartz (SiO_2) at 20.9 and 26.6° 2θ, followed by kaolinite [$Al_2(Si_2O_5)(OH)_5$] at 22.0° 2θ and hematite (Fe_2O_3) at 31.4° 2θ. As observed by Okoroigwe et al. [8], the presence of quartz is an indication of silicate in PKS. The formation of kaolinite and hematite in PKS as proven by the XRF data in Table 2, further indicates that aluminium oxide and iron oxide are the significant compounds present in PKS.

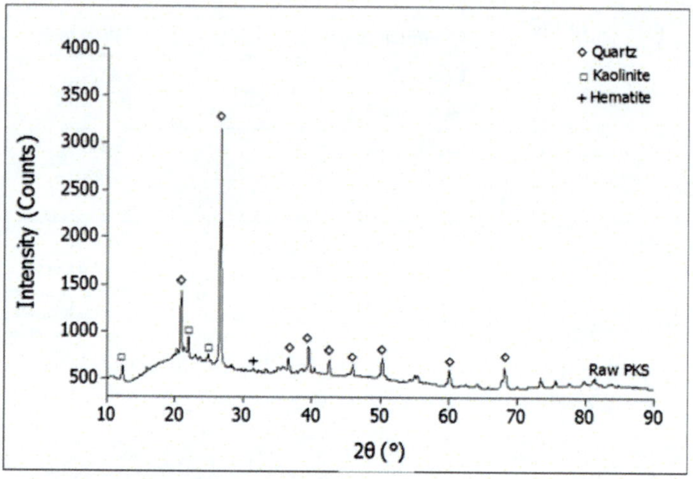

Fig. 4 XRD of palm kernel shell

3.2 *Microstructure and Thermogravimetric of Raw Materials*

The SEM image of clay soil magnified at 500 times is shown in Fig. 5. Clay soil was observed to have an angular soil particle with irregular shape. The surface of soil was also rough. From the analysis using ImageJ, the particle size of clay soil ranged

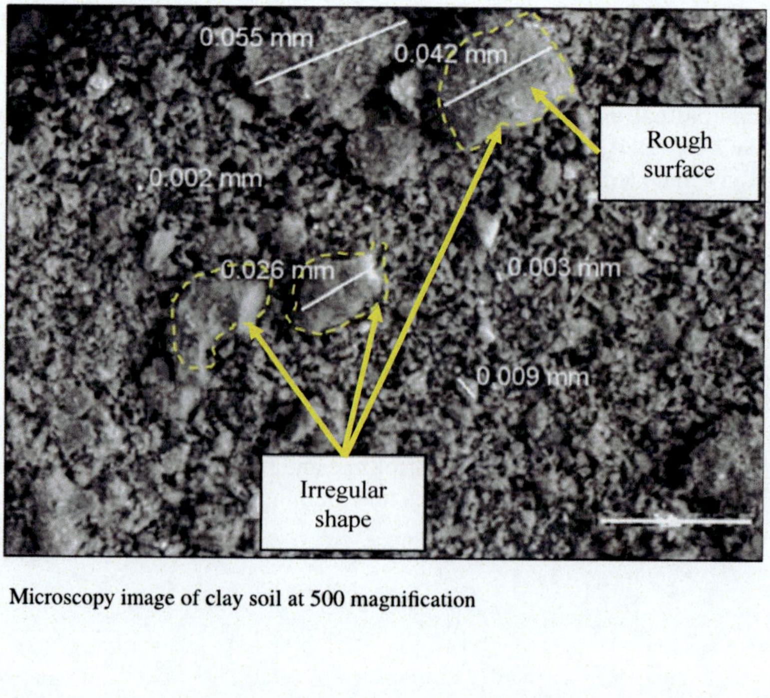

Fig. 5 Microscopy image of clay soil at 500 magnification

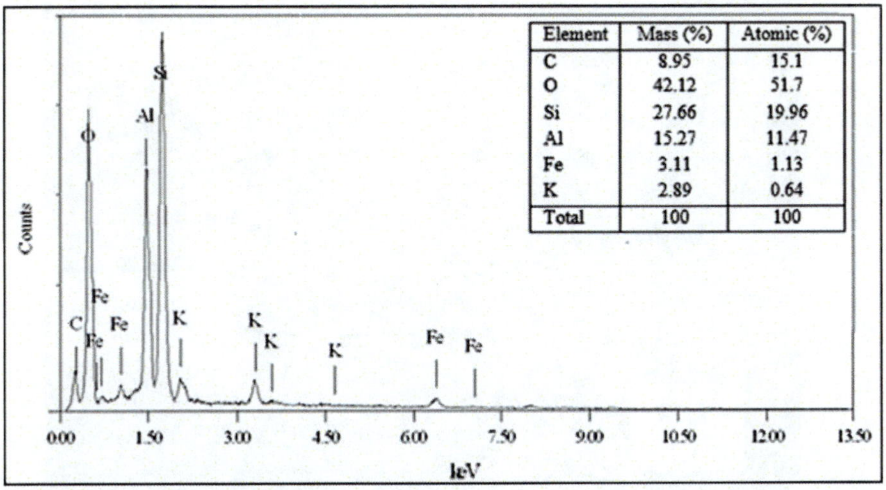

Fig. 6 EDX spectrum of clay soil at 500 magnification

from 0.002 to 0.055 mm. The particle size below 0.002 mm present in clay soil contributes to the brick strength. The particle sizes between 0.002 and 0.063 mm are beneficial in reducing the plasticity properties which prevent higher shrinkage during the drying process. This finding is in agreement with Mueller et al. [33] which shows the suitable sizes of clay soil particles would influence the workability during making brick while enhancing brick strength after firing. Meanwhile, the EDX spectrum in Fig. 6 shows carbon (C), oxygen (O), silica (Si), aluminium (Al), iron (Fe) and potassium (K) present in clay soil with. It appears that the elements obtained from the EDX spectrum are conforming that Si and Al are the highest elements obtained from XRF in Table 2.

Meanwhile, the morphology of PKS at 500 magnification is presented in Fig. 7. From the figure, it indicates that the surface of PKS was rough and non-uniform shapes, even with non-porous on the surface, similarly discovered by Bazargan et al. [11]. It was also observed that silica deposit in the morphology image of PKS, indicating that silicon dioxide is the major component in PKS during the chemical analysis. The same observation also observed by Anyika et al. [44] where PKS contains silica deposit due to impurities. According to the EDX spectrum in Fig. 8, major and minor elements that are detected in PKS such as carbon (C), oxygen (O), silica (Si), aluminium (Al) and iron (Fe). Note that Si content is higher than other elements showing that silica deposit abundantly attached on the PKS sample which contributes to the quartz formation.

The thermogravimetric of clay soil can be seen in Fig. 9. Apparently, total weight loss was observed to decrease by around 1.1 mg (5% decomposed). The first stage was represented by an endothermic activity via removal of hygroscopic water from the clay body, as previously observed by Monteiro et al. [45]. The weight loss in the first stage was expected to decrease by 0.4 mg in between 20 and 261.9 °C. From

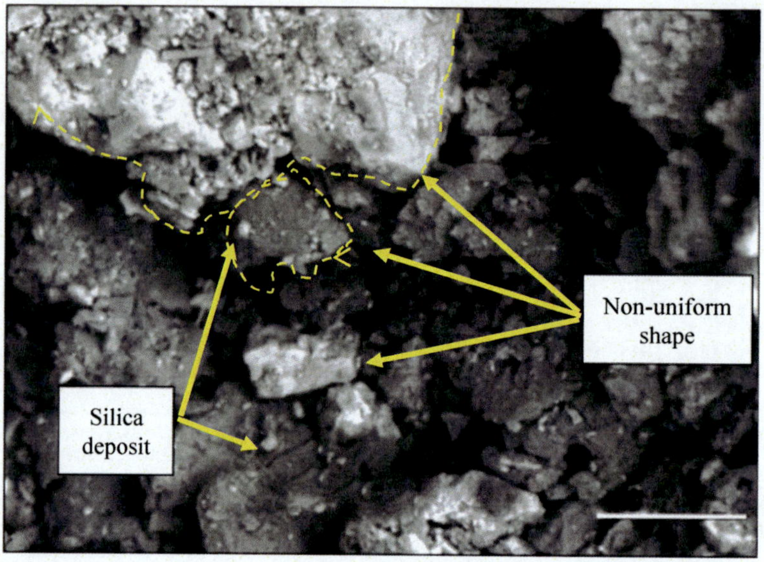

Fig. 7 EDX spectrum of palm kernel shell at 500 magnification

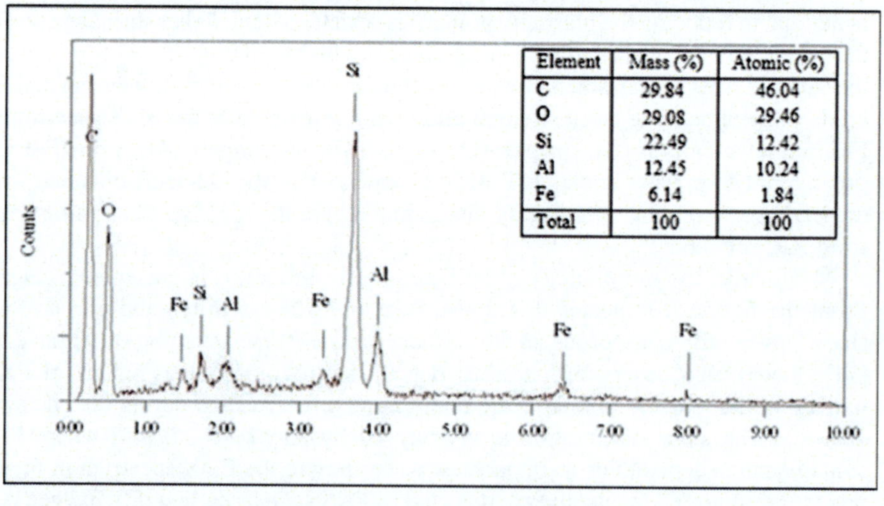

Element	Mass (%)	Atomic (%)
C	29.84	46.04
O	29.08	29.46
Si	22.49	12.42
Al	12.45	10.24
Fe	6.14	1.84
Total	100	100

Fig. 8 EDX spectrum of palm kernel shell at 500 magnification

261.9 to 685.9 °C, an exothermic reaction was detected at the peak of 404.1 °C. This reaction is fundamentally related to the oxidation of organic matter and dehydroxylation reaction of clay mineral (usually alumina or iron), which is consistent with Ramachandran et al. [46]. Due to these changes, about 0.7 mg weight loss was recorded. The exothermic reaction also indicates that organic matter begins to oxidize and emit significant pollutants such as carbon monoxide and carbon dioxide

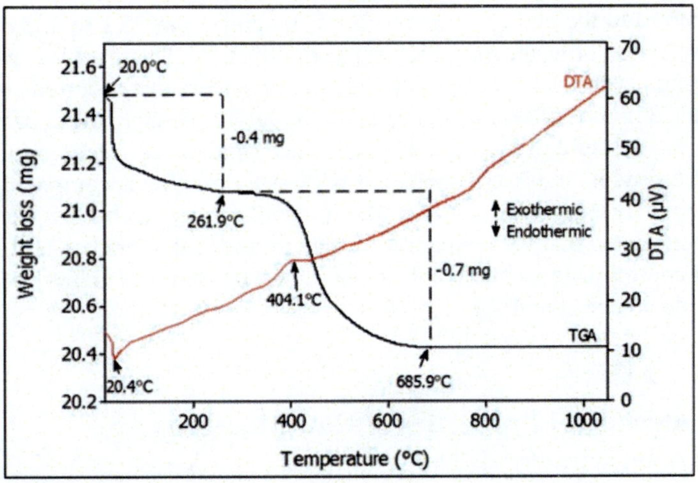

Fig. 9 DTA-TGA of clay soil

to the surrounding which corroborates the findings of Rathossi and Pontikes [47]. At the third stage, an exothermic and endothermic reaction occurred near 800 °C, indicated by the crystallization process of the current phase such as alumina which subsequently reached vitrification point (near 900 °C). The reaction, as observed in Fig. 9, is consistent with the previous studies due to the nature of clay minerals present in clay soil [48].

Figure 10 displayed the thermal behaviour of palm kernel shell (PKS). The total mass loss of PKS was recorded as 7.10 mg (63% decomposed). The first mass loss

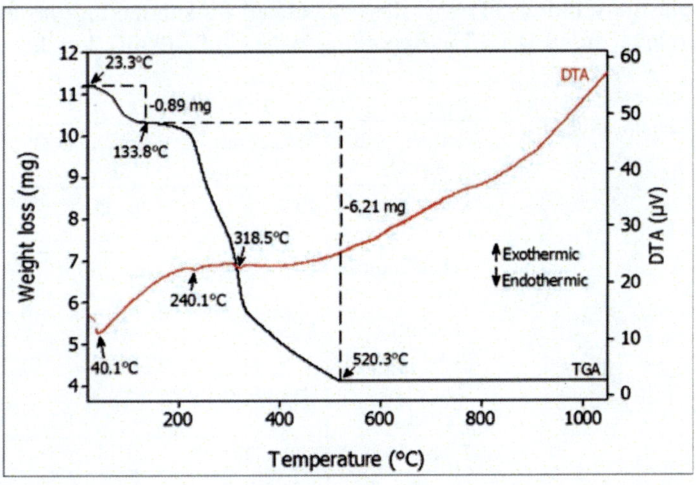

Fig. 10 DTA-TGA of clay soil of palm kernel shell

was referred to the endothermic reactions (ranging from 23.3 to 133.8 °C) due to the dehydration of water from PKS at peak of 40.1 °C. The mass loss was recorded at 0.89 mg. At the second stage, two endothermic reactions occurred at the peak of 240.1 and 318.5 °C. Noticeably, this reaction was attributed to the decomposition or devolatilization of cellulose and hemicellulose (mass loss of 6.21 mg). At the last stage, the residual lignin contained in PKS was thoroughly decomposed through an endothermic reaction (starting after 520 °C) until no weight changes were recorded. It has been noted that the temperature range for thermal degradation of PKS in this study is approximately comparable to the studies by Huang et al. [39], Ma et al. [49] and Ninduangdee et al. [50].

4 Geotechnical Properties of Raw Materials

4.1 Atterberg Limit of Clay Soil

The engineering behaviour of the soil is greatly influenced by the amount of water present in the sample. Therefore, the Atterberg limit test was performed to evaluate the soil behaviour of clay soil in terms of plastic limit and liquid limit. The soil classification is presented in Table 4.

The plastic limit (PL) at which the soil starts to crumble when rolled into a thread approximately 3 mm in diameter was 14.6%. Meanwhile, the liquid limit (LL) was obtained from the flow curve as in Fig. 11, in which a straight-line relationship between water content and penetration was plotted. From the graph, the water content was observed as 29.9% denoted by the liquid limit of the soil sample which corresponds to a penetration of 20 mm.

The plasticity index (PI) was then calculated by subtracting the PL from the LL, wherein recorded as 15.3%. According to the Unified Soil Classification System

Table 4 Classification of soil

Soil physical properties	Tests result	Mueller et al. [33]	ILO [34]
Liquid limit (%)	29.9	20–35	30–35
Plastic limit (%)	14.6	n.a	12–22
Plasticity index (%)	15.3	7–16	7–18
Degree of plasticity	Low plasticity	n.a	n.a
Type of soil	Silty clay	n.a	n.a
Specific gravity (SG)	2.56 (clay)	n.a	n.a

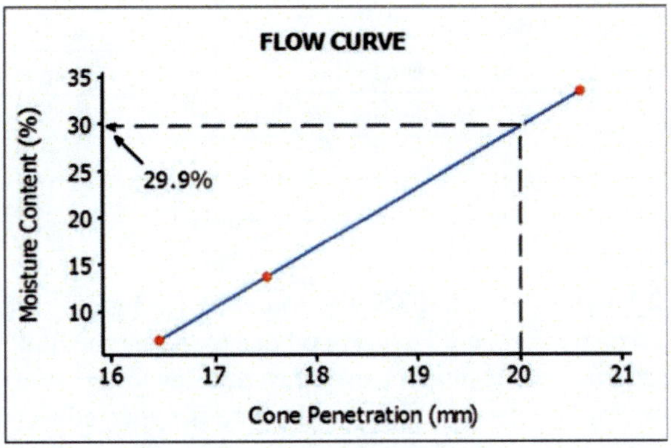

Fig. 11 Graph moisture content against cone penetration

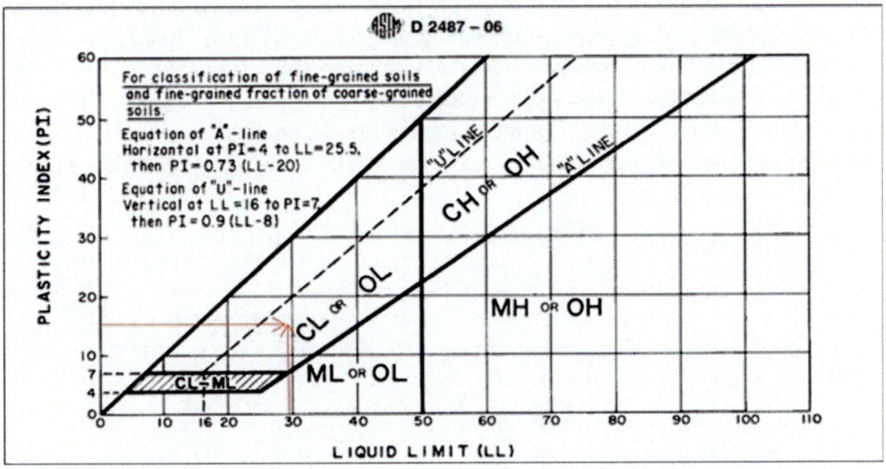

Fig. 12 Classified of soil according to unified soil classification system (USCS)

(USCS) in Fig. 12, the studied clay soil was classified as silty clay with low plasticity. The result suggested that the plastic limit, liquid limit and plasticity index of clay soil complied with standard requirements proposed by Mueller et al. [33] and ILO [34]. However, the soil studied in this study was categorized as a low plasticity index, thus decreasing the workability during extrusion, as reported by ILO [34].

4.2 Particle Density of Raw Materials

The particle density of clay soil and PKS was determined using small density bottles, and the data is recorded in Table 5. The result showed that the particle density of

Table 5 Particle density of the samples

Specimen reference	Experimental value	Fono-Tamo and Koya [52]
	Particle density (g/cm^3)	
Clay soil	2.56	n.a
PKS	1.62	1.26

clay soil was 2.56 g/cm^3, while PKS was found as 1.62 g/cm^3. PKS represents the lowest particle density compared to clay soil due to the lignocellulose characteristic, which also contains a great number of pores, similar to the study by Anuar et al. [51]. Therefore, the low particle density of waste is expected to reduce the density of bricks in sequence from the lightest to heaviest bricks; PKSB < Control brick.

Standard proctor test was conducted to obtain the moisture-density relationship. The first series of compaction tests were performed on control brick to determine the control brick compaction. Afterwards, the second series of compaction was made by mixing clay soil with a predetermined percentage of POMW (1, 5 and 10%). Accordingly, the maximum dry density (MDD) and optimum moisture content (OMC) were obtained from the peak point of the compaction curve as in Fig. 13, and the result is recorded in Table 6. From the compaction curve, dry density was increased with an increase in water content until they achieved MDD. Then, the water must be added

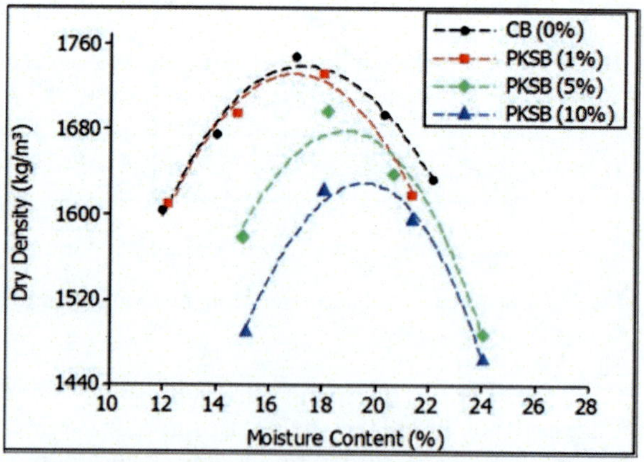

Fig. 13 Optimum moisture content of manufactured brick

Table 6 Maximum dry density and optimum moisture content of PKS brick

Mixture identification	Percentages (%)	Maximum dry density (g/cm^3)	Optimum moisture content (%)	Water content (mL)
CB	0.0	1.75	17.0	476
PKSB	1.0	1.74	17.6	493
PKSB	5.0	1.70	18.5	518
PKSB	10.0	1.64	19.0	532

in a proper amount to obtain the correct consistency for moulding while ensuring the homogeneity of the mixture. Therefore, the addition of water encouraged the formation of plastic bodies, thus reduced the workability during manufactured brick.

5 Physical and Mechanical Properties of Manufactured Brick

5.1 Firing Shrinkage of Manufactured Brick

Firing shrinkage is the parameter to measure the quality of brick. It was determined by measuring the length of the specimen before and after firing. Based on Fig. 14, firing shrinkage of manufactured brick was found to increase slightly with respect to the increasing amount of waste. The addition of waste content from 1 to 10% ultimately increased the firing shrinkage of PKSB from 0.6 to 0.9%.

From the result, control brick showed the lowest firing shrinkage since it contained an appropriate amount of fine-grained size of clay soil that eventually consumed lower water content during the mixing process. Hence, the correct amount of fine-grained in clay soil would increase the workability during moulding and decrease the firing shrinkage rate, as similarly proposed by Mueller et al. [33] and ILO [34]. The result also showed that incorporation of more than 5% POMW had produced a higher shrinkage compared to other bricks, parallel with the pattern observed by Eliche-Quesada and Leite-Costa [14], Kurovics et al. [53] and Barbieri et al. [36].

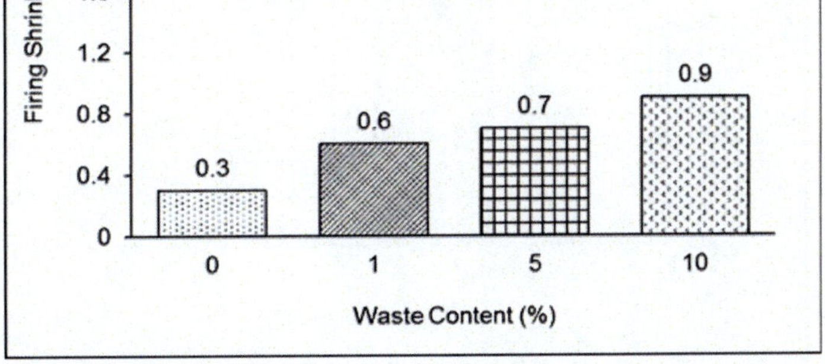

Fig. 14 Firing shrinkage of fired clay brick

Referring to the material properties of brick masonry, firing shrinkage usually falls in the range of 2.5–4% [54]. It was observed that firing shrinkage for all types of bricks was below the allowable limit, and size variations of the bricks were virtually homogeneous. As mentioned in the literature, firing temperature and heating rate should be done slowly and evenly in order to avoid clay stress as it will cause high shrinkage [14]. Therefore, it can be observed that a good pattern of firing shrinkage obtained in this study was due to the proper drying and firing shrinkage. In fact, the incorporation of POMW into fired clay bricks with a proper amount of clay soil particles may also help to reduce the firing shrinkage rate on the PKS brick, even though PKS was observed to contain a high amount of cellulose materials.

5.2 *Dry Density of Manufactured Brick*

Dry density was determined by weighing each brick sample to represent the ambient mass of brick. Figure 15 showed that the density of PKS brick is inversely proportional to the percentages of waste added in the mixture where dry density was decreased significantly with the increasing amount of waste. The dry density of control brick remains the heaviest compared to others brick with 1799 kg/m^3. In the meantime, the incorporation of 1–10% of PKS has significantly decreased dry density from 1794 to 1645 kg/m^3.

It was perceived that incorporation of up to 10% of PKS has significantly decreased the density of brick up to 9% lighter than control brick. Besides the reduction of weight through the decomposition of organic matter, the density of brick was also

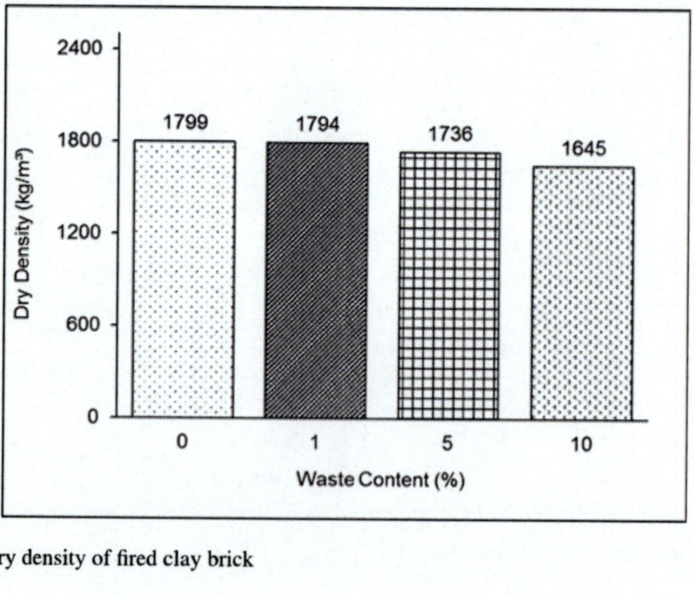

Fig. 15 Dry density of fired clay brick

decreased due to the removal of mechanical and chemical water from the brick body during the firing process, similarly as observed by Bories et al. [55].

Meanwhile, data of dry density obtained in this study was below the conventional brick density reported by Standard Australia [56]. As evident, through a thermal analysis of PKS, almost all organic material contained in PKS has been decomposed at below 500 °C thus leaving numerous pores inside the brick, in companion with the result obtained by Kizinievič et al. [57]. Therefore, lightweight fired clay brick was expected to develop. Apparently, lower density brick gives advantages in construction fields in terms of easier handling and higher production of brick [58–60].

5.3 Initial Rate of Absorption of Manufactured Brick

Initial Rate of Absorption (IRA) is used as an indicator of the expected bond strength of the brick surface and mortar. According to Fig. 16, the addition of PKS has produced a significant increase in the IRA. Additionally, the IRA of control brick was the lowest compared to other bricks at 1.3 kg/m^2 min. The incorporation of 1–10% of PKS into fired clay bricks has eventually increased the IRA value from 3.7 to 7.9 kg/m^2 min. It can be understood that the incorporation of up to 10% of PKS has significantly increased IRA values due to the interconnected pores, voids and capillaries presence in the brick body. The results collected in this study are relatively close to the previous study in which the increasing waste content into clay bricks will potentially increase the suction rate [57].

Furthermore, BS EN 771-1 recommended that IRA should be lower than 2 kg/m^2 min [61]. Unfortunately, all the bricks manufactured in this study are above the acceptable limit of IRA. Hence, this might affect the bonding between brick and mortar. However, the standard had suggested that bricks with higher IRA should be immersed for 3 to 24 h before. Higher IRA value may also cause defects in the ceramic body, which in turn lower the durability as pointed by Ukwatta et al. [62]. Thus, for the mentioned reasons, higher values of IRA should be avoided.

5.4 Water Absorption of Manufactured Brick

Water absorption is fundamental in influencing the properties of bricks, such as the resistance of brick to freeze and thaw, and their chemical stability [63, 64]. Figure 17 shows the addition of POMW has demonstrated the increase of water absorption. Following this, the water absorption of control bricks remained lower at 3%. However, the incorporation of 1–10% of PKS into fired clay bricks has ultimately increased the water absorption value. The increasing of PKS content into fired clay bricks tends to increase the water absorption from 8 to 12%. Water absorption may happen due to the elimination of organic matter from the brick bodies during sintering

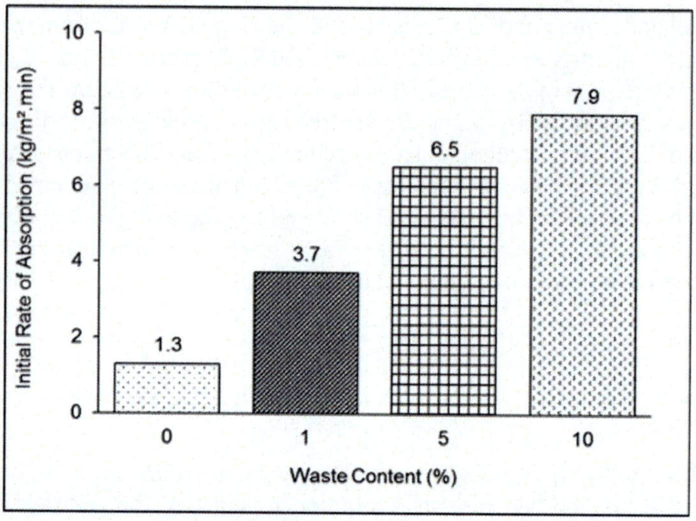

Fig. 16 IRA of fired clay brick

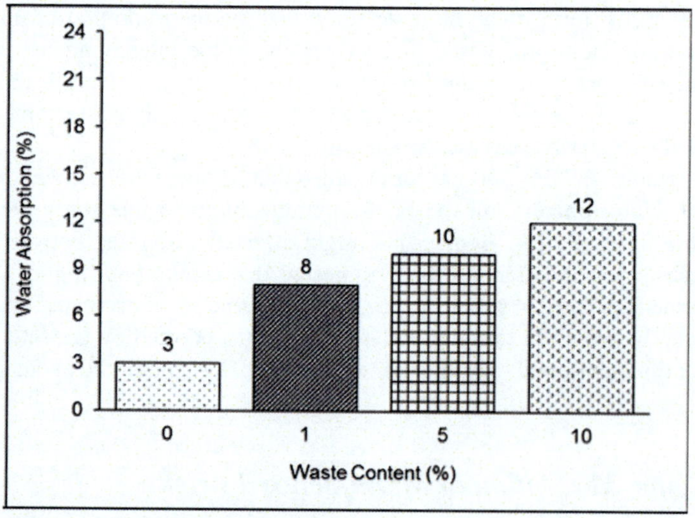

Fig. 17 Water absorption of fired clay brick

process which then created numerous pores inside the brick as found to be consistent with the trend by previous studies [65, 67].

Water absorption may happen due to the elimination of the organic matter from the brick bodies during sintering process which thus created numerous pores inside the brick as found to be consistent with the trend by previous studies [65–67]. Maximum water absorption can be referred to British Standard, whereby the water absorption is classified with low water absorption ≤4.5% by mass as Engineering Brick A, while brick with high water absorption ≤7.0% by mass as Engineering Brick B. Nevertheless, the results were found to be beyond the limit of ≤4.5% and ≤7.0% by mass. Thus, these manufactured bricks could not be used as a high loading application, but merely applicable for non-loading applications. However, all the results comply with the ASTM standard, where maximum water absorption was set not to exceed 22% for moderate-weather-resistant bricks [68]. Henceforth, minimum water absorption is required since water penetration is the key factor that leads to the durability of the brick [62, 69].

5.5 Compressive Strength of Manufactured Brick

The compressive strength is a fundamental property to ensure the engineering quality of bricks to withstand loads. Brick which is fired to high temperature can easily withstand high loads and suitable for almost all structural applications. It can be seen in Fig. 18, the increase of PKS content into fired clay bricks has eventually decreased the compressive strength of brick.

Control brick is capable of withstanding the loads for up to 24.6 MPa. Meanwhile, the incorporation of 1–10% of PKS into fired clay bricks subsequently decreased the compressive strength. As presented, the trend shows that the compressive strength of PKSB decreased from 19.8 to 11 MPa. According to the standard limit, the results obtained in this experiment are applicable to be used as a moderate-weather-resistant

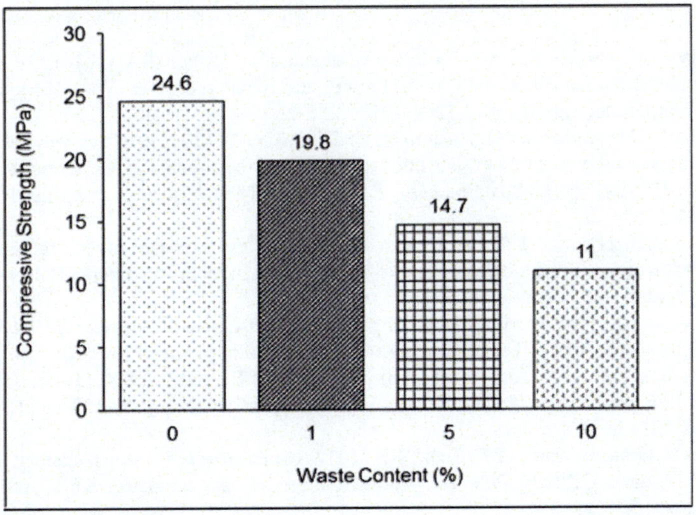

Fig. 18 Compressive strength of fired clay brick

brick (≥17.2 MPa), loading bearing 1 and 2 (≥7 and ≥14 MPa), non-loading bearing partition (≥1.4 MPa) and a loadbearing internal wall (≥5.2 MPa) [68, 70].

6 Conclusion

Considering the huge volume of palm oil mill waste which is produced yearly around the world, and the demand of bricks fabricated as a building material, recycling palm oil mill waste into fired clay bricks is a practical and possibly important contribution towards a sustainable solution to one of the genuine environmental pollution problems in the world. Based on the study, different percentages of PKS (1, 5 and 10%) were incorporated into fired clay bricks while the samples without PKS were considered as control bricks. Following this, preliminary examinations on the brick appearance found out that there are insignificant defects such as minor crack and bloating observed after firing. In any case, a certain percentage of PKS was observed to have black core on the fracture section of the PKS brick. Accordingly, the results indicated that it is possible to incorporate 1 and up to 5% of PKS into fired clay brick, which satisfies several limit standards and ensures the physical and mechanical properties are similar to those of control brick. However, the incorporation of 20 up to 30% of PKS has been found to fail due to major disintegration effects during the sintering stage.

References

1. Onoja E, Chandren S, Abdul Razak FI, Mahat NA, Wahab RA (2018) Oil palm (Elaeis guineensis) biomass in Malaysia: the present and future prospects. Waste Biomass Valorizat 1–19. https://doi.org/10.1007/s12649-018-0258-1
2. Awalludin MF, Sulaiman O, Hashim R, Nadhari WNAW (2015) An overview of the oil palm industry in Malaysia and its waste utilization through thermochemical conversion, specifically via liquefaction. Renew Sustain Energy Rev 50:1469–1484. https://doi.org/10.1016/j.rser.2015.05.085
3. Hansen UE, Nygaard I (2014) Sustainable energy transitions in emerging economies: the formation of a palm oil biomass waste-to-energy niche in Malaysia. Energy Policy 66:666–676. https://doi.org/10.1016/j.enpol.2013.11.028
4. Malaysian Palm Oil Board (2019a) Oil palm planted area as at December 2018 (Hectares). In: Oil palm plant. Area 2018. https://bepi.mpob.gov.my/images/area/2017/Area_summary.pdf
5. Malaysian Palm Oil Board (2019b) Monthly FFB yield 2018 (Tonnes/Hectare). In: Mon. FFB yield. https://bepi.mpob.gov.my/index.php/statistics/yield/129-yield-2014/690-yieldfiles-2014.html
6. Yunus R, Omar R, Abidin ZZ, Biak DRA (2012) Oil Palm as bioenergy feedstock. In: Lai O-M, Tan C-P, Akoh CC (eds) Palm oil: production, processing, characterization, and uses. AOCS Press, pp 653–692
7. Faizal HM, Shamsuddin HS, Heiree MHM, Hanaff MFMA, Rahman MRA, Rahman MM, Latiff ZA (2018) Torrefaction of densified mesocarp fibre and palm kernel shell. Renew Energy 122:419–428. https://doi.org/10.1016/j.renene.2018.01.118

8. Okoroigwe EC, Saffron CM, Kamdem PD (2014) Characterization of palm kernel shell for materials reinforcement and water treatment. J Chem Eng Mater Sci 5(1):1–6. https://doi.org/10.5897/JCEMS2014.0172
9. Uemura Y, Omar WN, Tsutsui T, Yusup SB (2011) Torrefaction of oil palm wastes. Fuel 90(8):2585–2591. https://doi.org/10.1016/j.fuel.2011.03.021
10. Poku K (2002) Small-scale palm oil processing in Africa. Food and Agricultural Organization of the United Nations
11. Bazargan A, Kostić MD, Stamenković OS, Veljković VB, McKay G (2015) A calcium oxide-based catalyst derived from palm kernel shell gasification residues for biodiesel production. Fuel 150:519–525. https://doi.org/10.1016/j.fuel.2015.02.046
12. Kazmi SMS, Munir MJ, Wu YF, Hanif A, Patnaikuni I (2018) Thermal performance evaluation of eco-friendly bricks incorporating waste glass sludge. J Clean Prod 172:1867–1880. https://doi.org/10.1016/J.JCLEPRO.2017.11.255
13. Adazabra AN, Viruthagiri G, Shanmugam N (2017) Management of spent shea waste: an instrumental characterization and valorization in clay bricks construction. Waste Manag 64:286–304. https://doi.org/10.1016/j.wasman.2017.03.006
14. Eliche-Quesada D, Leite-Costa J (2016) Use of bottom ash from olive pomace combustion in the production of eco-friendly fired clay bricks. Waste Manag 48:323–333. https://doi.org/10.1016/j.wasman.2015.11.042
15. Luna-cañas LM, Ríos-reyes CA, Quintero-Ortíz LA (2014) Recycling of agroindustrial solid wastes as additives in brick manufacturing for development of sustainable construction materials. DYNA 81(188), 34–41. https://doi.org/10.15446/dyna.v81n188.39717
16. Velasco PM, Ortíz MPM, Giró MAM, Velasco LM (2014) Fired clay bricks manufactured by adding wastes as sustainable construction material—a review. Constr Build Mater 63:97–107. https://doi.org/10.1016/j.conbuildmat.2014.03.045
17. Kadir AA, Maasom N (2013) Recycling sugarcane bagasse waste into fired clay brick. Int J Zero Waste Gener 1(1):21–26. https://doi.org/10.1016/j.jenvman.2012.01.032
18. Eliche-Quesada D, Pérez-Villarejo L, Iglesias-Godino FJ, Martínez-García C, Corpas-Iglesias FA (2011) Incorporation of coffee grounds into clay brick production. Adv Appl Ceram 110(4):225–232. https://doi.org/10.1179/1743676111Y.0000000006
19. Mohammed MS, Ahmed AE-SI, Osman RM, Khattab I (2008) Combinations of organic and inorganic wastes for brick production. Polym Polym Compos 16(2), 101–113. https://doi.org/https://doi.org/10.1002/pc.22647
20. Murmu AL, Patel A (2018) Towards sustainable bricks production: an overview. Constr Build Mater 165:112–125. https://doi.org/10.1016/j.conbuildmat.2018.01.038
21. Shakir AA, Mohammed AA (2013) Manufacturing of bricks in the past, in the present and in the future: a state of the art review. Int J Adv Appl Sci 2(3):145–156. https://doi.org/10.11591/ijaas.v2i3.1751
22. British Standard Institution (1990) Method of test for soil for civil engineering purposes—Part 2: Classification tests. London, BS 1377-2
23. British Standard Institution (1990) Methods of test for soils for civil engineering purposes—Part 4: Compaction-related tests. London, BS 1377-4
24. British Standard Institution (1990) Methods of test for soils for civil engineering purposes—Part 3: Chemical and electro-chemical tests. London. BS 1377–3
25. Phonphuak N, Kanyakam S, Chindaprasirt P (2016) Utilization of waste glass to enhance physical–mechanical properties of fired clay brick. J Clean Prod 112(112):3057–3062. https://doi.org/10.1016/j.jclepro.2015.10.084
26. Kadir AA, Mohajerani A (2012) Properties improvement of fired clay bricks incorporating with cigarette butts. Adv Mater Res 535–537:1723–1730. https://doi.org/10.4028/www.scientific.net/AMR.535-537.1723
27. Baspinar MS, Demir I, Orhan M (2010) Utilization potential of silica fume in fired clay bricks. Waste Manag Res 28(2):149–157. https://doi.org/10.1177/0734242X09104385
28. Muñoz VP, Morales OMP, Letelier GV, Mendívil GMA (2016) Fired clay bricks made by adding wastes: assessment of the impact on physical, mechanical and thermal properties. Constr Build Mater 125:241–252. https://doi.org/10.1016/j.conbuildmat.2016.08.024

29. Taha Y, Benzaazoua M, Hakkou R, Mansori M (2016) Natural clay substitution by calamine processing wastes to manufacture fired bricks. J Clean Prod 135:847–858. https://doi.org/10.1016/j.jclepro.2016.06.200
30. Bennour A, Mahmoudi S, Srasra E, Boussen S, Htira N (2015) Composition, firing behavior and ceramic properties of the Sejnène clays (Northwest Tunisia). Appl Clay Sci 115:30–38. https://doi.org/10.1016/j.clay.2015.07.025
31. Alonso-Santurde R, Coz A, Viguri JR, Andrés A (2012) Recycling of foundry by-products in the ceramic industry: green and core sand in clay bricks. Constr Build Mater 27(1):97–106. https://doi.org/10.1016/j.conbuildmat.2011.08.022
32. British Standard Institution (1998) Methods of test for masonry units-part 7: determination of water absorption of clay masonry damp proof course units by boiling in water. London. BS EN 772-7
33. Mueller H, Maithy S, Prajapati S, Bhatta AD, Shrestha BL (2008) Green brick making manual. Nepal
34. ILO (1984) Small scale brickmaking. International Labour Office
35. El Ouahabi M, Daoudi L, Hatert F, Fagel N (2015) Modified mineral phases during clay ceramic firing. Clays Clay Miner 63(5):404–413. https://doi.org/10.1346/CCMN.2015.0630506
36. Barbieri L, Andreola F, Lancellotti I, Taurino R (2013) Management of agricultural biomass wastes: preliminary study on characterization and valorisation in clay matrix bricks. Waste Manag 33(11):2307–2315. https://doi.org/10.1016/j.wasman.2013.03.014
37. Monazam ER, Breault RW, Siriwardane R (2014) Kinetics of magnetite (Fe_3O_4) oxidation to hematite (Fe_2O_3) in air for chemical looping combustion. Ind Eng Chem Res 53(34):13320–13328. https://doi.org/10.1021/ie501536s
38. Monteiro SN, Vieira CMF (2004) Influence of firing temperature on the ceramic properties of clays from Campos dos Goytacazes, Brazil. Appl Clay Sci 27:229–234. https://doi.org/10.1016/j.clay.2004.03.002
39. Huang Y, Liu H, Yuan H, Zhan H, Zhuang X, Yuan S, Yin X, Wu C (2018) Relevance between chemical structure and pyrolysis behavior of palm kernel shell lignin. Sci Total Environ 633:785–795. https://doi.org/10.1016/j.scitotenv.2018.03.238
40. Nizamuddin S, Shrestha S, Athar S, Ali BS, Siddiqui MA (2016) A critical analysis on palm kernel shell from oil palm industry as a feedstock for solid char production. Rev Chem Eng 32(5):489–505. https://doi.org/10.1515/revce-2015-0062
41. Arsenovic M, Pezo L, Radojevic Z (2012) Response surface method as a tool for heavy clay firing process optimization: roofing tiles. Process Appl Ceram 6(4):209–214. https://doi.org/10.2298/pac1204209a
42. Akinshipe O, Kornelius G (2017) Chemical and thermodynamic processes in clay brick firing technologies and associated atmospheric emissions metrics—a review. J Pollut Eff Control 5(2):1–12. https://doi.org/10.4172/2375-4397.1000190
43. Ingham J (2013) Bricks, terracotta and other ceramics. In: Geomaterials under the microscope. Academic Press, pp 163–170
44. Anyika C, Asri NAM, Majid ZA, Yahya A, Jaafar J (2017) Synthesis and characterization of magnetic activated carbon developed from palm kernel shells. Nanotechnol Environ Eng 2(1):16. https://doi.org/10.1007/s41204-017-0027-6
45. Monteiro SN, Alexandre J, Margem JI, Sánchez R, Vieira CMF (2008) Incorporation of sludge waste from water treatment plant into red ceramic. Constr Build Mater 22(6):1281–1287. https://doi.org/10.1016/J.CONBUILDMAT.2007.01.013
46. Ramachandran VS, Paroli RM, Beaudoin JJ, Delgado AH (2002) Clay-based construction product. In: Handbook of thermal analysis of construction materials (Chap. 12). Noyes Publications, United States of America, pp 488–530
47. Rathossi C, Pontikes Y (2010) Effect of firing temperature and atmosphere on ceramics made of NW Peloponnese clay sediments. Part I: Reaction paths, crystalline phases, microstructure and colour. J Eur Ceram Soc 30(9), 1841–1851. https://doi.org/10.1016/j.jeurceramsoc.2010.02.002

48. Tsozué D, Nzeugang AN, Mache JR, Loweh S, Fagel N, Nzeukou A, Richard J, Loweh S, Fagel N (2017) Mineralogical, physico-chemical and technological characterization of clays from Maroua (Far-North, Cameroon) for use in ceramic bricks production. J Build Eng 11(February):17–24. https://doi.org/10.1016/j.jobe.2017.03.008
49. Ma Z, Chen D, Gu J, Bao B, Zhang Q (2015) Determination of pyrolysis characteristics and kinetics of palm kernel shell using TGA-FTIR and model-free integral methods. Energy Convers Manag 89:251–259. https://doi.org/10.1016/j.enconman.2014.09.074
50. Ninduangdee P, Kuprianov VI, Cha EY, Kaewrath R, Youngyuen P, Atthawethworawuth W (2015) Thermogravimetric studies of oil palm empty fruit bunch and palm kernel shell: TG/DTG analysis and modeling. Energy Procedia 79:453–458. https://doi.org/10.1016/j.egypro.2015.11.518
51. Anuar MAM, Anting N, Shahidan S, Lee YY, Md Din MF, Khalid FS, Nazri WMHW (2017) Preliminary evaluation of physical and chemical characterization of waste palm oil shell as cool material replaced in asphaltic concrete as fine aggregate. In: IOP Conference Series: Materials Science and Engineering, 271, 01254
52. Fono-Tamo RS, Koya OA (2013) Characterisation of pulverised palm kernel shell for sustainable waste diversification. Int J Sci Eng Res 4(4):2229–5518
53. Kurovics E, Buzimov AY, Gömze LA (2016) Influence of raw materials composition on firing shrinkage, porosity, heat conductivity and microstructure of ceramic tiles. In: 3rd International conference on competitive materials and technology processes. In: IOP conference series: materials science and engineering, p 012058
54. Brick Industry Association (1992) Technical notes 3A—Brick masonry material properties. Reston, Virginia
55. Bories C, Borredon M-EE, Vedrenne E, Vilarem G (2014) Development of eco-friendly porous fired clay bricks using pore-forming agents: a review. J Environ Manage 143:186–196. https://doi.org/10.1016/j.jenvman.2014.05.006
56. Standard Australia (2011) AS 3700. Masonry structure. https://infostore.saiglobal.com/en-au/Standards/AS-3700-2011-99150_SAIG_AS_AS_267881/
57. Kizinievič O, Kizinievič V, Malaiškienė J (2018) Analysis of the effect of paper sludge on the properties, microstructure and frost resistance of clay bricks. Constr Build Mater 169:689–696. https://doi.org/10.1016/J.CONBUILDMAT.2018.03.024
58. Goel G, Kalamdhad AS (2017) An investigation on use of paper mill sludge in brick manufacturing. Constr Build Mater 148:334–343. https://doi.org/10.1016/j.conbuildmat.2017.05.087
59. Islam A, Mizan A, Ahmed T, Juel MAI, Mizan A, Ahmed T (2017) Sustainable use of tannery sludge in brick manufacturing in Bangladesh. Waste Manag 60:259–269. https://doi.org/10.1016/j.wasman.2016.12.041
60. Ukwatta A, Mohajerani A (2017) Effect of organic content in biosolids on the properties of fired-clay bricks incorporated with biosolids. J Mater Civ Eng 29(7):1–11. https://doi.org/10.1061/(ASCE)MT.1943-5533.0001865
61. British Standard Institution (2015) Specification for masonry units—Part 1: Clay masonry units. London. BS 1377-1
62. Ukwatta A, Mohajerani A, Eshtiaghi N, Setunge S (2016) Variation in physical and mechanical properties of fired-clay bricks incorporating ETP biosolids. J Clean Prod 119:76–85. https://doi.org/10.1016/j.jclepro.2016.01.094
63. Haque MO, Sharif A (2014) Utilization of open pit burned household waste ash—a feasibility study in Dhaka. Waste Manag Res 32(5):397–405. https://doi.org/10.1177/0734242X14526465
64. Demir I (2009) Reuse of waste glass in building brick production. Waste Manag Res 27(6):572–577. https://doi.org/10.1177/0734242X08096528
65. De Silva GHMJS, Surangi MLC (2017) Effect of waste rice husk ash on structural, thermal and run-off properties of clay roof tiles. Constr Build Mater 154:251–257. https://doi.org/10.1016/j.conbuildmat.2017.07.169
66. Sicakova A, Draganovska M, Kovac M (2017) Water absorption coefficient as a performance characteristic of building mixes containing fine particles of selected recycled materials. Procedia Eng 180:1256–1265. https://doi.org/10.1016/j.proeng.2017.04.287

67. Victoria AN (2013) Characterisation and performance evaluation of water works sludge as bricks material. Int J Eng Appl Sci 3(3):69–79
68. American Society for Testing and Materials (ASTM) (2017) Standard specification for building brick (solid masonry units made from clay or shale). https://www.astm.org/Standards/C62.htm
69. Martínez-Martínez S, Pérez-Villarejo L, Eliche-Quesada D, Carrasco-Hurtado B, Sánchez-Soto PJ, Angelopoulos GN (2016) Ceramics from clays and by-product from biodiesel production: processing, properties and microstructural characterization. Appl Clay Sci 121–122:119–126. https://doi.org/10.1016/j.clay.2015.12.003
70. Department of Standard Malaysia (2007) Methods of test for masonry units-part 1: determination of compressive strength. https://www.standardsmalaysia.gov.my

Characterization of Petroleum Sludge Prior to Be Treated in Cement Plants

Ali Benlamoudi, Aeslina Abdul Kadir, and Mohamed Khodja

Abstract Researchers are trying to find a suitable method to treat petroleum sludge (PS) waste and they proposed different methods. In this study, PS was characterized to understand their suitability to be treated in cement plants. Therefore, PS was analyzed for their physical, chemical and gas emission properties. The results revealed that PS is a hazardous waste that could pose serious problems to the environment and living things if it is disposed of in the environment without treatment. In addition, the treatment of PS is limited to incineration method in closed space in order to ensure the complete burning of this waste, which could be executed within the cement kiln. Thus, one of the alternative methods is to treat PS within the cement production plants, which may provide the advantages of using this waste as ingredient to decrease the consumption of raw materials and fuel within the cement plants.

Keywords Petroleum sludge · Incineration · Cement · Physical · Chemical · Gas emission

1 Introduction

Nowadays, Petroleum sludge (PS), which is the waste generated throughout oil drilling operations, is accumulated in nature year by year and causes various environmental problems. It is estimated that 0.37 kg of PS is generated per barrel of petroleum

A. Benlamoudi · A. A. Kadir (✉)
Faculty of Civil Engineering and Built Environment, Universiti Tun Hussein Onn Malaysia, 86400 Parit Raja, Batu Pahat, Johor, Malaysia
e-mail: aeslina@uthm.edu.my

A. Benlamoudi
e-mail: aliptbe2008@gmail.com

M. Khodja
Division Research & Development SONATRACH, Central Direction of Research and Development, Algiers, Algeria
e-mail: Mohamed.Khodja@Sonatrach.dz

A. Abdul Kadir et al. (eds.), *Sustainable Waste Utilization in Bricks, Concrete, and Cementitious Materials*, Lecture Notes in Civil Engineering 129,
https://doi.org/10.1007/978-981-33-4918-6_8

produced, resulting in around 35 thousand tons per day [1]. PS is a thick, viscous intractable mixture encountered during the crude petroleum exploration, production, transportation, storage and refining processes [2, 3]. PS consists mainly of water, solids of the crossed geological formations during the drilling operation, and hydrocarbons deduced from both the drilled petroleum itself and the substances used for the drilling operation. Researchers have tried to treat PS using different methods to reduce their quantity from the environment, eliminate its source of danger, recover valuable fuel, reuse for different purposes and dispose of unwanted PS after ensuring their recalcitrance [4]. However, by reviewing the various methods examined, PS was found to be correlated with advantages and disadvantages that drive one approach to be applied for certain purposes but not for another.

In this chapter, PS samples were collected and analyzed for their physical, chemical, mineralogical and other properties to determine whether PS treatment in cement plants is appropriate to remove this waste from the environment.

2 Materials and Method

Petroleum sludge was collected and characterized for their physical, chemical, mineralogical properties, calorific value and gas emissions. Tests conducted were triplicated to ensure the accuracy of the results obtained.

2.1 *Samples Collection*

Petroleum sludge (PS) samples were collected from the dumping area of the drilling oil well-coded OMF-46 in the drilling field of Hassi Messaoud (a desert region in Algeria) (Fig. 1). Hassi Messaoud is the largest oil field in Algeria, with a production of 350 thousand b/d. Figure 2 shows the PS sample, which has a brown colour and a semi-solid physical status due to the water and the oil content. The difference in

Fig. 1 Petroleum sludge from the oil drilling field of Hassi Messaoud

Fig. 2 Petroleum sludge sample before drying

colour between Figs. 1 and 2 is due to the sand covering PS because it is situated in a windy area.

2.2 *Physical Characteristics*

Physical characterization tests include moisture content, loss on ignition, volatile matter and ash content.

2.2.1 Moisture Content

This test was conducted according to ASTM E949-88 [5]. Empty containers were dried and covered; then 100 g of each PS sample was put inside the container and dried within the preheated oven under a temperature of 105 °C for one hour (Fig. 3).

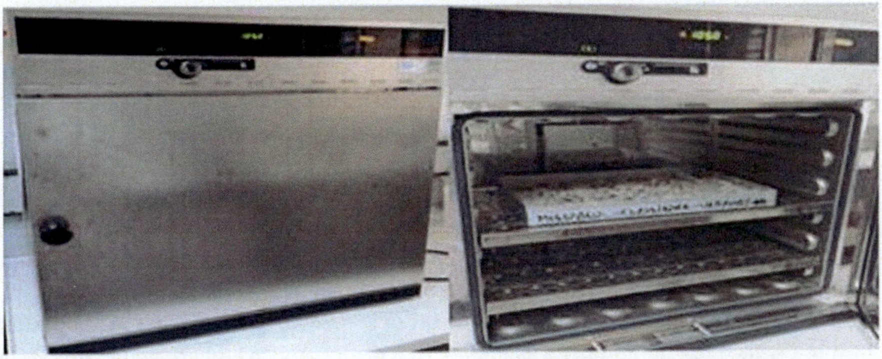

Fig. 3 Petroleum sludge sample before drying

After that, the containers were quickly placed in the desiccator to reach room temperature. The samples were weighed, and the moisture content values were calculated according to Eq. 1 [5].

$$\mathrm{M} = \frac{A - B}{C} * 100 \tag{1}$$

where

M Moisture content (%);
A weight of wet sludge + container (g);
B weight of dried sludge + container (g);
C weight of container (g).

2.2.2 he Loss on Ignition (LOI)

This test was conducted according to ASTM D7348 [6] by weighing 1 g of the dried PS in a crucible, placing it in a furnace and heating it under a temperature of 950 °C for 2 h. The fired sample was then cooled down to room temperature in a desiccator and weighed again. The weight loss associated with firing is known as the loss on ignition (LOI). The LOI values were calculated according to Eq. 2 [7].

$$\mathrm{LOI}(\%) = \frac{A - B}{C} * 100 \tag{2}$$

where

LOI Loss on ignition (%);
A weight of wet sludge + crucible (g);
B weight of fired sludge + crucible (g);
C weight of crucible (g).

2.2.3 Volatile Content Test

The volatile content test was determined according to Standard Method Agency 2540 G by using several apparatus that include an evaporating dishes of 100 ml made of porcelain, 90 mm diameter, a platinum or High-silica glass, a muffle furnace for operation at 550 °C, a steam bath, a desiccator, which is provided with a desiccant containing a colour indicator of moisture concentration or an instrumental indicator, a drying oven, for operation at 103 to 105 °C, an analytical balance with a sensibility of up to 10 mg, a graduated cylinder and a low-form beaker [8].

The process began with igniting a clean evaporating dish at 550 °C for 1 h in a muffle furnace; then, the dish was cooled in a desiccator, weighed, and stored in the desiccator until it was ready for use. Secondly, 25 g of the sample was placed in the evaporating dish and weighed, then, it was evaporated to dry inside the oven, dried at 103–105 °C for 1 h, cooled in an individual desiccator containing fresh desiccant, and weighed. Thirdly, the dried residue was transferred to the cool muffle furnace, heated to 550 °C, and ignited for 1 h. Then, the sample was cooled in a desiccator and weighed. Igniting for 1 h, cooling, desiccating and weighing steps were repeated until the weight change was less than 4%. The volatile content was calculated according to Eq. 3 [8]:

$$\text{Volatile content}(\%) = \frac{(m_1 - m_2) * 100}{m_1 - m_3} \quad (3)$$

where

- m_1 weight of dried residue + crucible (g);
- m_2 weight of residue + crucible after ignition (g);
- m_3 weight of crucible (g).

2.2.4 Ash Content

This test was conducted according to ASTM D3174 [9] by placing 6 g of sludge (m_1) into a dried pre-weighed porcelain crucible; then, the crucible with its content was burned in an oven at temperatures above 550 °C, for 4 h (Fig. 4). After that, the crucible was weighed after it was cooled to room temperature in a desiccator. The dried sludge was then recorded and labeled (m_2). Contrasting volatile content test, for the ash content test, ignition was continued for 4 h. The total ash content is calculated by dividing the difference in weight between m_1 and m_2 by the weight of m_1, and the total was multiplied by 100 (Eq. 4) [9]. The operation was continued until obtaining a constant weight of the residue.

Fig. 4 Furnace used to determine petroleum sludge ash

$$\text{Ash}(\%) = \frac{100 * (m_1 - m_2)}{m_1} \tag{4}$$

where

m_1 Weight of crucible with residues after smouldering (g);
m_2 Weight of empty crucible (g);

2.3 Chemical Characteristics

The main chemical characterizations determined in this research include leaching test, total petroleum hydrocarbons (TPH) and organic pollutant characterization.

2.3.1 Leaching Test

The leaching was conducted in two main steps: leaching by using the Toxicity Characteristics Leaching Procedure (TCLP) and analysis by using Inductively Coupled Plasma Mass Spectrometry (ICP-MS) Agilent 7700—2009 manufactured by Agilent Technologies, Inc. Hewlett-Packard. The TCLP is intended to be a laboratory test designed to simulate leaching in a municipal landfill. It was designed to determine the mobility of both organic and inorganic compounds presented in liquid, solid as well as multiphasic samples. In this test 100 g of the sample was crushed before the leaching test to pass a 2 mm sieve and retain on a 0.3 mm sieve. Samples were extracted according to USEPA Method 1311 [10]. Extractions were carried out with 20:1 liquid to solid ratio in an acetate solution at pH 2.88 ± 0.05 depending on the sediment pH, in 2 L bottles and to be rotary agitated at 30 rpm for 18 h using a rotary system, model Reax20/12, P/N541-20,012-00—2016, manufactured by Heidolph Company (Fig. 5) [10].

After the samples were extracted in the acetic fluid, the solid and liquid phases were separated by filtration through 0.45-mm-pore-size membrane filters. The pH

Fig. 5 The rotary system for TCLP procedure (P/N541-20012-00, Reax20/12)

was measured. For the inorganic analysis, the TCLP extract was digested (acidified to 2% HNO_3) prior to analysis for metals using ICP-MS. Then, the quantities of the heavy metal in the solutions obtained were compared with the regulatory cited by the United States Environmental Protection Agency (USEPA). Figure 6 shows the detail of the procedure. After the sludge components were extracted, samples were subjected to be analyzed using ICP-MS, which is chosen because of its excellent detection limits for most of the elements [11].

2.3.2 Total Petroleum Hydrocarbons (TPH)

This test was conducted according to USEPA 3540C [12] by involving the Soxhlet extraction (Fig. 7). A total of 26 g of PS was weighed and placed in a cartridge. This cartridge was introduced into a "Soxhlet apparatus" that is placed between a refrigerant circulating water and a glass flask filled with dichloromethane (Cl_2H_2), which is introduced as a solvent. The extraction of the organic matter was carried out by dissolving in dichloromethane using a balloon heater. The extraction time was about 72 h. The extraction was terminated when the dichloromethane was becoming colourless in the Soxhlet system. The weight of the hydrocarbons was then determined after the evaporation of the solvent according to Eq. 5.

$$\mathrm{TPH}(\%) = \frac{m_2}{m_1} * 100 \tag{5}$$

where

m_1 The initial weight of the sample (contained in the cartridge)
m_2 The weight of the extracted organic matter.

2.3.3 Organic Pollutant Characterization

This characterization involves two steps: extraction and analysis. The extraction was conducted according to the Algerian standard [13], which is equivalent to USEPA 1311 [10]. The analysis was conducted according to ASTM D1945-14 using Gas chromatography-mass spectrometry (GC–MS) (Fig. 8) that is attached to a computer. Firstly, the injection was performed using an autosampler with a 10 μL syringe injection; then, eight solvent washes from each wash bottle were performed. After that, a 1.0 μL sample with a fast plunger speed, three sample washes, and three sample pumps were injected. The inlet was run in split mode at 240 °C with a 50:1 split ratio and the carrier gas was the hydrogen. Secondly, the oven was run starting at 80 °C for 0 min; then, it was ramped at 20 °C/min until 280 °C was obtained. After that, the temperature was held at 280 °C for 2 min. The column used was HP-5 ms that has 0.25 mm of diameter with a 0.25 μm film thickness. The column was kept at a constant flow of 1.6 ml/min. The detector temperature was set at 300 °C, the flow

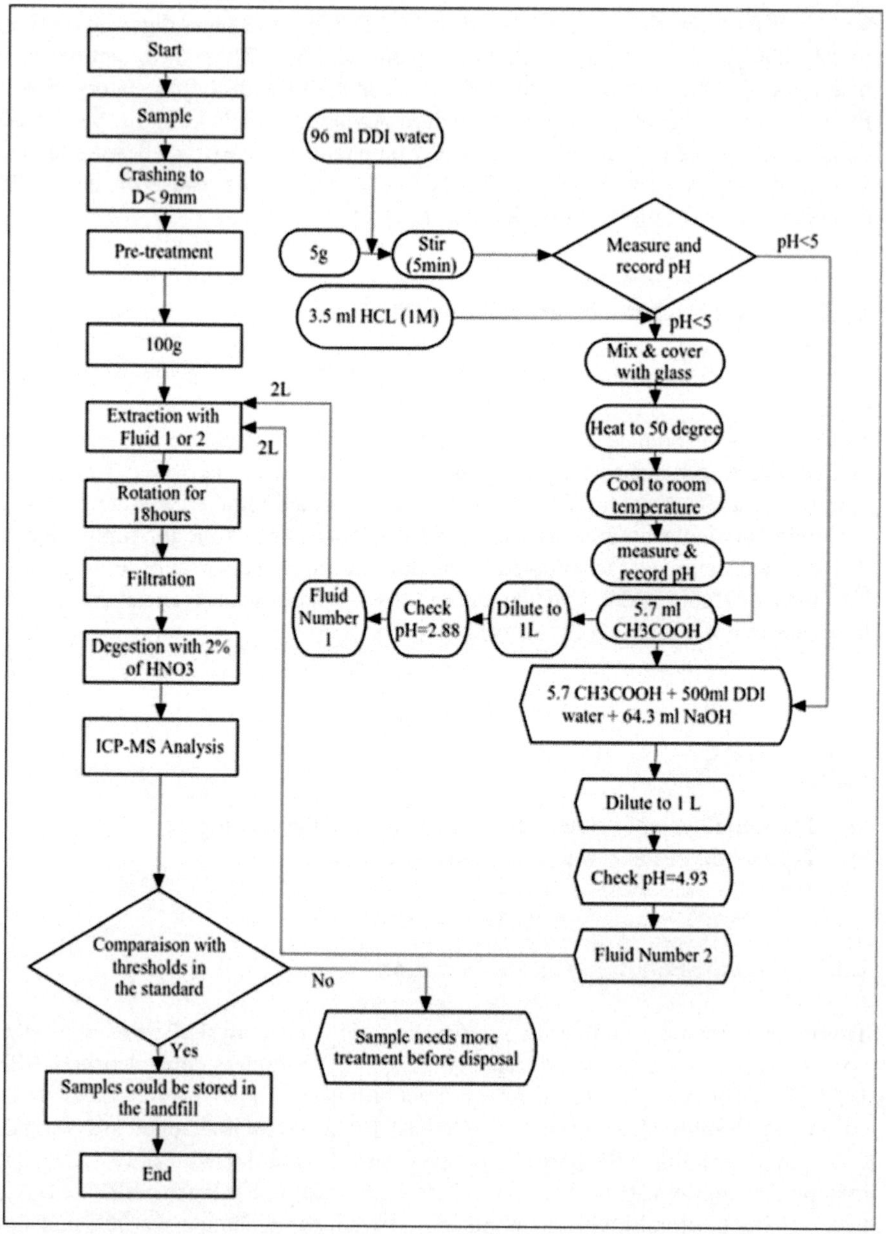

Fig. 6 Steps of TCLP procedure

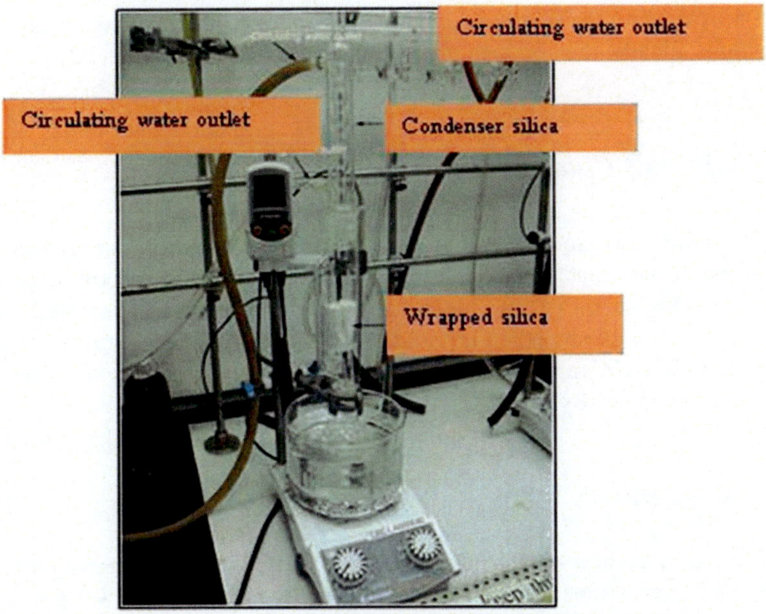

Fig. 7 Soxhlet extraction method (CBE-000645, Heidolph)

Fig. 8 GC–MS system (Agilent 7890A, Agilent Technologies, Inc. Hewlett-Packard)

rate for hydrogen was 30 ml/min, air was 400 ml/min, and the flow of nitrogen was 25 ml/min. Finally, the obtained graph was analyzed and interpreted.

2.4 *Minerals Characteristics*

The PS sample was ground into 0.002–0.005 mm cross-section carefully to obtain a smooth flat surface that may facilitate close alignment with the spectrometer center. It was then packed flat on a glass slide and confined within a thin plastic ring cemented to the glass. The sample was then exposed to Bruker D8 Advance X-ray diffractometer (XRD) with Cu Kα radiation and 2θ scanning ranging from 10 to 70° that eases drawing of the intensity peaks diagram [14].

2.4.1 Gas Emission

A standardless method was performed and 17 kg of PS (the high quantity was chosen for better interpretation of the results as advised by the company) was put in a small industrial kiln made to burn different types of wastes. Then, the PS was burned under a temperature of 600 °C in order to ensure the disappearance of the total organic matter, as reported by the United Nations Environment Program [15]. A few minutes after placing PS, a measurement of the gases emitted was performed using an electric probe detector. Then, the registered gases were compared to the emission thresholds allowed for cement plants set by the Algerian standard 06-138 [16] in order to determine the maximum percentages of PS that could be incorporated into cement production [17] according to Eq. 6.

$$CO_{permitted}(\%) = \frac{CO_{threshold}}{CO_{obtained}} * 100 \tag{6}$$

2.5 *Calorific Analysis*

According to ASTM D5468, this test was conducted by weighing 1 g of sample in a combustion crucible [18]. The pressed tablet was weighed and placed in the combustion crucible. The ignition thread was fixed to the ignition wire. 5 ml of distilled water was poured into the calorimeter bomb, which was filled with 30 bar oxygen and placed in the calorimeter. After combustion, the calorimeter bomb was unscrewed and rinsed with distilled water. Some drops of 'mixed indicator 5' were added to the solution. Afterwards, the solution was titrated with 0.1 n caustic soda until the colour changed to green.

The calorific value was calculated according to Eq. 7:

$$GCV_{(ad)} = \frac{\mathrm{C} * \mathrm{T} - Q_{NS} - Q_{Z} - Q_{S}}{\text{sample weigh}} \tag{7}$$

where as

GCV	Gross calorific value (the value after the complete combustion) (kJ/kg);
C	heat capacity of device (J/K);
T	Increase of temperature (K);
Q_{NS}	Consumption ml 0.1 n NaOH (b) * 6;
Q_Z	Ignition thread = 50 J + ignition energy (70 J), according to manufacturer data;
Q_S	Fuel sample weigh * 10 * % sulfur (ad) * 5.7.

3 Results and Discussions

Results of different characteristics of PS were obtained and studied.

3.1 *Physical Characteristic*

In this section, results of the different physical tests were obtained, discussed and compared to previous studies.

3.1.1 Analysis of Moisture Content

The drying of PS has given a moisture content of 6.5%, which has taken a long time to be obtained due to the oil content within the samples that prevent its drying. This low value is due to PS's oily origin and the geological structure of the oil wells situated in dry areas. The benefit of this low value for cement plants is that it does not require high energy to get rid of the water content. The value obtained of the moisture content is considered very low compared to the values obtained by Fang et al. [19] and Xu et al. [20] who have studied the incorporation of sewage sludge into cement production and obtained moisture contents of 78–83% and 86.6%, respectively.

However, in Fang et al. [19] study, the cement plant has to build a drying plant for the sewage sludge with a temperature of 300 °C since it contains high moisture content. This temperature (300 °C) is high and is considered as an additional cost for the whole cement production. Nevertheless, the current study does not need any plant for the drying process and this will be detailed in the calorific value section.

3.1.2 Analysis of Loss on Ignition (LOI)

Loss on ignition of PS was conducted within a muffle furnace at a temperature of 950 °C; the results gave an average loss value of 20.83%, which is considered low. The reason for this low value is mainly because of the nature of the sludge itself, which is coming only from the drilling operation of the crude oil and not from other processes like refining. However, the PS used in this study, which is generated during the drilling operation, is associated with the rocks of the geological structure of the oil well and not from the oil itself, which means that the oil is not the main matter associated with the sludge.

This low value, which is a simulation of the mass loss of PS once placed in a cement kiln, indicates the low organic matter content within the PS samples. On the other hand, it describes the water quantity, the organic content like onentalkanes, aeromatics, asphaltenes and resin [21] and even some percentages of carbonate matters. It also contains volatile organic carbons (VOCs) and semi-volatile organic carbons (SVOCs) like polycyclic aromatic hydrocarbons (PAH) [19, 22]. The organic content of PS is the source of heat that offers supplementary energy to the kiln, and this may be determined by calorific value test. VOCs and SVOCs are significant sources of air contamination; however, they would be eliminated under a temperature of 550 °C [15]. From another study, Huang et al. [23] used PS to produce sustainable cement and obtained a loss on ignition of 60.91%, which is explained by the high organic content of PS samples used.

3.1.3 Analysis of Volatile Content

From the analysis, the volatile content of PS was 14.8%. This value is very low compared to the value obtained by Lechtenberg and Parter [1], which is due to the different sources of the PS studied. In fact, there are very limited researches on PS that showed its volatile content. Lechtenberg and Parter [1] has characterized the PS before being used as an alternative fuel to feed the cement plants. In his study, he declared that the PS samples contain more than 61% as an average for the volatile matter.

In this research, it is ensured that injection of PS into the cement kiln could eliminate the whole volatile content since similar condition in temperature was settled compared to the studies of Pipilikaki et al. [24] and Kara [25] who have studied the incorporation of tire-derived fuel (TDF) and refused derived fuel (RDF), respectively into cement production. They concluded that even the obtained volatile contents were very high, which were 72 and 92.3%, respectively, but the disappearance of the total volatile content was ensured inside the cement kiln.

3.1.4 Analysis of Ash Content

This test was conducted to determine the effective mineral content within the PS that will substitute some percentages of the cement raw materials in general terms. However, placing PS in the cement kiln provided energy from its organic content and raw materials from its mineral content. The results showed that the PS samples had an ash value of 81.3% of the sample's total weight. This value is of great importance since its incorporation will be high in the cement plant. The high value obtained in this study was mainly due to the high mineral content within the PS samples and the low organic content, which was confirmed with the LOI results obtained. As explained before, high mineral content was due to the nature of PS samples that was originated from the geological structure of the drilling well and not from the oil. Nevertheless, this test should be followed by the chemical analysis of PS so as to know the exact percentages of the different minerals of this waste that will be in use for the substitution of the cement raw materials by PS samples.

The ash value obtained in this research was higher than that of Lechtenberg [1] and Conesa et al. [26] who have studied the incorporation of PS and solid recovered fuel (SRF), respectively and obtained an ash content of 43 and 10.2% respectively. This difference in ash values was due to the difference in geological structure and difference in PS source because each PS has their own characterization.

3.2 Chemical Characteristic

In this section, results obtained from different chemical analyses were obtained and discussed.

3.2.1 Analysis of Heavy Metals Content

PS was subjected to a leaching procedure using the Toxicity Characteristic Leaching Procedure (TCLP), followed by the leachate analysis using Inductively Coupled Plasma Mass Spectrometry (ICP-MS). The leachate analysis of PS samples and the maximum concentrations of heavy metals in cement production are presented in Table 1.

This analysis shows that the heavy metals did not exceed the cement's regulatory limits for all the elements except for lead (Pb), which is 149.02 ppm compared to the maximum percentage in cement production, which is 75 ppm. Thus, incorporation of PS in cement plants should consider Pb's value to prevent any leachability from this waste. The results obtained are better than the study conducted by Xu et al. [20], who have studied the incorporation of sewage sludge into cement production and obtained very high concentration for almost all the heavy metals. As for their results, the sewage sludge studied contains 202.76 ppm, 523.59 ppm, 30.33 ppm, 125.45 and 41.34 ppm for Cu, Zn, Ni, Cr and Pb, respectively [27].

Table 1 Heavy metals content in PS compared to the maximum concentrations in cement production

Heavy metals	Content within PS samples (ppm)	Maximum concentrations in cement production (ppm)
Manganese (Mn)	25.77	2366
Nickel (Ni)	0.34	129
Zinc (Zn)	0.20	321
Iron (Fe)	37.40	N/A
Copper (Cu)	3.43	N/A
Lead (Pb)	149.02	75
Cadmium (Cd)	0.44	1.12
Chromium (Cr)	0.48	422

3.2.2 Analysis of the Total Petroleum Hydrocarbons (TPH)

TPH of PS has been extracted according to USEPA method 3540C (Soxhlet extraction) [12] by using an excellent extractor, which is methylene chloride (CH_2Cl_2). The results showed a value of 8.5% of organic matter contained in this waste. This value was mainly due to the oil content associated with PS waste during the drilling operation. The result obtained was relatively high and considered very harmful to the environment and living things reported by Tang et al. [28]. However, Tang et al. [28] have studied the contamination of soil by petroleum hydrocarbons and concluded that content of only 1.5% of TPH is a critical value for the germination of plants and the living of the worms. In addition to that, they also mentioned that only 0.5% of TPH might influence bacteria activities.

In this research, the burning of PS samples within the cement kiln was chosen as the treatment method to eliminate PS from the environment, as reported by many researchers. However, Fang et al. [19], Xu et al. [20] and Kara [25] have studied the incorporation of sewage sludge, lime dried sludge (LDS) and refuse-derived fuel (RDF), respectively, into cement production and concluded the efficiency of this method to eliminate wastes from the environment. It was also reported that incineration of waste within the cement kiln ensures the devastation of the total organic polluted matters due to the high temperature at sufficiently long retention time [15], and it also offers complete destruction of dioxins and furans [29].

3.2.3 Analysis of PS Alkanes

The analysis of the different n-alkanes (C11–C31) of PS has been conducted by using the gas chromatography-mass spectrometry (GC-MS), and the results are shown in Fig. 9 for the chromatogram, and Table 2 for the weight percentages of the n-alkanes contained in PS samples.

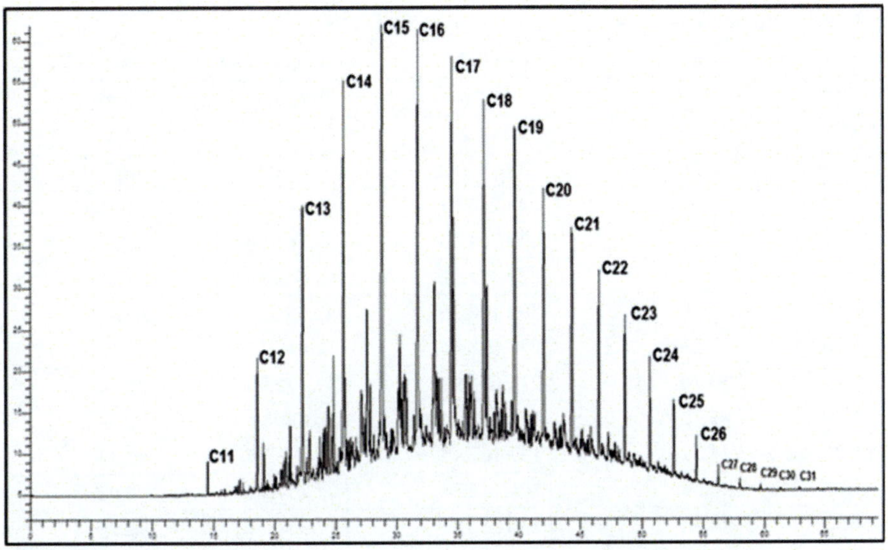

Fig. 9 The chromatogram of petroleum sludge

Almost 50% of the weight concentrations were between C14 and C17 and only around 1% of the weights were from C26 to C31. The high concentrations of the elements between C14 and C17 indicate the harmful impact of this waste. This negative impact is due to the association of oil to the PS samples during the drilling process, which is a mixture of different hydrocarbons and most commonly the n-alkanes.

However, it was reported that the harmful impact of petroleum hydrocarbon is related mainly to the fractions that have lower boiling points, especially to the range of C10 to C19, among these compounds, many of them are considered as toxic, mutagenic and carcinogenic [28, 30, 31]. Consequently, PS is considered highly harmful and should be treated with high precautions if it is not eliminated from the environment.

Nevertheless, the chromatogram did not show the n-alkanes from C1 to C10 with temperatures below 174 °C since some of them are gaseous and most likely disappeared during of drilling operation, where others have very low boiling points that make them difficult to be detected [28]. Organic components from C1 to C4 are toxic gaseous and from C5 to C10 are toxic liquids that pose serious impacts to the human being through exposure [28]. Thus, serious precautions should be taken when handling the PS.

Table 2 Percentages in weight of different n-alkanes of PS

Peak	Time (min)	Area (µV s)	Height (µV)	Area (%)
C11	14.49	1,185,104	3992.32	0.48
C12	18.58	57,144.22	16,008.15	2.31
C13	22.29	15,699,182	32,371.43	6.33
C14	25.69	249,248.67	42,661.96	10.06
C15	2.86	329,764.73	50,166.66	13.31
C16	31.81	329,151.92	48210M7	13.28
C17	34.60	292,108.26	42,658.29	11.79
C18	37.23	240,018.64	38,619.21	9.68
C19	39.74	216,670.93	37,172.95	8.74
C20	42.12	163,287.34	30308J87	6.59
C21	44.39	137,162.48	26,786.79	5.53
C22	46.58	104,614.34	22,257.15	4.22
C23	48.66	73,165.28	17,547.84	2.95
C24	50.67	50,323.03	13,401.30	2.03
C25	52.59	32,344.63	9025.47	1.31
C26	54.45	17,774.33	5425.06	0.72
C27	56.24	7933.16	2494.02	0.32
C28	57.98	4285.14	1291.32	0.17
C29	59.67	2124.87	670.90	0M9
C30	61.31	1063.53	329.17	0.04
C31	62.91	1412.78	301.14	0.06
Sum		2,478,445.15	441,700.08	100.00

3.3 Mineralogical Characteristics

Figure 10 presents the diffractogram of petroleum sludge. The diffractogram shows that PS contains different components that are anhydrite ($CaSO_4$), calcite ($CaCO_3$), dolomite ($CaMg(CO_3)_2$), quartz (SiO_2), barite ($BaSO_4$) and halite (NaCl). It is known that $CaCO_3$ and SiO_2 are the main minerals that enter in the production of cement clinker since they are the primary sources of calcium oxide (CaO) and silica (SiO_2), respectively [15]. Thus, it is worth to study the possibility of using this waste as a substituent to the cement raw materials. Rodríguez et al. [29] conducted a study on the sewage sludge as an alternative fuel on Portland cement clinker production and concluded that the quartz and calcite were the main crystalline phases contained in the sludge. They also concluded the ability and the efficiency to incorporate this sludge into cement clinker. On the other hand, the diffractogram of PS shows a high content of $CaSO_4$, which is the main composition of gypsum, followed by the other compositions; thus, it is worth also to study the possibility of PS to substitute gypsum, which is considered as an additive material that develops the strength of the cement.

Following the highlight on the composition of petroleum sludge, semi-quantitative analysis was also performed on this waste using the same equipment of XRD, and the results are presented in Table 3. Table 3 shows a percentage of 54% of anhydrite, which is relatively high and confirms the diffractogram analysis, followed by barite, dolomite and calcite, then; a very low percentage of quartz. The high content of $CaSO_4$ is due to the geological structure of the oil well from where the sludge was coming out. Nevertheless, the $BaSO_4$ content is due to the drilling mud that is used for the oil drilling operation. Furthermore, the other components are associated with both the geological structure and the drilling mud. As explained above, this waste could substitute some percentages of limestone due to the content of calcite ($CaCO_3$) and dolomite ($CaMg(CO_3)_2$), which contain high amounts of calcium (Ca). It can also substitute gypsum due to its high anhydrite content.

3.4 *Gas Emission*

For PS, Table 4 presents the gases emitted from the burning of this waste within a small industrial kiln and compared to the permitted thresholds. From Table 4, an important observation through this analysis is the zero content of carbon dioxide (CO_2). This result is highly appreciated mainly for the incorporation of PS samples into the cement production. This is because it was reported that 1 kg of petroleum coke used in the cement production emits almost 70% of CO_2 [25]. Also, it was reported that 1 tonne of cement clinker resulted in the emission of more than 900 kg of CO_2 [32], which is a considerable quantity that contributes directly to the greenhouse gases (GHG). In Algeria, the cement company of Ain EL-Kebira, Algeria (SCAEK) reported an emission of 0.5 tonnes of CO_2 per tonne of cement produced [33]. As a consequence, the incorporation of any percentage of PS to produce cement in this company should result in a decrease of the CO_2 emission to the environment. From another study, Abdul-Wahab et al. [34] have incorporated petroleum sludge as a partial substitution of limestone in cement plants and concluded that PS could reduce CO_2 emission into the atmosphere during cement production. The same results were obtained by Scrivener et al. [35] who declared that partial substitution of fuel in cement plants reduces the global emission of CO_2.

In addition to that, the results also showed 0 ppm for sulfur dioxide (SO_2) and nitrogen oxides (NO_x), which has a benefit in lowering their emission when the PS is incorporated with the cement raw materials to produce cement. This result is in line with the studies conducted by Fang et al. [19], Kara [25] and Conesa et al. [26], who have studied the incorporation of sewage sludge, refuse-derived fuel (RDF) and solid recovered fuel (SRF), respectively into the cement production. Kara [25] and Conesa et al. [26] concluded that incorporating those substances has contributed to the decrease of SO_2 and NO_x emissions. Other than that, Fang et al. [19] have declared that the NO_x decrease was more than 65%. Nakomcic-Smaragdakis et al.

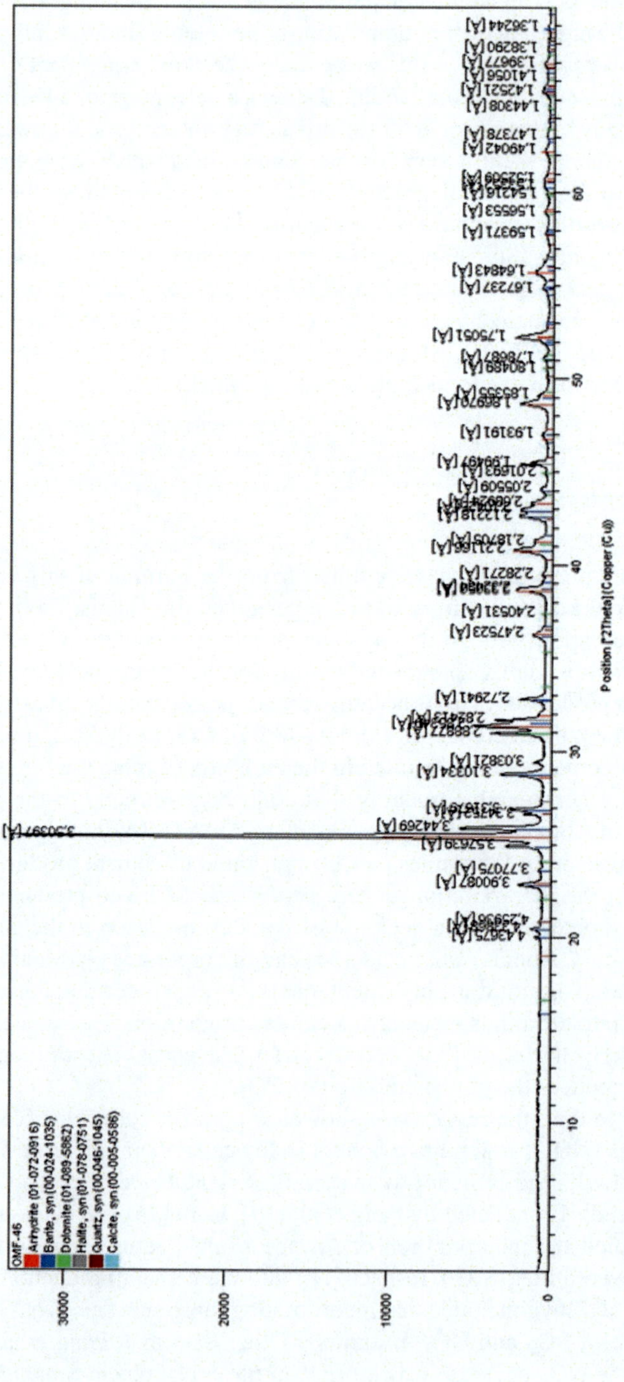

Fig. 10 Diffractogram of petroleum sludge

Table 3 Semi-quantitative composition of petroleum sludge

Mineral compositions	Percentages (%)
Quartz	05
Calcite	09
Dolomite	13
Anhydrite	54
Barite	13
Others	06

Table 4 The gases emitted from burning of PS [13]

Gases	Values	Thresholds
CO_2	[a]0.0%	[b]N/A
NO_x	0.0 ppm	878.0 ppm
SO_2	0.0 ppm	266.0 ppm
CO	392.0 ppm	162.0 ppm

[a]The value of CO_2 in the standard is measured in % not in ppm
[b]Not detected

[36] have also concluded that SO_2 and NO_x emissions were slightly decreased by incorporating scrap tires into cement plants.

Nevertheless, the carbon monoxide (CO) value shown in Table 4 is 392.0 ppm. This value is of concern since it exceeds the threshold emission for cement plants, which is 162.0 ppm [37]. Thus, preliminary precaution should be taken prior to burning this waste within cement factories. According to Eq. 8, the incorporation of PS should not exceed 41.33% in order to ensure that the CO emissions will not surpass the emission permitted limit.

$$CO_{permitted}(\%) = \frac{CO_{threshold}}{CO_{obtained}} * 100 \tag{8}$$

3.5 *Calorific Analysis*

Calorific analysis of petroleum sludge samples gave a value of 2.8 MJ/kg. This value has a substantial benefit on the cement plant, but it is dependent on the percentage of PS samples incorporated together with the cement raw materials. The calorific value obtained in this study is close to the average associated energy consumption level for cement plants, which is 2.9–3.3 MJ for 1 kg of produced clinker (2.9–3.3 MJ/kg) [24]. This referential average (2.9–3.3 MJ/kg) was set by the reference documents on the Best Available Techniques (BREFs). The same document (BREFs) also indicated that the energy required to dry the cement raw materials is 0.150–0.599 MJ for 1 kg of clinker [38]. However, based on the energy required to dry the cement raw materials that were set by BREF, the incorporation of only 5.4% of PS as a raw material can substitute the whole minimum energy required for drying, which is 0.150 MJ/kg. This percentage of PS (5.4%) can also decrease the amount of 4.6–5.2% of the total fuel consumption required to produce cement. The percentage of fuel substituted by introducing PS into the cement plant even it seems low, but it is considered necessary regarding the enormous energy required for the total cement production.

The result obtained is supported by previous researchers who have studied incorporating different types of wastes into cement plants. Fang et al. [19] has concluded that sewage sludge used in their study could provide 3–10% of energy required for cement production. Pipilikaki et al. [24] also got beneficial results for using tire-derived fuel (TDF), which has offered 6% of energy to produce cement. Other than that, Kara [25] used the refused derived fuel (RFD) as an additional fuel source for the cement industry and concluded that it could offer 15% as an alternative fuel to the cement plant.

However, Lechtenberg and Parter [1] has conducted a study on using PS in cement plants. As a result, calorific values of more than 16.2 MJ/kg were obtained in the study. This value is far higher than the result obtained in the current research, which is simply due to the nature of the sludge generated that differs from one source to another. Also, the calorific value depends on the organic matter associated with the PS samples, for Lechtenberg study, the organic matter associated with PS was around 60%, nevertheless, in the present study, the organic matter is lower than 20%. This lower organic percentage is a direct impact on the lower calorific value obtained.

4 Conclusion

The results revealed that petroleum sludge is a hazardous waste that could pose serious problems to the environment and living things if disposed of in the environment without treatment. However, the mineralogical analysis of PS indicated that it could substitute gypsum due to the high anhydrite content. Also, it can substitute some percentages of limestone due to its content in calcite ($CaCO_3$) and dolomite ($CaMg(CO_3)_2$). The high carbon monoxide (CO) emitted through the burning of this waste, limits the choice of using the incineration treatment of PS in open spaces in order to prevent environmental pollution. This incorporation is possible only in a closed space process like the use of cement kiln to decrease the emission of CO_2, NO_x and SO_2 gases. Moreover, the treatment of PS into cement plants may provide a support fuel due to its calorific value.

Overall, this method may ensure the complete elimination of petroleum sludge from the environment on the one hand and reduce cement production cost by offering alternative sources that support cement raw materials on the other hand.

References

1. Lechtenberg MVW Parter (2010) Waste management and cement industry in Arab countries. Cement Build Mater Rev 39:27–34 (Unpublished)
2. Prakash V, Saxena S, Sharma A, Singh S, Singh S (2015) Treatment of oil sludge contamination by composting. J Bioremediat Biodegrad 6(3). https://doi.org/10.4172/2155-6199.1000284
3. Organization of the Petroleum Exporting Countries (OPEC) (2015) Monthly oil market report. Feature article: review of 2015, outlook for 2016, Economic Brochure, Vienna, Austria
4. Hu G, Li J, Zeng G (2013) Recent development in the treatment of oily sludge from petroleum industry: a review. J Hazard Mater 261:470–490. https://doi.org/10.1016/j.jhazmat.2013.07.069
5. American Society for Testing and Materials (ASTM) (2004) Standard test methods for total moisture in a refused-derived fuel laboratory sample. West Conshohocken, USA, pp E949–E988
6. American Society for Testing and Materials (ASTM) (2013) Standard test methods for Loss on Ignition (LOI) of solid combustion residues. West Conshohocken, USA, p D7348
7. American Society for Testing and Materials (ASTM) (2000) Standard test methods for specific gravity of soil solids by water pycnometer. ASTM, West Conshohocken, USA, pp D854–00
8. Standard Method Agency (SMC) (1997) Total, fixed and volatile solids in solid and semisolid samples. G2540
9. American Society for Testing and Materials (ASTM) (2002) Standard test methods for ash in the analysis of sample of coal and coke from coal. ASTM, West Conshohocken, The USA. D3174–02
10. United States Environmental Protection Agency (1992) Test method 1311—Toxicity characteristic leaching procedure (TCLP). USEPA 1311
11. Johnson O, Napiah M, Kamaruddin I (2015) Encapsulation of petroleum sludge in building blocks. Constr Build Mater 78:281–288. https://doi.org/10.1016/j.conbuildmat.2014.12.122
12. United States Environmental Protection Agency (USEPA) (2009) Soxhlet extraction method EPA 3540. USA. 3540C
13. Normalisation Francaise (1992) Waste. Leaching test. AFNOR X 31-210, Paris
14. Lin Y, Zhou S, Li F, Lin Y (2012) Utilization of municipal sewage sludge as additives for the production of eco-cement. J Hazard Mater 213–214:457–465. https://doi.org/10.1016/j.jhazmat.2012.02.020
15. United Nations Environment Program (UNEP) (2010) Open-ended working group of the basel convention on the control of transboundary movements of hazardous wastes and their disposal: Seventh session. Draft technical guidelines on co-processing of hazardous waste in cement kilns. Industrial Technical Guidelines, Geneva, Switzerland
16. Algerian Standard (2006) Cement for work in highly aggressive environments (French language).XP X32–210, Algeria
17. Benlamoudi A, Kadir AA, Khodja M (2017) Petroleum Sludge as gypsum replacement in cement plants: its impact on cement strength. In: International research and innovation summit (IRIS). IOP conference series: materials science and engineering, p 12059
18. American Society for Testing and Materials (ASTM) (2007) Standard test method for gross calorific and ash value of waste materials. ASTM, West Conshohocken, USA, pp D5468–02
19. Fang P, Tang ZJ, Huang JH, Cen CP, Tang ZX, Chen XB (2015) Using sewage sludge as a denitration agent and secondary fuel in a cement plant: A case study. Fuel Process Technol 137:1–7. https://doi.org/10.1016/j.fuproc.2015.03.014
20. Xu W, Xu J, Liu J, Li H, Cao B, Huang X-F, Li G (2014) The utilization of lime-dried sludge as resource for producing cement. J Clean Prod 83:286–293. https://doi.org/10.1016/j.jclepro.2014.07.070
21. Diallo M, Cagin T, Faulon J-L, Goddard W (2000) Thermodynamic properties of asphaltenes: a predictive approach based on computer assisted structure elucidation and atomistic simulations. Asph Asph II Dev Pet Sci Ser 40(B40):103–124. https://doi.org/10.1016/S0376-7361(09)70276-6

22. Rahman A, Rasul M, Khan MM, Sharma S (2015) Recent development on the uses of alternative fuels in cement manufacturing process. Fuel 145. https://doi.org/10.1016/j.fuel.2014.12.029
23. Huang M, Ying X, Shen D, Feng H, Li N, Zhou Y, Long Y (2017) Evaluation of oil sludge as an alternative fuel in the production of Portland cement clinker. Constr Build Mater 152:226–231. https://doi.org/10.1016/j.conbuildmat.2017.06.157
24. Pipilikaki P, Katsioti M, Papageorgiou D, Fragoulis D, Chaniotakis E (2005) Use of tire derived fuel in clinker burning. Cem Concr Compos 27(7):843–847. https://doi.org/10.1016/j.cemconcomp.2005.03.009
25. Kara M (2012) Environmental and economic advantages associated with the use of RDF in cement kilns. Resour Conserv Recycl 68:21–28. https://doi.org/10.1016/j.resconrec.2012.06.011
26. Conesa JA, Rey L, Egea S, Rey MD (2011) Pollutant formation and emissions from cement kiln stack using a solid recovered fuel from municipal solid waste. Environ Sci Technol 45(13):5878–5884. https://doi.org/10.1021/es200448u
27. Kadir AA, Benlamoudi A (2017) Petroleum sludge treatment and reuse for cement production as setting retarder. In: IOP conference series materials science and engineering, vol 203, p 12010. https://doi.org/10.1088/1757-899X/203/1/012010
28. Tang J, Wang M, Wang F, Sun Q, Zhou Q (2011) Eco-toxicity of petroleum hydrocarbon contaminated soil. J Environ Sci 23(5):845–851. https://doi.org/10.1016/S1001-0742(10)60517-7
29. Rodríguez N, Martínez-Ramírez S, Blanco-Varela MT, Donatello S, Guillem M, Puig J, Fos C, Larrotcha E, Flores J (2013) The effect of using thermally dried sewage sludge as an alternative fuel on Portland cement clinker production. J Clean Prod 52:94–102. https://doi.org/10.1016/j.jclepro.2013.02.026
30. Ololade I (2010) Occurrence and toxicity of hydrocarbon residues in crab (*Callinectes sapidus*) from contaminated site. J Appl Sci Environ Manag 11(4). https://doi.org/10.4314/jasem.v11i4.55183
31. Ogunfowokan AO, Asubiojo OI, Fatoki OS (2003) Isolation and determination of polycyclic aromatic hydrocarbons in surface runoff and sediments. Water Air Soil Pollut 147(1):245–261. https://doi.org/10.1023/A:1024573211382
32. European Integration Pollution Prevention Control (IPPC) Bureau (EIPPCB) (2000) Reference document on best available techniques in the cement and lime manufacturing industries. Seville, Spain
33. Company of the Cements of Ain El Kebira, Algeria (SCAEK) (2014) Atmospheric releases. Annual report (Unpublished)
34. Abdul-Wahab SA, Al-Rawas GA, Ali S, Al-Dhamri H (2016) Impact of the addition of oil-based mud on carbon dioxide emissions in a cement plant. J Clean Prod 112(5):4214–4225. https://doi.org/10.1016/j.jclepro.2015.06.062
35. Scrivener KL, John VM, Gartner EM (2018) Eco-efficient cements: Potential economically viable solutions for a low-CO_2 cement-based materials industry. Cement Concr Res 114:2–26. https://doi.org/10.1016/j.cemconres.2018.03.015
36. Nakomcic-Smaragdakis B, Cepic Z, Senk N, Doric J, Radovanovic L (2016) Use of scrap tires in cement production and their impact on nitrogen and sulfur oxides emissions. Energy Sources Part A Recover Util Environ Eff 38(4):485–493. https://doi.org/10.1080/15567036.2013.787473
37. Chen W, Hong J, Xu C (2015) Pollutants generated by cement production in China, their impacts, and the potential for environmental improvement. J Clean Prod 103:61–69. https://doi.org/10.1016/j.jclepro.2014.04.048
38. European Integrated Pollution Prevention Control Bureau (EIPPCB) (2015) Best available techniques (BAT) reference document for the production of cement, lime and magnesium oxide. Technical Catalogue, Publications Office of the European Union, Luxembourg

Fabrication of Lightweight Ceramic Materials Using Geopolymer Technology

Romisuhani Ahmad, Mohd Mustafa Al Bakri Abdullah, Kamarudin Hussin, and Wan Mastura Wan Ibrahim

Abstract Depending on density, lightweight ceramic can be used in several applications such as construction, cutting tools, amour system, wear lining refractories, and wastewater treatment. The growth of lightweight ceramics in such applications becomes a motivation in developing the materials using geopolymer technology. With the uniqueness of geopolymer composition and the help of geopolymerization, high-temperature routes are no longer required to produce ceramic materials. The geopolymer, so called inorganic polymer will convert the amorphous and semi-crystalline phase into a crystalline phase during sintering. The synthesis at low temperatures is energy efficient and more environmentally friendly than older materials. The formation of geopolymer is a polymerization process similar to the polycondensation of an organic polymer. Therefore, the overall process is termed as geopolymerization. The geopolymerization can transfer Al and Si containing wastes into geopolymers with high mechanical strength and highly durable construction materials. The use of kaolin and fly ash as source materials in the formation of geopolymer ceramics could achieve good economic and environmental benefits. This chapter highlights the role of geopolymer systems in the production of ceramic materials for lightweight applications. It covers basic knowledge on ceramics, the formation of geopolymer, fabrication of light-weight ceramic materials using the geopolymer method, and the characterization and properties of the products.

Keywords Geopolymer · Geopolymer ceramics · Lightweight ceramics · Kaolin geopolymer · Fly ash geopolymer

---

R. Ahmad (✉) · M. M. A. B. Abdullah · K. Hussin · W. M. W. Ibrahim
School of Materials Engineering, Center of Excellence Geopolymer and Green Technology, Universiti Malaysia Perlis, 01000 Kangar, Perlis, Malaysia
e-mail: romisuhani@unimap.edu.my

M. M. A. B. Abdullah
e-mail: mustafa_albakri@unimap.edu.my

K. Hussin
e-mail: kamarudin@unimap.edu.my

W. M. W. Ibrahim
e-mail: wanmastura@unimap.edu.my

A. Abdul Kadir et al. (eds.), *Sustainable Waste Utilization in Bricks, Concrete, and Cementitious Materials*, Lecture Notes in Civil Engineering 129,
https://doi.org/10.1007/978-981-33-4918-6_9

1 Introduction

The superior character of ceramic materials gives growth to huge applications in this modern industry. Primarily, ceramics usually having high melting points, therefore there are generally labeled as refractory. Different types of ceramics will provide different properties. In general, ceramic materials are high in modulus, high in strength, low thermal conductivity, high hardness, and chemically inert [1]. The two types of bonding in ceramic materials that are ionic and covalent from the crystal structure influence the bonding stability. However, the fragility of the ceramic materials will increase caused by the strong bonds that limit the applications [2]. In this study, ultra-high molecular weight polyethylene (UHMWPE) has been used as an additive to assist the primary function of providing lightweight properties to the ceramic green body. Additionally, in recent times, increasing demand for stronger, stiffer, and lightweight ceramic materials creates a great interest in the industry. Therefore, materials used selection, appropriate fabrication methods, and sintering profile can all affect the texture of the materials to meet the required properties and performance.

Several possible methods can be used in producing ceramic materials and the selected method is one of the crucial factors as it can affect the properties of the products. Some sequence of steps has been outlined for the ceramic fabrications which start with raw material, proceed with batch preparation, forming and complete with firing. An alternate way of producing ceramic materials is by using the geopolymer method since the amorphous to semi-crystalline phases of geopolymer will transform into crystalline ceramic phases upon heating [3–5]. With the help of geopolymer composition and geopolymerization reactions, high temperature procedures or processes are no longer necessary to achieve materials of ceramic-like structure with comparable properties [6]. Besides, geopolymer can also be directly transformed into the final structural ceramic part of interest and practically design the chemical compositions of the final product [7].

In this study, geopolymer ceramics were fabricated by using powder metallurgy methods. Kaolin and fly ash were used as source materials and ultra-high molecular weight polyethylene was added as a binder. The geopolymer was pulverized into powder form, followed by mixing with the binder, compressing and sintering. The mechanical and physical properties, microstructural, and phase analysis were investigated. This study could be potentially used as a basis for understanding the effect of the geopolymer system toward producing lightweight ceramic materials.

2 Ceramics

Production of ceramic materials consists of compacting and sintering, becoming progressively essential in the modern industry due to their outstanding mechanical and physical properties. The crystallinity of ceramic materials ranges from crystalline to partly crystalline structure, vitrified, and maybe completely amorphous (e.g., glass) [8]. The definition of ceramics is often restricted to inorganic crystalline materials as opposed to the non-crystalline glasses. In general, crystalline ceramics are polycrystalline, fabricated by an excessive number of small crystals or grains separated from one another by grain boundaries [9]. Atomic and crystal structures are two types of structures that have been concerned as valuable factors in determining the ceramic material properties. For the atomic scale, the crystal structure and the type of bonding while for the large scale are the microstructure, referring to the distribution of the structural elements or phases, nature, and the amount of the ceramic [10].

2.1 Ceramics Fabrication

A varied range of processing routes has been recommended for ceramics production, engaging various raw materials from nature or artificial, prominent to a variation of microstructure and pore morphologies. The main methods for fabricating ceramic materials can be classified as in the gaseous phase, liquid phase, or solid-phase fabrication depending on the initial material used as in Table 1. The fabrication of solid (powders) materials was chosen in this study since the geopolymer material is in solid form. This powder metallurgy has the potential in making an almost unlimited variety of materials and products, which places greater significance on the applicable utilization of experimental techniques.

The powder Metallurgy method consists of three main steps which are the production of powders, pressing the powders in dies, and subsequent or simultaneously heating (sintering) the compacted powder in a controlled atmosphere and temperature to obtain the required density and strength [12]. The powder flows into a die cavity at room temperature and is densified by deformation with the deformation of weak friction bonds under pressure [12]. The structure is then heated to a temperature and the particles sinter to one another to form strong bonds thus improves the strength. The main concerns to make this method to be practical are strength, stiffness, toughness, and cost of the production products.

Table 1 Typical methods used to fabricate ceramic materials [11]

Initial phase	Method	Product
Gases	Chemical vapor deposition	Films, monoliths
Gas-liquid	Directed metal oxidation	Monoliths
Gas-solid	Reaction bonding	Monoliths
Liquid-solid	Reaction bonding	Monoliths
Liquids	Sol-gel process Polymer pyrolysis	Films, fibers Fibers, films
Solids (powder)	Melt casting Sintering of powder	Monoliths Monoliths, films

The different mechanisms in the sintering process will promote the material densification and high temperature needed to facilitate the process while diffusion is the matter transport mechanism that promotes densification and grain growth [13]. The effect on the sintering method on the microstructure of ultrafine alumina was studied by Chinelatto et al. [14]. Heating the samples to 1050 °C results in smaller final grain were observed due to the elimination of the finest particles and the narrower distribution of particle size. While Vogiatzis et al. [15] found that the two-step sintering can avoid any thermal shock that leads to cracking and deteriorating the structure of the samples. As a conclusion, the temperature of sintering depends on the melting point of the major components of the raw material used. Therefore, the sintering temperature and method are other variables affecting the mechanical properties of the ceramic materials.

2.2 *Ceramics Binder*

The properties of ceramic materials critically depend on the powder packing and the additives such as binder, plasticizers, surfactants, dispersants, and lubricants play a crucial role, irrespective of the various processing techniques [16]. The term binder system is used to define those additives that remain in the as-formed ceramic body after drying. Binder is regularly added in ceramics fabrication since the ceramics powder particles are very hard and having minimal plastic deformation capability at room temperature that causes difficulty in the compaction of ceramic powders [17]. The main purpose of the binder usage is to plastically deform the ceramic particles to transmit sufficient strength to the green compact, thus maintaining the integrity of the green part during ejection and presintering treatment. Though, the binder's existence can also affect the relative density [18].

Generally, a long chain of polymers was used as binders to assist the primary function of providing strength to the green ceramics body by developing bridges between the particles. Polymers that are categorized into two groups which are thermoplastic and thermoset typically consist of a backbone of covalently bonded carbon–carbon (C–C) linkages with side groups attached at frequent intervals along the length of the molecule. The polymers that heated above their respective decomposition temperatures will decompose to form volatile degradation products [19]. The binder burnout is an important part of ceramics processing and requires an extended heating time which reduces the production rate and releases toxic gases hence causing air pollution and global warming [20]. However, the presence of binder offers remarkably high green strength due to the polymer to polymer and ceramics to polymer interaction and the decreasing in green density with the increasing of a binder concentration were observed due to the separations of ceramic particles occupied by the volume of binder phase [21].

As per review on the usage of binder, polymeric species are the primary constituent of most binder systems. For that, the usage of ultra-high molecular weight polyethylene (UHMWPE) probably could enhance the strength of the green bodies by binding the geopolymer ceramic particles due to their characteristic features includes the dimensionality of the covalently bonded network, the type of functional group, and their molecular chain length, weight, and distribution [20]. These features affect the determination of the properties as well as their interaction with the ceramic surfaces.

2.3 Lightweight Ceramics

Ceramic materials have a broad range of applications, especially in construction, industry, chemical, environmental protection, and agriculture due to their advantageous properties of low density [22]. The specific range of the properties is not fixed depending on each sector requirement since the application of lightweight ceramics is dependent on the suitability properties of each sector [23].

Depending on the density, lightweight ceramic can be used in a number of applications such as amour systems, cutting tools, wear lining refractories, construction, and wastewater treatment [24]. Medvedovski [25] reported that the range density for the lightweight ceramics using alumina, alumina-mullite, and silicon carbide composites for amour application is in between 3.00 and 3.82 g/cm^3. Kitouni and Harabi [26] stated that the bulk densities for conventional porcelain products are 2.45 g/cm^3. Meanwhile, Cetin, Marangoni and Bernardo [27] found that the target density of lightweight ceramic for tiles application is at 2.00 g/cm^3. De Santis, Rossignolo and Morelli [28] studied the fabrication of calcined clay lightweight ceramics using wood sawdust and sodium silicates as additives. The wood sawdust was added to reduce the specific mass from 1.92 to 1.76 g/cm^3 while the alkaline activator reduces the water absorption and increases the compressive strength. The influence of nanopowders and pore-forming agents used in the fabrication of porous alumina–zirconia ceramics were investigated by Khabas et al. [29]. Al-Marahle [30] investigated the feasibility of lightweight ceramics clay bricks production by adding various types of combustible materials to the clay. The substantial ratio of bagasse, sawdust, rice husk, and straw were mixed with clay and the results found that the combustible materials lower the density of the bricks between 0.80 and 0.90 g/cm^3.

The fabrication of lightweight ceramics involves consideration of several factors including the properties of the components. The selection of materials, proper method and selection of additive plays could play such an important role in lightweight ceramic materials fabrication to achieve the lower density with superior mechanical strength.

3 Geopolymer for Lightweight Ceramic Material Applications

Inorganic polymers denoted as geopolymers are aluminosilicate materials that display superior physical and chemical properties with a diverse range of possible potential applications such as cement for concrete, coating, high-temperature ceramics, and binder for fireplace resistant fiber composite [31]. Hardening at ambient temperatures, geopolymer possesses amorphous to semi-crystalline phase contain a random tetrahedral network of silicon and aluminium atoms [32]. Theoretically, any alkali and alkaline earth cation can be used as an alkali element in the geopolymerization process.

High alkali solutes, such as sodium silicate (Na_2SiO_3) and sodium hydroxide (NaOH) are integrated into source materials with a high content of SiO_2 and Al_2O_3, yielding Si–O–Al–O bonds. Porous structures are commonly found in the geopolymer system, and water molecules along with the positive ion are trapped within the three-dimensional cross-linked Si–O–Al structure. The complex geopolymer structure involves chains, three-dimensional networks, and sheet-like made up of several Q unit types of connected SiO_4 and AlO_4 tetrahedra. Types of geopolymers depend on the Si/AL ratios namely poly (sialate), poly (sialate-siloxo) and poly (sialate-disiloxo).

An exothermic reaction called geopolymerization is carried out through oligomers (dimer, trimer) which are the fundamental unit structures for three-dimensional macromolecular edifice [33]. Generally, geopolymerization involves several processes including dissolution, reorientation, and solidification, as shown in Fig. 1 [34, 35]. In the dissolution step, Si–Al raw materials contact the alkali solution producing both Si and Al species. While during the reorientation step, the dissolved Si and Al species (eg., Al^{3+}, Si^{4+}) are diffused into oligomers. The oligomers in the aqueous phase form relatively large networks by condensation, resulting in the formation of a gel. Meanwhile, reactive Al and Si species will further leach from the raw materials which occur when the dissolved Al^{3+} and Si^{4+} on the surface of source Si–Al materials are removed [36]. At the solidification step, the gelation system continues to rearrange and reorganize as the connectivity of the gel network increases. The amorphous or semi-crystalline three-dimensional alumino-silicate network is commonly accredited to geopolymer. Two major factors determining the properties of the final geopolymeric products are temperature and air circulation [36]. It needs to be specifying that there is no specific order for these three major steps. In other words, all steps occur simultaneously [37].

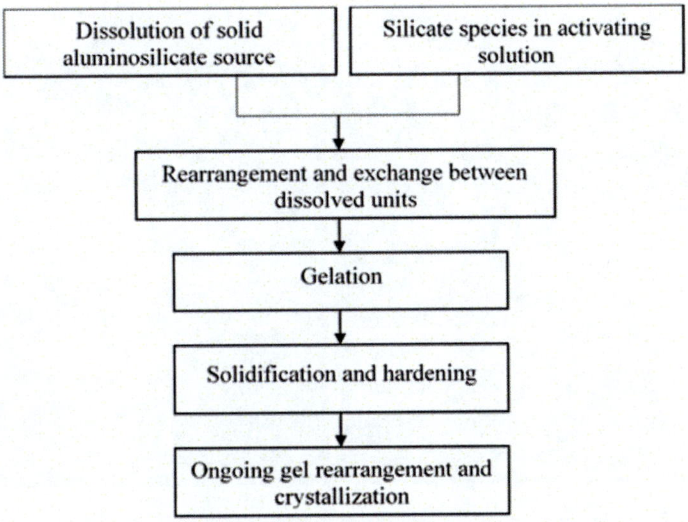

Fig. 1 Summary of geopolymerization process

3.1 Constituents of Geopolymer for Lightweight Ceramic Material Applications

The development of geopolymer material for lightweight ceramics applications consists of several main constituents which are source materials (kaolin and fly ash) alkaline activator solutions (sodium hydroxide solution and sodium silicate solution) and binder (ultra-high molecular weight polyethylene).

3.1.1 Source Materials—Kaolin and Fly Ash

Any materials that contain mostly silica (Si) and alumina (Al) can be used as source material in the formation of geopolymer. However, based on the nature of the source material, it was identified that the calcined source materials, such as calcined kaolin, fly ash, and slag revealed a higher final strength compared to the non-calcined materials such as kaolin clay and natural minerals [36, 38]. The formation of geopolymers using calcined materials was found to have higher early strength, while those formed from non-calcined materials possessed higher strength during the later curing [36]. The calcined materials such as metakaolin are made from calcining the kaolin clay at 600–800 °C, thus using non-calcined kaolin as raw materials in producing geopolymer would be a significant achievement to reduce the energy consumption.

Kaolin is natural clay in one of our natural world resources with the potential to be used as geopolymer sources materials since it has high content of Si and Al. Kaolinite with a chemical formula of $Al_2O_3 \cdot 2SiO_2 \cdot 2H_2O$ has 1:1 uncharged dioctahedral layer

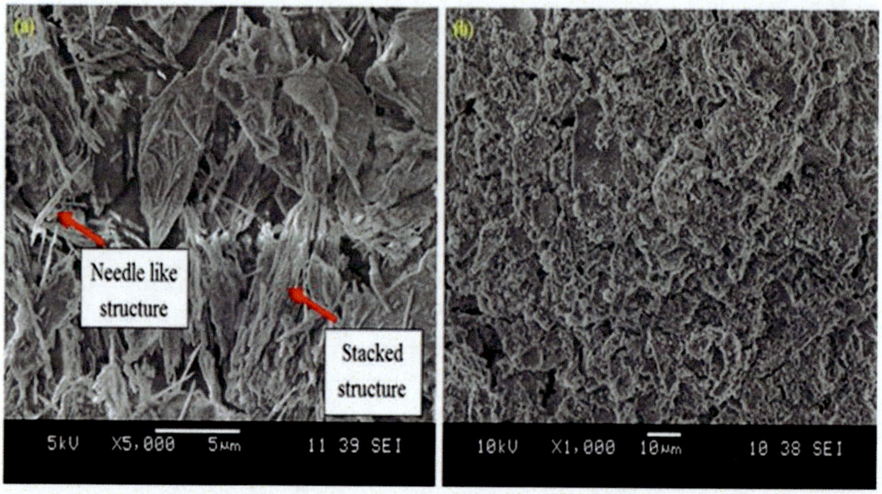

Fig. 2 SEM micrograph of **a** raw material kaolin and **b** kaolin geopolymer

structure. This layer comprises of $Al(OH)_3$ sheet and the $(Si_2O_5)n^{2-}$ sheet linked together by sharing oxygen atoms and held together by van der Waals and hydrogen bonds [39]. Extensive research was conducted by Xu and Van Deventer [36] on the use of kaolinite with the addition of sixteen different alumino-silicate minerals in order to synthesis the geopolymers. From the observation, the addition of kaolinite is necessary for the formation of gel. A weak structure is formed if kaolinite is used without the presence of aluminosilicates, thus the cooperation between different alumino-silicates seems to be relatively important [36]. The SEM micrograph of kaolin and kaolin geopolymer is illustrated in Fig. 2.

While for fly ash, it is an acidic material containing acidic oxides such as Al_2O_3, SiO_2 and Fe_2O_3 which provide the potential for alkali reaction [40]. The chemical composition of fly ash depends on the type of fuel burned and the combustion method, concerning the chemical composition classified into class C (high Ca and low Al) and class F (low Ca and high Al). Class F is usually used in low calcium geopolymer [41]. Between those types, Class F is available in the biggest quantities to compare to Class C. The chemical composition of both class C and class F Fly ash tabulated in Table 2.

Kong and Sanjayan [44] investigated geopolymer paste after being exposed to an elevated temperature of 800 °C. The major change of color was detected at the temperature exposure of 800 °C. There was no surface crack and the compressive strength of fly ash-based geopolymer increased with the increase of temperature. More significantly, the source materials play an important role in the geopolymeric reaction and affect the mechanical properties of the final hardened product.

Table 2 Chemical composition of class C and class F fly ash

Compound	Class C	Class F
	Mass (wt%) [42]	Mass (wt%) [43]
SiO_2	37	55.90
Al_2O_3	19.59	28.1
Fe_2O_3	9.94	6.97
CaO	23.88	3.84
TiO_2	1.42	2.21
K_2O	1.06	1.55
SrO	1.2	0.37
RuO_2	1.32	0.24
LOI	4.59	0.23

3.1.2 Alkali Activator

Alkali activator is used to activate the aluminosilicate materials as well as prompt the crystallization and precipitation of the silica and alumina species present in the solution. The hydroxyl radical (OH^-) execute as a reaction catalyst and alkali metal (Na^+) as a structural forming element in balancing the negative framework carried by the tetrahedral aluminium [45]. The alkali activator solution affects the properties of geopolymer and geopolymer ceramics significantly. It also promotes the solubility and dissolution of aluminosilicate besides accelerating the geopolymerization reaction [4, 46].

Common alkali activator used is a mixture of potassium hydroxide (KOH) or sodium hydroxide (NaOH) and potassium silicate or sodium silicate with different mass ratios [36, 47, 48]. Palomo et al. [37] reported that the type of alkali liquid plays an important role in the polymerization process. Alkali activator comprises soluble silicate either potassium or sodium silicate, leading to a higher rate of reaction than the use of only alkali hydroxides. According to Xu and Van Deventer [48], the reaction between the source materials and the solution was boosted with the addition of sodium silicate solution to the sodium hydroxide solution as the alkali activator. Sodium silicate (Na_2SiO_3) will rapidly dissolve and starts to bond the particles of the raw materials.

3.2 Geopolymer Ceramics for Lightweight Applications

Geopolymers have the advantage to be used in various fields due to better properties than those of ceramics and cement-based materials. By owing excellent physical and mechanical properties, the feasibility of forming ceramics from geopolymer has been verified by few researchers as geopolymer are a class of amorphous, aluminosilicate materials that harden hydrothermally at ambient temperatures and can be converted into ceramics by heating [49, 50]. Bell et al. [51] proved that it was possible to produce pollucite ceramic from the natural metakaolin-based geopolymer in the formation of pollucite ($Cs_2O{\cdot}Al_2O_3{\cdot}4SiO_2$). Geopolymer that consists of a mixture of metakaolin and cesium silicate solution was prepared using an overhead mixer. The resultant slurry was then cured at 50 °C for 24 h. The Cs-based geopolymer was found to be more refractory compared with K-based and Na-based geopolymer. The pollucite crystallization occurred gradually during heating at temperatures in the range of 900–1250 °C.

Based on the study by He et al. [52], the ultrasonic vibration technique was used in order to produce geopolymer ceramic composites. A geopolymer resin was obtained by mixing metakaolin powder with a potassium silicate solution with the molar ratios of $SiO_2/Al_2O_3 = 3$, $K_2O/SiO_2 = 0.33$, and $H_2O/K_2O = 11$. Unidirectional carbon fibers were used as reinforcement in the geopolymer matrix. The composite was prepared by infiltrating geopolymer resin into the unidirectional, continuous Pan-based carbon fiber preform with the help of the ultrasonic vibration treatment. Then, the samples were heated in the temperature range of 1000–1400 °C to produce leucite ($K_2O{\cdot}Al_2O_3{\cdot}4SiO_2$) ceramics.

The mechanical properties of leucite ceramic derived from a geopolymer were studied by Xie et al. [53]. Leucite ceramic was fabricated by grinding the $K_2O{\cdot}Al_2O_3{\cdot}4SiO_2{\cdot}11H_2O$ geopolymer to a powder, followed by drying, isotactic pressing, and sintering at 950–1200 °C. Samples sintered at 1200 °C for 3 h provide a maximum density, fracture toughness, and biaxial flexure strength. In the study by He et al. [52], a leucite ceramic disk was developed by sintering the geopolymer without first grinding it into a powder. The relative density was 93% after it was soaked at 1000 °C for 2 h. The geopolymer with a composition of $SiO_2/Al_2O_3 = 5$, $K_2O/SiO_2 = 2$, and $H_2O/K_2O = 11$ (molar ratio) was formed by mixing metakaolin with potassium silicate solution. After curing the geopolymer at 70 °C for 48 h, the samples were sintered at different temperatures with 120 min of soaking. As a result, a substantial amount of leucite was formed at a temperature of 800 °C.

He and Jia [54] studied the crystallization, sintering, and thermal expansion behaviors of a synthetic-metakaolin cesium-based geopolymer. A Cs-based geopolymer with the composition (mole ratios) of $SiO_2/Al_2O_3 = 4$, $Cs_2O/Al_2O_3 = 0.9$, and $H_2O/Cs_2O = 11$ (molar ratio) was obtained by mixing synthetic metakaolin with an alkaline silicate solution and curing the mixture at 50 °C for 48 h. Based on crystal nucleation and growth mechanism, the amorphous Cs-geopolymer completely crystallized into pollucite on heating at 1200 °C for 2 h. Due to attractive refractory properties and the ability to convert to leucite ($KAlSi_2O_6$) on heating, K-based geopolymer ($K_2O{\cdot}Al_2O_3{\cdot}4SiO_2{\cdot}11H_2O$) and the comparisons between clay and bentonite to obtain leucite at low potassium silicate solution was investigated by El-Maghraby et al. [49]. The kaolinite and bentonite were calcined at 750 °C for 1 h and mixed with the caustic silicate solution with the composition of $SiO_2/Al_2O_3 = 4$, $K_2O/ SiO_2 = 0.3$, and $H_2O/K_2O = 11$ (molar ratio). After curing, the sample was ground using gad mortar, sieved and sintered from 800 to 1100 °C at a heating rate of 200 °C/h, soak time for 3 h and cooling rate of 10 °C. The quartz phase decreased and disappeared and a small crystal for nepheline appeared as the temperature increased to 1100 °C.

Meanwhile, Kuenzel et al. [55] explored the properties of metakaolin de-rived geopolymer mortars that contained 50% by weight of silica sand, after exposure to temperature up to 1200 °C with the heating rate of 20 °C/minute and 2 h soaking time. The sodium aluminate crystalline phase was found forms at above 800 °C and the geopolymer phase transforms to nepheline ($NaAlSiO_4$) and carnegieite ($NaAlSiO_4$) on heating at 900 °C. The mortar samples were transformed into polycrystalline nepheline/quartz ceramics after sintered to 1000 °C. Thus, the capability of using geopolymer technology in fabricating ceramic materials at low temperature creates an innovative ceramic application due to the forming of various crystalline types depends on the raw material used.

3.2.1 Fabrication Method of Geopolymer Ceramics

The difficulty of fabricating ceramic materials using geopolymer for various applications with superior properties and performance required can be obtained reliably and cost-effectively. Both goals of appropriate reliability and are cost dependent on the impact of components composition as well as size, shape and dimensional-surface finish requirement on fabrication routes and the parameters of the processing technique within the routes [11]. Previous work has primarily focused on the microstructural and physical evolution on heating of as-cast geopolymers. By heating of as-cast, the monolithic bodies typically encourage cracking due to large capillary forces acting on capillary channels during drying [56]. He and Jia [50] have developed leucite ceramic directly derived from the geopolymer and the ceramic products produced showed a low mechanical strength.

The successful method to overcome the problem is by sinter ground and compressed geopolymer powder [53]. By using this method, the researchers found that the use of geopolymer precursor provides a simple, low-cost approach to form relatively high strength leucite ceramics compared with the conventional method. Die pressing of dried geopolymer prevents cracking on heating due to capillary stresses created from the evaporation of water [56], thus resulting in better strength compared to the direct heating method. Therefore, the fabrication method plays an important role in producing ceramic with optimum mechanical properties and relative densities.

3.2.2 Sintering Temperature of Geopolymer Ceramic

Sintering is referred to a process of heat treatment in which powder or porous material is formed into a required shape and converted to a useful solid. As polycrystalline ceramics' properties are controlled by the microstructure, the density, grain size, and occurrence of heterogeneities [57], the sintering condition plays an important role in order to archive the desired microstructure thus improve the properties of the ceramic product. Besides, the sintering temperature strongly influences the phase transition of geopolymer from amorphous to crystalline.

According to Xie et al. [53], the sintering of geopolymer above 1000 °C was sufficient to consolidate the sample and begin leucite crystallization. Heating to 1200 °C for 3 h resulted in a maximum density, fracture toughness, and bi-axial flexure strength. The grain size and leucite concentration were influenced by the toughness and biaxial flexure. The higher sintering temperature was reported to produce the appropriate size and quantity of leucite grains required to reinforce the glassy matrix with the stronger high temperature, leucite ceramic phase. While in the study carried out by He et al. [52] found that the crystallization peak temperature of $K_2O{\cdot}Al_2O_3{\cdot}5SiO_2$ was 986.3 °C, and the temperature range of sintering for and 700–954.3 °C.

Contrarily, the crystallization peak temperature and the temperature range of sintering for $K_2O{\cdot}Al_2O_3{\cdot}4SiO_2$ found by Bell et al. [51] was 1130 and 850–1100 °C, respectively, which higher compared to He et al. [52]. The researchers discovered that the quartz introduced from metakaolin affect the reaction and facilitated the crystallization process. In another study by He et al. [54], the soaking time was increased from 60 to 120 min and soaking temperature from 800 to 1000 °C in order to make sure the quartz was fully dissolved.

3.3 Lightweight Geopolymer Ceramics Properties

The impact addition of ultra-high molecular weight polyethene (UHMWPE) as binder on geopolymer ceramics to be used for lightweight applications will be elaborated and discussed in this section. The optimum weight percentage of UHMWPE in the geopolymer ceramics has been determined by the strength performance. Geopolymer ceramics with UHMWPE addition were also tested under flexural strength test, density test, shrinkage, and water absorption test. The mechanical testing, phase analysis, morphology analysis, and functional group identical to the performance of UHMWPE on the geopolymer ceramics will be discussed.

3.3.1 Flexural Strength Analysis

In order to produce the desired lightweight ceramic materials with an improvement of the mechanical properties, the organic phase (UHMWPE) was used as a binder. The effects of the addition of different amounts of UHMWPE to the geopolymer ceramics with heating at 1200 °C on flexural strength is shown in Fig. 3. The result shows that the flexural strength of geopolymer ceramics has been improved after the integration of UHMWPE. The flexural strength of geopolymer ceramics without the addition of UHMWPE was 85.83 MPa, while the addition of 4 wt% of UHMWPE achieved the highest strength of 92.1 MPa. The increasing of flexural strength with the addition of different content of UHMWPE as binder confirmed that the decomposition of UHMWPE at high temperature resulted in the increasing of carbon content in the geopolymer system. The well-distributed of excess carbon can be beneficial for

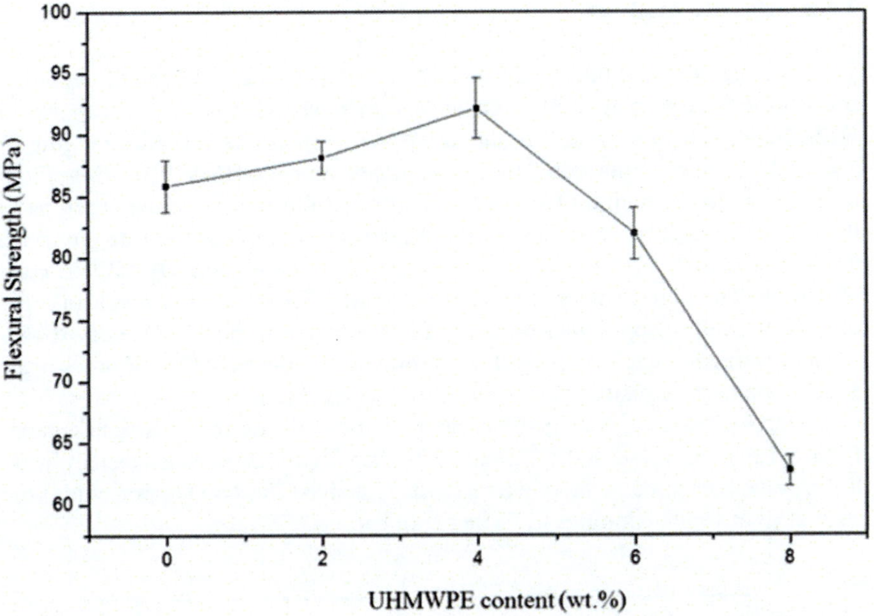

Fig. 3 Flexural strength of geopolymer ceramics with various content of UHMWPE sintered at 1200 °C

the resistance against crystallization up to high temperatures and simultaneously generates a complex microstructure for carbon-rich ceramic materials hence improve in flexural strength [58].

The addition of 2 wt% of UHMWPE proves the increasing of flexural strength of the geopolymer ceramics with the addition of binder. Despite, the addition of 2 wt% UHMWPE was considered as an introduced amount in the reaction between the binder and the geopolymer, which leads to involvement and increases the reaction rate. While the addition of 4 wt% of UHMWPE giving the highest flexural strength corresponds to the sufficient amount of binder to react with the geopolymer matrix. This also could be explained that the addition amount of UHMWPE was fully introduced in the reaction of the geopolymer system. However, the flexural strength decreased after the optimum value of UHMWPE content at 4 wt% was due to the high existence of UHMWPE. An excessive amount of UHMWPE inhibits the reaction between binder and polymer thus lead to the decomposition of UHMWPE that leads to the increasing of the amount and size of pores during the sintering process [59].

3.3.2 Density Analysis

The effects of the addition of different amounts of UHMWPE to the geopolymer ceramics with heating at 1200 °C on density is shown in Fig. 4. The incorporation of UHMWPE leads to decreases in the density of geopolymer ceramics ranging from 1.49 to 2.013 g/cm^3 contributed by the decomposition of UHMWPE during sintering resulting in the formation of pores hence affected the density of the green part. At the optimum weight percent of UHMWPE, the density recorded at 1.88 g/cm^3 while the lowest density of 1.49 g/cm^3 was achieved at the highest UHMWPE content of 8 wt%. The high in weight percent of UHMWPE added will result in a lower density of kaolin geopolymer ceramic. The formation of pores is clearly shown in the microstructure analysis in Sect. 3.3.6 proves that the existence of pores leads to a reduction of the density value of the geopolymer ceramic.

Generally, the density of geopolymer ceramic at the optimum weight content of UHMWPE is achieved at 1.88 g/cm^3. The density obtained considerably reaches lower value compared to the previous finding, lead to offer good opportunities to use such as lightweight ceramics in further applications.

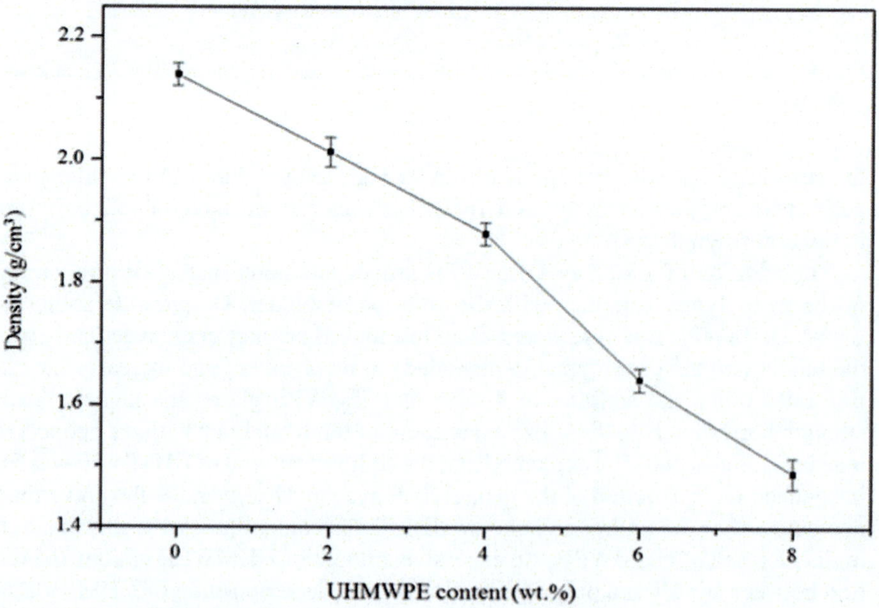

Fig. 4 Density of geopolymer ceramics at various content of UHMWPE sintered at 1200 °C

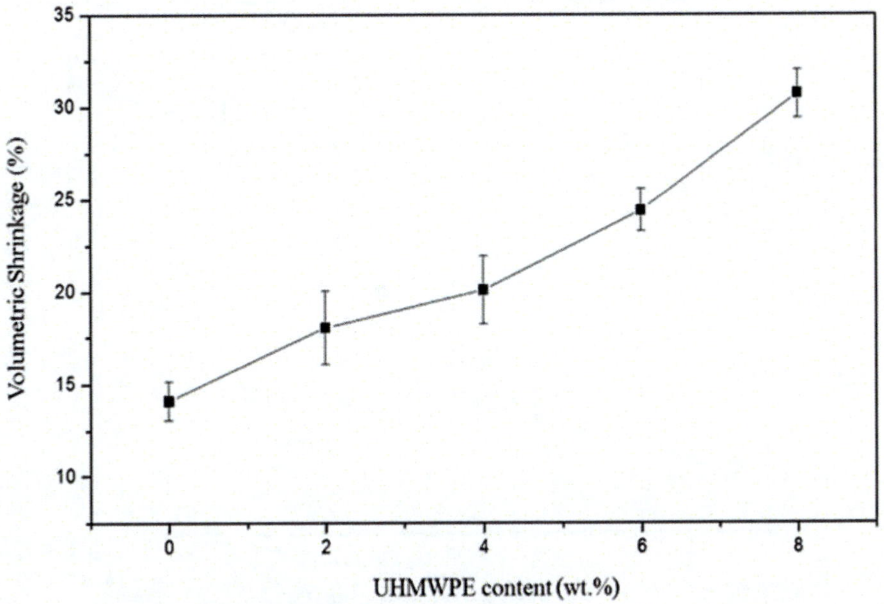

Fig. 5 Volumetric shrinkage of geopolymer ceramics at various content of UHMWPE sintered at 1200 °C

3.3.3 Volumetric Shrinkage

The volume shrinkage as a function of various content of UHMWPE of geopolymer ceramic that sintered at 1200 °C is presented in Fig. 5. The overall result indicates that the volumetric shrinkage increased with the increase of UHMWPE content. From the result, geopolymer ceramic without the addition of UHMWPE attained the lowest volumetric shrinkage value at 14.1% while at the optimum weight content of UHMWPE (4 wt%), the volumetric shrinkage observed at 20%. Shrinkage occurs as the binder initiates to decompose and the volume of the gases that evaporates leaving the system quantitatively accounted for the volume losses resulting from the decomposition of UHMWPE during sintering.

3.3.4 Water Absorption

The water absorption of the geopolymer ceramic with different UHMWPE percentages is shown in Fig. 6. The graph shows that the geopolymer ceramic with the addition of UHMWPE has a higher percentage of water absorption compared to the geopolymer ceramic without UHMWPE. The percentage of water absorption increased with the increase of weight content of UHMWPE due to the formation of pores that resulted from the degradation of polymer while sintering. Moreover, the sintering of geopolymer is also one of the causes of the formation of small pores

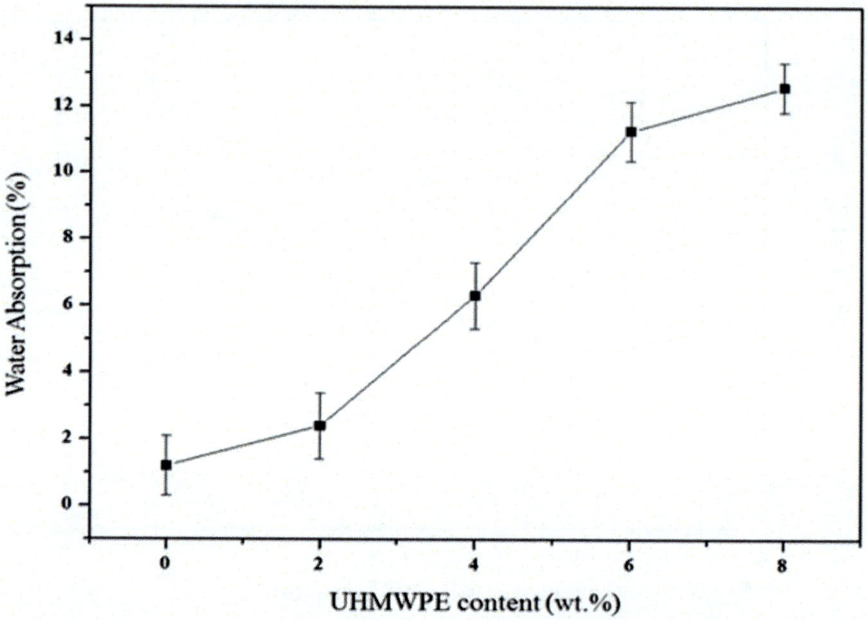

Fig. 6 Water absorption of geopolymer ceramic at various content of UHMWPE

throughout the samples. The existence of pores permits absorbing more water which increased the water absorption value [60]. Water absorption is primarily dependent on the porosity of the ceramic materials.

Khan et al. [60] found that the water absorption was initially decreased and increased for further increasing binder content. The initial decrease in the percentage of water absorption was due to the densification process and further increases were due to the formation of micropores following the dehydroxylation of the binder.

3.3.5 Phase Analysis

The phase analysis of geopolymer ceramics with and without the addition of various content of UHMWPE sintered at 1200 °C is illustrated in Fig. 7. It can be observed that the phase analysis for each sample with incorporation amount of 2, 4, 6 and 8 wt% of UHMWPE are quite similar to the geopolymer ceramics without UHMWPE. The appearance of the major phase, nepheline ($NaAlSiO_4$) was observed in both XRD patterns of geopolymer ceramics with and without the addition of UHMWPE. The crystallization of nepheline ($NaAlSiO_4$) from geopolymer gradually occurred on heating. However, in the XRD pattern geopolymer ceramics with the addition of various content of UHMWPE, the appearance of the extra peak of carbon (C) was observed at 2θ values of 17.7°, 31.0°, 35.9°, 37.5°, 41.7°, 45.7°, and 56.2°. The existence of carbon (C) peak accredited to the decomposition of UHMWPE while

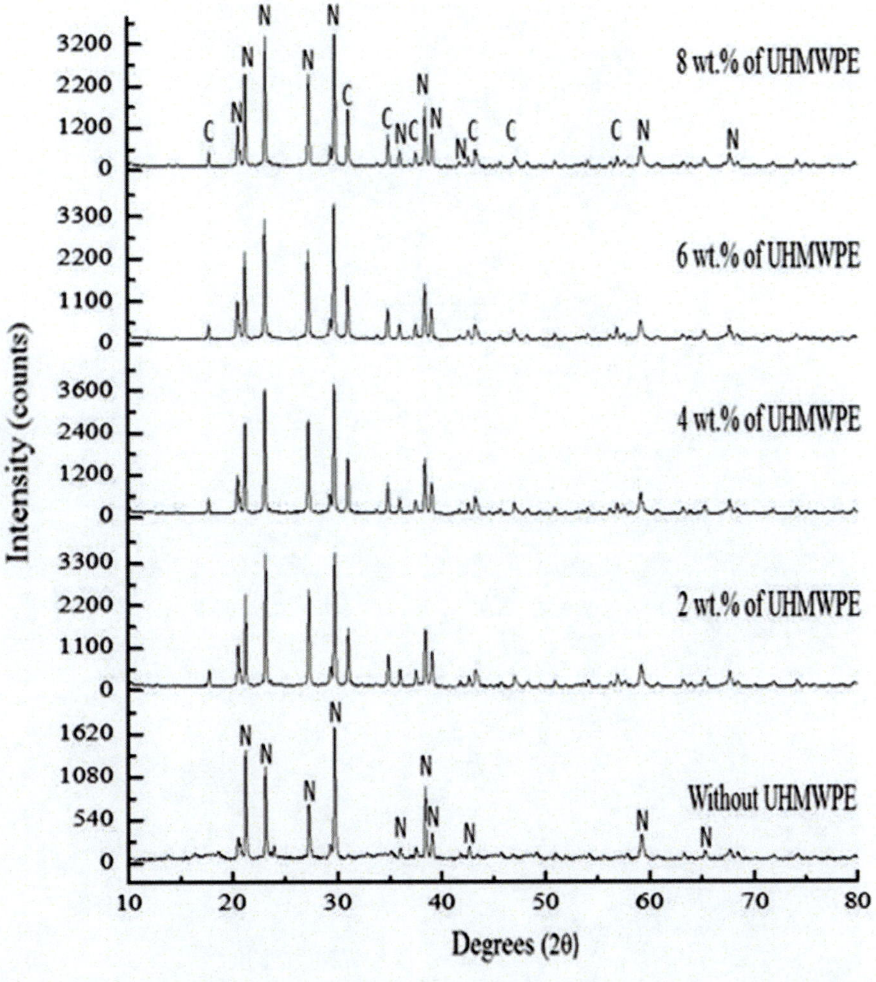

Fig. 7 Phase analysis of geopolymer ceramics with and without the addition of various content of UHMWPE sintered at 1200 °C (N = Nephaline, C = Carbon)

sintering at high temperatures. The molecule of carbon inside the geopolymer ceramic only has one source which is the UHMWPE. The intensities of nepheline peak also increased with the increasing weight percent of UHMWPE and started to decrease at 6 wt% of UHMWPE showed that adequate reaction between the raw material and alkali activator has occurred which contributes to better geopolymerization.

3.3.6 Morphology Analysis

Figure 8 displays the SEM micrograph of geopolymer ceramic with and without

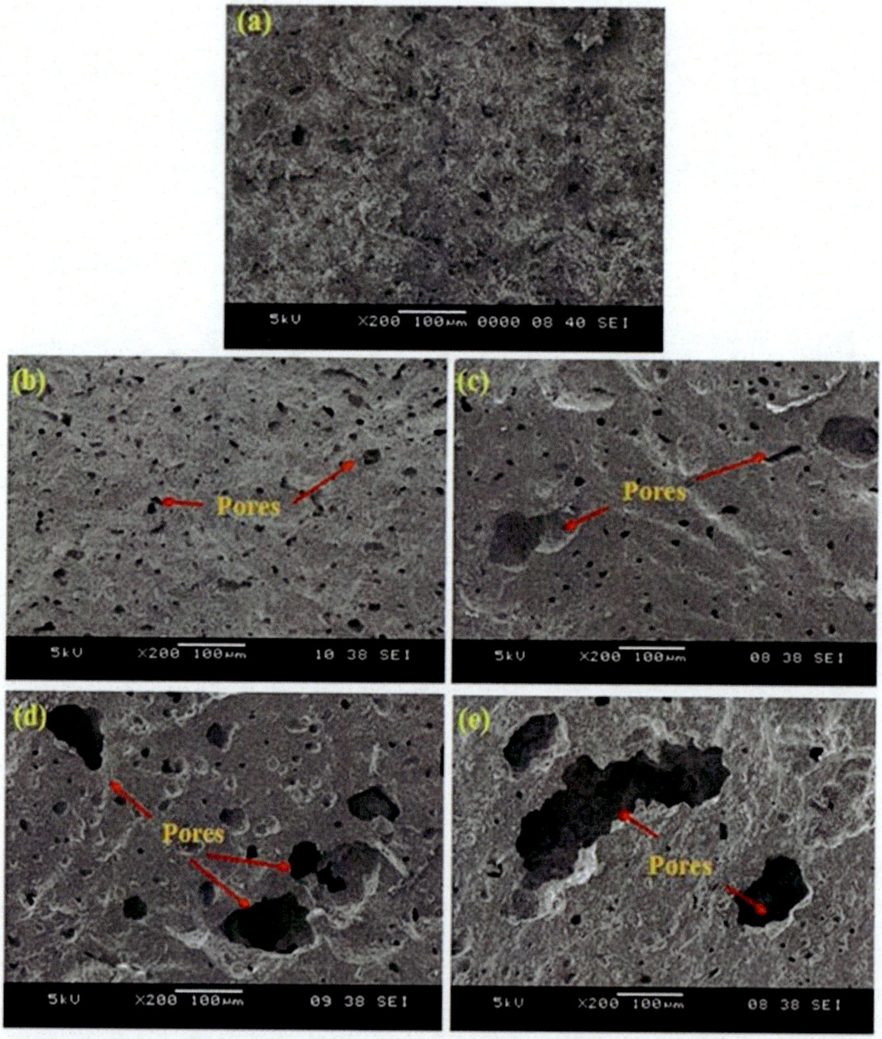

Fig. 8 SEM micrograph of geopolymer ceramic at various content of UHMWPE **a** 0 wt%, **b** 2 wt%, **c** 4 wt%, **d** 6 wt% and **e** 8 wt%

the addition of UHMWPE sintered at 1200 °C. The microstructure of fractured surfaces of geopolymer ceramics at various UHMWPE content is shown in Fig. 8b 2 wt%, (c) 4 wt%, (d) 6 wt%, and (e) 8 wt% showed that the microstructure of the investigated materials was considerably affected by the content of the UHMWPE. The smooth surfaces were observed in the microstructure of the samples with the addition of UHMWPE. The addition of UHMWPE on geopolymer ceramic increased the number of pores and connected pores compared to the sample without the addition of UHMWPE. The size of pores also increased with the increase of UHMWPE content due to the evaporation of different gas from the degradation of UHMWPE that sintered at high temperatures. The observation in microstructure is in agreement with the decreasing flexural strength of the samples after the addition of 4 wt% of UHMWPE due to the presence of more pores and interconnected pores. Furthermore, the decrease of density as well as an increase in the percentage of water absorption has been proven with the microstructural results.

3.3.7 Functional Group Analysis

The infrared spectra of geopolymer ceramics with and without the addition of various content of UHMWPE that sintered at 1200 °C in the range of 400–4000 cm^{-1} are presented in Fig. 9. The FTIR measurements are basically in accordance with the results of XRD and proved that the nepheline phase appeared when the samples were heated. The typical bands of geopolymeric materials remained in the spectra after the addition of UHMWPE. Nevertheless, several observations are relevant. Sintering the samples with and without the addition of UHMWPE shows a weak bond at around 3400 cm^{-1}. The samples with the addition of UHMWPE showed peaks at about 2842 and 2931 cm^{-1} corresponding to the C–H asymmetric stretching vibration due to the addition of UHMWPE as a binder in the samples. A solid polymer will break down into a smaller molecular fragment made up by a number of different chemical species and left residues [19]. As a result, the decomposition of UHMWPE converts the UHMWPE into hydrocarbon molecules simultaneously, reacted with the geopolymer, and formed a hydrogen bridge between the UHMWPE

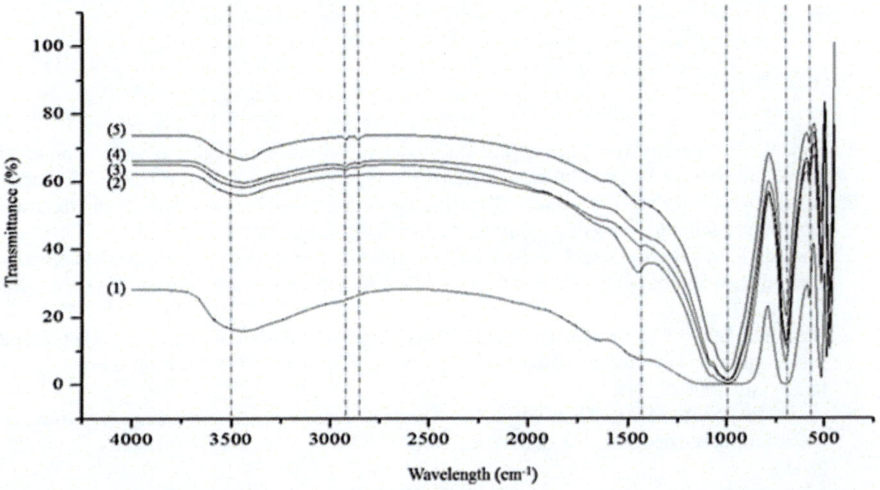

Fig. 9 Spectra of geopolymer ceramic with and without addition of various content UHMWPE sintered at 1200 °C (1) = Without UHMWPE, (2) = 2 wt% of UHMWPE, (3) = 4 wt% of UHMWPE, (4) = 6 wt% of UHMWPE and (5) = 8 wt% of UHMWPE

chain and geo-polymer. Due to the hydrogen bonding weakness, it could be broken at high temperatures and during a micro-reaction inside the system, which leads to the formation of water molecules. Those which get evaporate later and create micro-pores in the system. This affects the density and strength of the products. The appearance peak at about 700 and 1040 cm^{-1} corresponding to the Si–O–Si groups while the characteristic absorption band of O–H stretching vibration is revealed at about 3443 cm^{-1}.

4 Conclusion

The main conclusion of this experimental research study could be summarized in this section. It is shown that the optimum percentage, which could correspond to UHMWPE is 4 wt%, conducted to achieve a maximum flexural strength of 92.1 MPa. A homogenous and smooth surface has been observed by using SEM micrograph, with a density of 1.88 g/cm^3. However, the obtained density could be classified as low density compared to the conventional range of lightweight ceramics. Small and well-distributed pores are observable, due to the transformation of the amorphous phase to the crystalline one, during the progress of the sintering process at a high temperature. Further, the geopolymer ceramic exposed to high temperature drive to a decomposition of UHMWPE, which could be explained by the degradation of polymer and converting into chemical hydrocarbons and gases. A carbonaceous residue has remained inside the samples under high temperatures, which are shown in the phase and functional group analysis.

References

1. Popoola A, Olorunniwo O, Ige O (2014) Corrosion resistance through the application of anti-corrosion coatings. In: Aliofkhazraei M (ed) Developments in corrosion protection. IntechOpen
2. Iyasara A, Joseph MTT, Azubuike T (2014) Exploring ceramic raw materials in Nigeria and their contribution to nation's development. Am J Eng Res 3:127–134
3. Barbosa V, Mackenzie K (2003) Thermal behaviour of inorganic geopolymers and composites derived from sodium polysialate. Mater Res Bull 38:319–331. https://doi.org/10.1016/S0025-5408(02)01022-X
4. Duxson P, Lukey GC, van Deventer JSJ (2006) Thermal evolution of metakaolin geopolymers: part 1—physical evolution. J Non Cryst Solids 352(52):5541–5555. https://doi.org/10.1016/j.jnoncrysol.2006.09.019
5. He P, Jia D, Wang M, Zhou Y (2011) Thermal evolution and crystallization kinetics of potassium-based geopolymer. Ceram Int 37(1):59–63. https://doi.org/10.1016/j.ceramint.2010.08.008
6. Iwahiro T, Nakamura Y, Komatsu R, Ikeda K (2001) Crystallization behavior and characteristics of mullites formed from alumina–silica gels prepared by the geopolymer technique in acidic conditions. J Eur Ceram Soc 21(14):2515–2519. https://doi.org/10.1016/S0955-2219(01)00273-4

7. He P, Jia D, Wang M, Zhou Y (2010) Effect of cesium substitution on the thermal evolution and ceramics formation of potassium-based geopolymer. Ceram Int 36(8):2395–2400. https://doi.org/10.1016/j.ceramint.2010.07.015
8. Čtvrtlík R, Kulikovsky V, Bohác P, Slovance N (2010) Effect of structure on mechanical properties of covalent ceramics. In: NanoCon
9. Fragassa C (2015) Limits in application of international standards to innovative ceramic solutions. Int J Qual Res 9:279–298
10. Chen P, Kim G-Y, Ni J (2007) Investigations in the compaction and sintering of large ceramic parts. J Mater Process Technol 190(1):243–250. https://doi.org/10.1016/j.jmatprotec.2007.02.039
11. Rahaman MN (2007) Ceramic processing. CRC/Taylor & Francis
12. Al-Qureshi HA, Galiotto A, Klein AN (2005) On the mechanics of cold die compaction for powder metallurgy. J Mater Process Technol 166(1):135–143. https://doi.org/10.1016/j.jmatprotec.2004.08.009
13. Lóh NJ, Simão L, Faller CA, De Noni A, Montedo ORK (2016) A review of two-step sintering for ceramics. Ceram Int 42(11):12556–12572. https://doi.org/10.1016/j.ceramint.2016.05.065
14. Chinelatto A, Pallone E, Souza A, Manosso M, Chinelatto A, Tomasi R (2012) Mechanisms of microstructure control in conventional sintering. In: Lakshmanan A (ed) Sintering of ceramics—new emerging techniques. InTechOpen, pp 401–422
15. Vogiatzis CA, Tsouknidas A, Kountouras DT, Skolianos S (2015) Aluminum–ceramic cenospheres syntactic foams produced by powder metallurgy route. Mater Des 85:444–454. https://doi.org/10.1016/j.matdes.2015.06.154
16. Rajeswari K, Chaitanya S, Biswas P, Buchi Suresh M, Rao YS, Johnson R (2015) Binder burnout and sintering kinetic study of alumina ceramics shaped using methylcellulose. J Ceram Process Res 16(1):74–80
17. Lewis JA (1997) Binder removal from ceramics. Annu Rev Mater Sci 27(1):147–173. https://doi.org/10.1146/annurev.matsci.27.1.147
18. Baklouti S, Bouaziz J, Chartier T, Baumard J-F (2001) Binder burnout and evolution of the mechanical strength of dry-pressed ceramics containing poly(vinyl alcohol). J Eur Ceram Soc 21(8):1087–1092. https://doi.org/10.1016/S0955-2219(00)00305-8
19. Beyler C, Hirschler M (2002) Thermal decomposition of polymers. SFPE Handb Fire Prot Eng 2
20. Pizette P, Martin CL, Delette G, Sans F, Geneves T (2013) Green strength of binder-free ceramics. J Eur Ceram Soc 33(5):975–984. https://doi.org/10.1016/j.jeurceramsoc.2012.11.018
21. Bergström L (2002) Colloidal processing of ceramics. In: Holmberg K (ed) Handbook of applied colloid and surface chemistry. Ytkemiska institutet, Wiley, pp 201–218
22. Ducman V, Mirtič B (2009) The applicability of different waste materials for the production of lightweight aggregates. Waste Manag 29(8):2361–2368. https://doi.org/10.1016/j.wasman.2009.02.013
23. Bauer J, Hengsbach S, Tesari I, Schwaiger R, Kraft O (2014) High-strength cellular ceramic composites with 3D microarchitecture. Proc Natl Acad Sci 111(7):2453 LP–2458. https://doi.org/10.1073/pnas.1315147111
24. Zang W, Guo F, Liu J, Du H, Hou F, Guo A (2016) Lightweight alumina based fibrous ceramics with different high temperature binder. Ceram Int 42(8):10310–10316. https://doi.org/10.1016/j.ceramint.2016.03.170
25. Medvedovski E (2006) Lightweight ceramic composite armour system. Adv Appl Ceram 105:241–245. https://doi.org/10.1179/174367606X113537
26. Kitouni S, Harabi A (2011) Sintering and mechanical properties of porcelains prepared from Algerian raw materials. Cerâmica 57:453–460. https://doi.org/10.1590/S0366-69132011000400013
27. Cetin S, Marangoni M, Bernardo E (2015) Lightweight glass–ceramic tiles from the sintering of mining tailings. Ceram Int 41(4):5294–5300. https://doi.org/10.1016/j.ceramint.2014.12.049

28. de Santis B, Rossignolo JA, Morelli MR (2016) Calcined clay light-weight ceramics made with wood sawdust and sodium silicate. Mater Res 19(6):1437–1442. https://doi.org/10.1590/1980-5373-MR-2016-0249
29. Khabas TA, Maletina LV, Kamyshnaya KS (2014) Influence of nanopowders and pore–forming additives on sintering of alumma–zircorna ceramics. IOP Conf Ser Mater Sci Eng 66:12050. https://doi.org/10.1088/1757-899x/66/1/012050
30. Al-Marahle G (2005) Production of light weight ceramic from teils of local materials. Am J Appl Sci 2(4):778–783. https://doi.org/10.3844/ajassp.2005.778.783
31. Liew YM, Heah CY, Li L, Jaya NA, Abdullah MMAB, Tan SJ, Hussin K (2017) Formation of one-part-mixing geopolymers and geopolymer ceramics from geopolymer powder. Constr Build Mater 156:9–18. https://doi.org/10.1016/j.conbuildmat.2017.08.110
32. Duxson P, Lukey GC, Deventer van JSJ (2007) Physical evolution of Na-geopolymer derived from metakaolin up to 1000 °C. J Mater Sci 42(9):3044–3054. https://doi.org/10.1007/s10853-006-0535-4
33. Davidovits J (1994) Properties of geopolymer cements. In: First international conference on alkaline cements and concretes, pp 131–149
34. Duxson P, Provis JL, Lukey GC, Mallicoat SW, Kriven WM, Van Deventer JSJ (2005) Understanding the relationship between geopolymer composition, microstructure and mechanical properties. Colloids Surf Physicochem Eng ASP 269:47–58. https://doi.org/10.1016/j.colsurfa.2005.06.060
35. Duxson P, Fernández-Jiménez A, Provis J, Lukey G, Palomo A, Van Deventer J (2007) Geopolymer technology: the current state of the art. J Mater Sci 42:2917–2933. https://doi.org/10.1007/s10853-006-0637-z
36. Xu H, Van Deventer JSJ (2002) Geopolymerisation of multiple minerals. Miner Eng 15(12):1131–1139. https://doi.org/10.1016/S0892-6875(02)00255-8
37. Palomo A, Grutzeck MW, Blanco MT (1999) Alkali-activated fly ashes: a cement for the future. Cem Concr Res 29(8):1323–1329. https://doi.org/10.1016/S0008-8846(98)00243-9
38. Barbosa VFF, MacKenzie KJD, Thaumaturgo C (2000) Synthesis and characterisation of materials based on inorganic polymers of alumina and silica: sodium polysialate polymers. Int J Inorg Mater 2(4):309–317. https://doi.org/10.1016/S1466-6049(00)00041-6
39. Zhang Z, Yao X, Zhu H, Hua S, Chen Y (2009) Activating process of geopolymer source material: Kaolinite. J Wuhan Univ Technol Sci Ed 24(1):132–136. https://doi.org/10.1007/s11595-009-1132-6
40. Hardjito D, Rangan B (2005) Development and properties of low-calcium fly ash based geopolymer concrete. Res Rep Geopolymer Concr
41. Provis JL (2009) Activating solution chemistry for geopolymers. In: Provis JL, van Devender JSJ (eds) Geopolymers: structure. processing, properties and industrial applications, Woodhead Publishing and CRC Press, Cambridge and Boca Raton, pp 50–71
42. Rakngan W, Williamson T, Ferron RD, Sant G, Juenger MCG (2018) Controlling workability in alkali-activated Class C fly ash. Constr Build Mater 183:226–233. https://doi.org/10.1016/j.conbuildmat.2018.06.174
43. Shahedan N, Abdullah MMAB, Mahmed N, Kusbiantoro A, Bouaissi A (2018) Characterization of fly ash geopolymer concrete with glass bubble for thermal insulation application. In: 4th international conference on green design and manufacture. AIP conference proceedings, p 20294
44. Kong D, Sanjayan JG (2008) Damage behavior of geopolymer composites exposed to elevated temperatures. Cem Conc Comp 30(10):986–991. https://doi.org/10.1016/j.cemconcomp.2008.08.001
45. Fernández-Jiménez A, Palomo A, Criado M (2005) Microstructure development of alkali-activated fly ash cement: a descriptive model. Cem Concr Res 35(6):1204–1209. https://doi.org/10.1016/j.cemconres.2004.08.021
46. Komnitsas K, Zaharaki D (2007) Geopolymerisation: a review and prospects for the minerals industry. Miner Eng 20(14):1261–1277. https://doi.org/10.1016/j.mineng.2007.07.011

47. Swanepoel JC, Strydom CA (2002) Utilisation of fly ash in a geopolymeric material. Appl Geochem 17(8):1143–1148. https://doi.org/10.1016/S0883-2927(02)00005-7
48. Xu H, Van Deventer JSJ (2000) The geopolymerisation of alumino-silicate minerals. Int J Miner Process 59(3):247–266. https://doi.org/10.1016/S0301-7516(99)00074-5
49. El-Maghraby A, Khaled K, Elsabawy K (2013) Formation of leucite crystals from metakaolin-based geopolymer using kaolin and bentonite. Int J Chem Sci 11:740–750
50. He P, Jia D (2013) Low-temperature sintered pollucite ceramic from geopolymer precursor using synthetic metakaolin. J Mater Sci 48(4):1812–1818. https://doi.org/10.1007/s10853-012-6944-7
51. Bell J, Driemeyer P (2009) Formation of ceramics from metakaolin-based geopolymers. Part II: K-based geopolymer. J Am Ceram Soc 92:607–615. https://doi.org/10.1111/j.1551-2916.2008.02922.x
52. He P, Jia D, Lin T, Wang M, Zhou YU (2010) Effects of high-temperature heat treatment on the mechanical properties of unidirectional carbon fiber reinforced geopolymer composites. Ceram Int 36:1447–1453. https://doi.org/10.1016/j.ceramint.2010.02.012
53. Xie N, Bell J (2010) Fabrication of structural leucite glass–ceramics from potassium-based geopolymer precursors. J Am Ceram Soc 93(9):2644–2649. https://doi.org/10.1111/j.1551-2916.2010.03794.x
54. He P, Jia D, Wang S (2013) Microstructure and integrity of leucite ceramic derived from potassium-based geopolymer precursor. J Eur Ceram Soc 33(4):689–698. https://doi.org/10.1016/j.jeurceramsoc.2012.10.019
55. Kuenzel C, Grover LM, Vandeperre L, Boccaccini AR, Cheeseman CR (2013) Production of nepheline/quartz ceramics from geopolymer mortars. J Eur Ceram Soc 33(2):251–258. https://doi.org/10.1016/j.jeurceramsoc.2012.08.022
56. Gordon M, Bell J, Kriven WM (2006) Thermal conversion and microstructural evaluation of geopolymers or "alkali bonded ceramics" (ABCs). In: Bansal JS, Kriven WM (eds) Advances in ceramic matrix composites XI. Wiley, pp 215–224
57. Lin FJ, de Jonghe L, Rahaman MN (1997) Microstructure refinement of sintered alumina by a two-step sintering technique. J Am Ceram Soc 80:2269–2277. https://doi.org/10.1111/j.1151-2916.1997.tb03117.x
58. Mera G, Riedel R, Poli F, Müller K (2009) Carbon-rich SiCN ceramics derived from phenyl-containing poly(silylcarbodiimides). J Eur Ceram Soc 29(13):2873–2883. https://doi.org/10.1016/j.jeurceramsoc.2009.03.026
59. Lu Z, Lu J, Li X, Shao G (2016) Effect of MgO addition on sinterability, crystallization kinetics, and flexural strength of glass–ceramics from waste materials. Ceram Int 42(2, Part B):3452–3459. https://doi.org/10.1016/j.ceramint.2015.10.142
60. Khan J, Amritphale S, Chandra N, Patel M (2012) A novel binder-free and energy-efficient process for making ceramic tiles using red mud and sericitic pyrophyllite. Indian J Chem Technol 19:420–426

Performance of Sintered Pozzolanic Artificial Aggregates as Coarse Aggregate Replacement in Concrete

Rafiza Abd Razak, Mohd Mustafa Al Bakri Abdullah, Kamarudin Hussin, Subaer, Zarina Yahya, Rosnita Mohamed, and Alida Abdullah

Abstract The abundant increase in waste quantities year by year and the depletion of natural resources worldwide are the major concerns among researchers. Recycling and reusing various types of natural or by-product material waste from industry become highlighted in the recent study. The growing demand for concrete and the production of artificial aggregate become crucial in the construction industry. Artificial aggregate can be produced either by sintering, auto-clave or cold bonding method. Although auto-clave and cold bonding methods can minimize energy consumption, the selection of the sintering method always leads to high quality and better properties of artificial aggregates produced. The use of sintering methods to produce artificial aggregate made from pozzolanic material focuses on the use of volcanic ash as raw material activated by geopolymerization process to produce artificial

R. Abd Razak (✉)
Faculty of Civil Engineering Technology, Universiti Malaysia Perlis, 01000 Kangar, Perlis, Malaysia
e-mail: rafizarazak@unimap.edu.my

R. Abd Razak · M. M. A. B. Abdullah · K. Hussin · Z. Yahya · R. Mohamed · A. Abdullah
Geopolymer and Green Technology, Centre of Excellence (CEGeoGTech), Universiti Malaysia Perlis, 01000 Kangar, Perlis, Malaysia
e-mail: mustafa_albakri@unimap.edu.my

K. Hussin
e-mail: kamarudin@unimap.edu.my

Z. Yahya
e-mail: zarinayahya@unimap.edu.my

R. Mohamed
e-mail: rosnitamohamed21@gmail.com

A. Abdullah
e-mail: alida@unimap.edu.my

Subaer
Geopolymers and Green Material Group, Physics Department, FMIPA, Universitas Negeri Makassar, Makassar, Indonesia
e-mail: subaer@unm.ac.id

A. Abdul Kadir et al. (eds.), *Sustainable Waste Utilization in Bricks, Concrete, and Cementitious Materials*, Lecture Notes in Civil Engineering 129,
https://doi.org/10.1007/978-981-33-4918-6_10

lightweight aggregate-based geopolymer will be summarized in this chapter. This chapter discusses the result of the physical and mechanical properties of artificial lightweight aggregate or known as sintered geopolymer volcanic ash artificial lightweight aggregate (SGVA). The interfacial transition zone (ITZ) of sintered geopolymer volcanic ash artificial lightweight aggregate (SGVA) is presented. The performance of sintered geopolymer volcanic ash artificial lightweight aggregate (SGVA) in concrete is also discussed.

Keywords Geopolymer · Artificial lightweight aggregate · Volcanic ash · Palletizing · Aggregate impact value (AIV) · Aggregate crushing value (ACV)

1 Introduction

Since concrete is the primary construction material in civil engineering, concrete production is considered one of the most economic indicators to the countries. Concrete production is always regarded as environmental pollution due to excessive resource consumption. Pozzolanic materials are the best alternative to solve this problem and always lead to the preservative environment, greenhouse and converting waste into a value-added product as construction materials.

Generally, the artificial aggregate can be produced through the agglomeration process of fine ash or powdered waste materials towards the hydration process for hardening mechanism to form pallets of artificial aggregate in the desired size. The hardening process of fresh pallets can be achieved through autoclaving, cold-bonding or sintering processes. The sintering process is the best method to produce high-quality artificial lightweight aggregates in terms of the mechanical properties of aggregates and concretes. The sintered artificial lightweight aggregates can be used in various applications, including structural and non-structural types.

Aardelite, Lytag, Leca, Trefoil, Pollytag are some of the commercially available sintered artificial lightweight aggregates around the world. The inclusion of lightweight artificial aggregate in concrete will be beneficial, such as reducing the total of dead loads, steel reinforcement and also reduction in transportation costs. At present, sintered artificial aggregate showed good properties in terms of mechanical properties compared with the artificial aggregate produced with different methods (auto-clave or cold-bonding method). Geopolymer is become highlighted by many researchers due to better properties produced, including high strength and durability in terms of concrete. Thus, research on developing artificial aggregate with the inclusion of geopolymer system should be carried out. In this study, volcanic ash has been selected due to the pre-heated process of raw materials collected from the volcanic mud that contribute to the pore system that leads to lightweight properties, thus making it suitable to be used as a lightweight artificial aggregate. Introducing the geopolymer system in artificial aggregate production using volcanic ash as raw material will give better performance in terms of physical and mechanical properties.

2 Geopolymer

Geopolymer is a combination of two words, geo and polymer that refers to the combination of polymer-based natural resources or by-product materials. Geopolymer describes inorganic polymers based on aluminosilicates which can be produced by synthesizing pozzolanic compounds (high in silica, Si and alumina, Al) or aluminosilicate source materials from natural resources or by-product materials with highly alkaline solutions [1, 2]. Pozzolanic materials consist of silica and alumina compounds. Meanwhile, alkaline solutions are referred to as combinations of sodium hydroxide (NaOH) and sodium silicate (Na_2SiO_3). Geopolymer concrete is proved as an excellent property in terms of its high durability, high strength including early-age strength and high sustainability due to the use of waste materials. The hydration process in geopolymer is referred to as geopolymerization process is claimed as environmentally friendly due to very little $CaCO_3$ as a by-product but replaced with the N–A–S–H (N refer to Si or Al) or C–A–S–H (C refer to Ca) product that leads to the hardening mechanism of geopolymer. Ca-rich that is usually aligned with by-product materials such as ground granulated blast furnace slag (GGBS) and Class C fly ash will produce C–A–S–H gel as a product of the geopolymerization process. Otherwise, Si-rich and Al-rich of natural resources such as kaolin, metakaolin, Class F fly ash, volcanic ash will lead to the N–A–S–H gel as a product of the geopolymerization process.

Figure 1 shows the geopolymerization process towards hardening pozzolanic

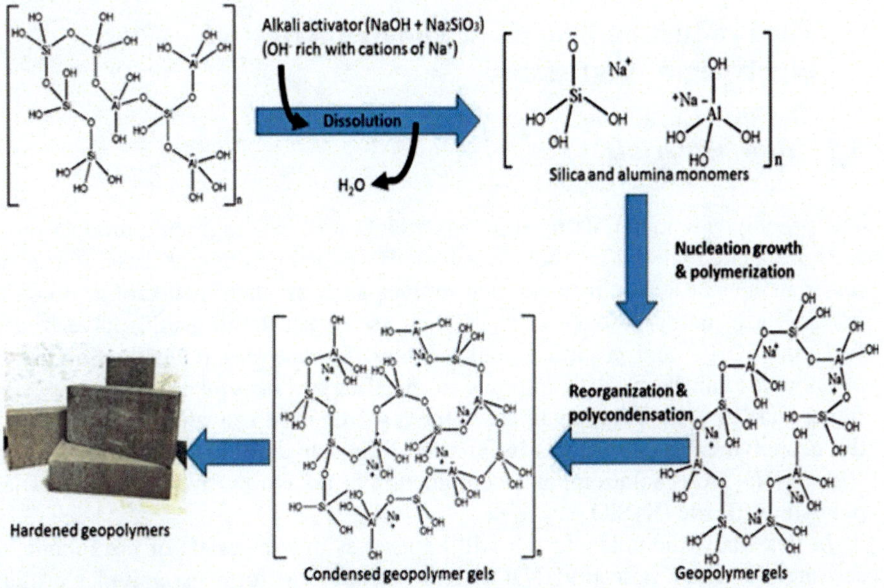

Fig. 1 Geopolymerization process for hardening mechanism

material mechanisms to form the construction material products. The first process leads to break down of the Si–O–Si and Si–O–Al covalent bonds into a colloid phase once contact with an alkali activator solution through the chemical attack at high alkalinity medium which then known as the dissolution process. The interaction among dissolved products will be created to form the coagulated structure as silica and alumina monomers. The intermediate product has high aluminium content, and when the reaction progress continuous, the geopolymer gels will be formed as a by-product and transformed for the polymerization process, forming more geopolymer gels with more silicon content. In the last process, the gel will grow through the reorganization and polycondensation process to form the three-dimensional frameworks of the new geopolymeric matrix of condensed geopolymer gels that contribute to strength development.

All the processes involved in the geopolymerization process, which is known as hydration process, will produce water and heat as a by-product that plays an important role in the hardening mechanism. The curing temperature plays a significant role in the geopolymerization process as it will accelerate the process of hydration to form a geopolymer matrix at a faster setting time compared to ordinary Portland cement (OPC). However, the performance of geopolymer produced at low and ambient temperature also shows comparable to the OPC. The mix design proportion should be optimized to achieve a high strength of the product. Different raw materials of pozzolanic used will have a different optimum ratio of mix design proportion due to differences in chemical composition and the particle size and shape.

3 The Production Process of Sintered Artificial Lightweight Aggregates

3.1 Raw Materials

The primary raw materials used in lightweight artificial aggregate production are the pozzolanic material and additives to activate the hydration process. Pozzolanic materials can be either by natural resources such as clay, metakaolin, dolomite, volcanic ash, or by-product material such as fly ash, bottom ash, pulverized oil fuel ash (POFA) and ground granulated blast furnace slag (GGBS). Some types of additives can be used to improve artificial aggregate properties such as silica fume, glass powder, or nano-silica. Some research used activator of alkali through the geopolymerization process by mixing sodium hydroxide (NaOH) and sodium silicate (Na_2SiO_3) solutions or a combination of potassium hydroxide (KOH) and potassium silicate (K_2SiO_3).

In this study, the volcanic ash will be used as raw materials of pozzolanic and activated by alkali activator (NaOH and Na_2SiO_3) to form geopolymer artificial aggregate. Table 1 presents the summary of the raw materials, the binder used, method

Table 1 Various materials used for manufacturing artificial lightweight aggregate using sintering process

Pozzolanic materials source	Binder of activator	Methods producing pellets of aggregate	Post-process
Fly ash [3]	Bentonite + powdered Limestone + glass powder	Pelletizing disk (20 min)	Sintered artificial aggregate at high temperature of 1100–1200 °C for 1 h
Fly ash [4]		Pelletizing disk (40 rpm)	Sintered artificial aggregate at high temperature of 1100 °C for 1 h
Bottom ash [5]	–	Rolled by hand manually	Sintered the aggregate in the rotary furnace at high temperature of 1050 °C rapidly
Sewage sludge ash [6]	Clay	Rolled by hand manually	Sintered the aggregate in the rotary furnace at high temperature of 1020–1080 °C
Clay [7]	–	Breaking the clay to the thickness slice of 2 mm, then conventional rolled by hand to form pellets of aggregate	Sintered the aggregate in the rotary kiln at high temperature of 1300 °C
Clay [8]	–		Sintered the aggregate in the chamber furnace at high temperature of 1120 °C for 8 min
Clay [9]	Ferrochrome sludge	Granulation process of aggregate pellets	Sintered the aggregate in the rotary kiln at high temperature in the range of 1050–1250 °C using typical dwell times for up to 20 min
Expanded clay [10]		Aggregates were produced by manually rolled, then dried at 105 °C, followed by pre-heated at 250 °C for 2 h	Sintered the aggregate in the rotary kiln at high temperature of 1120 °C
Volcanic ash [2]	Alkali activator	Palletizing process	Sintered the aggregate at temperature less than 1000 °C (950 °C) in the chamber furnace

processing, and post-processing in the production of sintered artificial lightweight aggregate.

3.2 *Mixing, Palletizing, and Sintering Process*

The fundamental production process of sintered artificial aggregate consists of three stages; the mixing process of raw materials, palletization or granulation process and hardening process through sintering. The methods mentioned are focused on the production of sintered geopolymer volcanic ash artificial lightweight aggregate (SGVA). The first step to producing artificial lightweight aggregates is by mixing all raw materials including pozzolanic and additives used with specified water/cement ratio (w/c ratio) or solid-to-liquid ratio to form the fresh pastes. The solid-to-liquid ratio used in this study is 1.7. The molarity of sodium hydroxide (NaOH) used is 12 M due to the optimum properties produced. Then the fresh pastes will be going through the agglomeration process. Palletization process is well-known as non-pressure agglomeration through the rolling movement either by hand manual or pelletizer disc that can be adjusted at a specific angle and speed to ensure good collision of all raw materials in the aggregates.

Some research also stated that curing in the water gives the best medium for accelerating the hydration process to harden aggregates. In this study, to gain the desired strength of artificial aggregate, the fresh pellet aggregates will just be cured in the ambient temperature, then proceed to the sintering process at high temperature as low as 500 until 1000 °C. Is should be noted that the performance of geopolymer cured at ambient temperature is comparable with the OPC that needs to be cured in the water. Figure 2 shows the flow chart of the summary process based on the previous study included producing sintered artificial light-weight aggregate using a pozzolanic-based geopolymer as raw material.

4 Physical Properties of Sintered Artificial Lightweight Aggregates

4.1 *Shape and Texture*

The shape of the aggregates produced is very significant as it influences the aggregate and cement matrix interlocking. The sintered artificial aggregate is mostly spherical and brown in colour with pores in the internal with black colour which is attributed to the carbon content and oxidation state of iron. The spherical shape of sintered artificial aggregates contributes to low shape index compared to angular shape. However, the performance of spherical aggregates sometimes is similar to the angular shape of aggregates. The surface texture of sintered artificial aggregate also may affect the

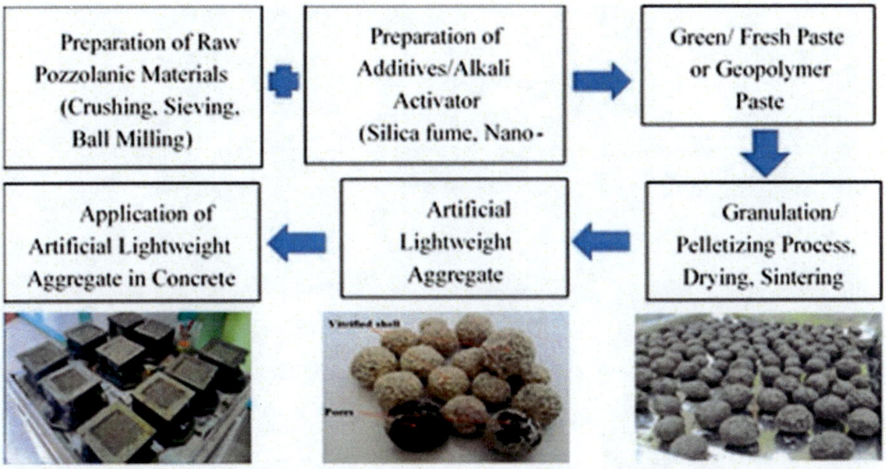

Fig. 2 The summary of steps on producing sintered artificial aggregate using pozzolanic based geopolymer

surface friction towards the harshness during the mixing process. The porous state in the sintered artificial aggregate combined with the rough surface may promote the penetration of cement hydration products into cavities or large pores on the aggregate surface. Thus, this phenomenon will lead to multiple hooks to bind the aggregate phase and the paste phase together [11]. Figure 3 shows the shape and surface of sintered geopolymer volcanic ash artificial lightweight aggregate (SGVA) at various sintering temperatures.

Based on Fig. 3, the colour of SGVA produced was various at different sintering temperatures. The SGVA is red-brownish when sintered at a temperature of 500 and 600 °C. The colour of SGVA then turns to grey when sintered at an increasing temperature of 700 and 800 °C. The inner part of SGVA turns to partially dark brown and black, starting at 910 °C, then completely turn to black colour in the inner part at 950 °C sintering temperature. The artificial aggregate produced at 950 °C shows the distribution of small pores uniformly, thus reducing the specific gravity value. However, the SGVA sintered at 1000 °C produces pores in bigger sizes and creates pores at the surface of aggregate, thus expanding the original size of aggregate for about 20%. The large pores produced will contribute to low specific gravity value to the lighter lightweight aggregate, but it will decrease the strength of the lightweight artificial aggregate produced.

The hard shell was naturally formed at a sintering temperature of 900 °C to strengthen the lightweight artificial aggregate produced. Thus, the treatment on the SGVA is unnecessary. Some of the production of lightweight artificial aggregate needs the treatment by coating the aggregate produced to act as a barrier and strengthen the aggregate produced, thus increasing the production cost. Figure 4 shows the close-up image of SGVA. The porous structure was developed well in

Fig. 3 Shape and textures of sintered geopolymer volcanic ash artificial lightweight aggregates (SGVA) at various sintering temperatures

Fig. 4 The sintered geopolymer volcanic ash artificial lightweight aggregate (SGVA)

the inner part of aggregate SGVA. The vitrified shell was naturally produced, thus reducing the water absorption value of the aggregate produced.

4.2 Specific Gravity

The specific gravity value increases as the sintering temperature increase. As the specific gravity decreases, the sintered artificial lightweight aggregates tend to experience the bloating effect due to gas generation during the sintering process at high temperatures (up to 1200 °C). However, another condition where the sintered artificial lightweight aggregate produced is formed through the dense structure; this leads to higher specific gravity that is believed due to excessive glass formation at higher sintering temperatures. The melting temperature of each type of pozzolanic material differs, thus affect the formation of pores in-side the aggregates that transform from solid to gas states. Most of the pozzolanic binders used in the sintered artificial lightweight aggregate are claimed to contain low melting point minerals than fly ash [11]. As the aggregates are sintered, the binder will be melted to form a liquid phase that filled the voids between the particles through capillary action. Due to this statement, the voids will be eliminated which existed in the aggregate at its fresh state, leading to the formation of the dense solid matrix. The surface of sintered artificial lightweight aggregate becomes more vitrified with smooth, porous and rough textures for increasing the binding force between aggregates and cement matrix.

The specific gravity for fly ash sintered artificial aggregate is in the range of 1.30–2.35 [4, 6]. Meanwhile, kaolin showed lower specific gravity that achieves lower than 1.00 (bloating effect), which sometimes leads to low properties that are not suitable for structural application. Sintered geopolymer volcanic ash artificial lightweight aggregate (SGVA) contributes to the specific gravity value lower as 1.10 [2]. Thus, different types of raw materials used and the additives used influence the value of specific gravity produced, which is in line with the performance of lightweight artificial aggregate in terms of physical and mechanical properties.

4.3 Water Absorption

It is well-known that the porous nature of sintered artificial lightweight aggregate is responsible for the high-water absorption. The high absorption is not encouraged in good quality of concrete. There are also conditions that the water absorption of sintered artificial lightweight aggregate decreases as the sintering temperature increases. This can be due to the formation of the glassy texture at the outer part of aggregates that can hinder the internal pore connectivity for water to be absorbed. Most of the previous research reported that high water absorption results up to 33%. The SGVA produced in this study showed low water absorption at 950 °C but then increased at 1000 °C, as shown in the graph of Fig. 5. This phenomenon is due to the formation of small pores on the surface of aggregates that promotes to increase water absorption at 1000 °C. The same trend also can be found for other types of pozzolanic materials used.

Fig. 5 Water absorption of sintered geopolymer volcanic ash artificial lightweight aggregates (SGVA)

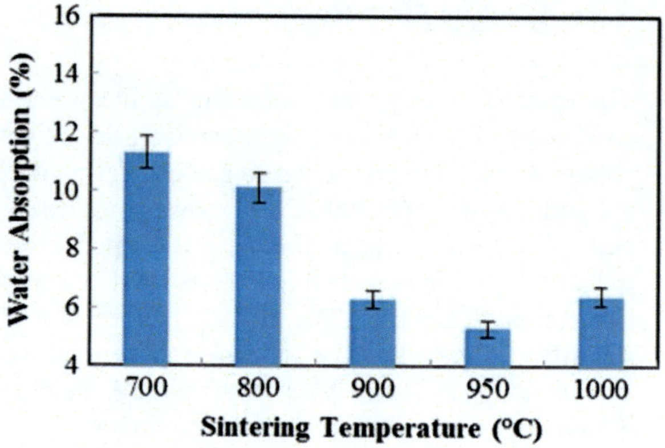

5 Mechanical Properties of Sintered Artificial Lightweight Aggregates

5.1 Aggregate Crushing Value (ACV)

The crushing strength always represents the strength of the aggregate. It is noted that for aggregate crushing value (ACV) less than 30, 'ten percentage fineness value' is preferred. That means the individual crushing strength of artificial aggregates with a smaller size is always higher than the large size. The ACV will increase depending on the sintering temperature. The ACV value for sintered geo-polymer volcanic ash artificial lightweight aggregate (SGVA) was achieved up to 1050 N at a sintering temperature of 950 °C. This value is comparable with the previous study that used different raw materials such as fly ash, bottom ash and GGBS as raw materials. However, these materials' artificial aggregate required a higher sintering temperature (mostly more than 1000 °C).

The sintering temperature below 900 °C is the recommended temperature for producing metakaolin artificial lightweight aggregate achieving good strength [12]. Sintered fly ash artificial aggregate can achieve up to 23.5% [11]. Additional additives material and alkali activator were proved to increase the ACV value of sintered artificial aggregate. A reduction in strength is influenced by the formation of large pores in the aggregate that is believed due to the bloating effect. Different materials will lead to different expansion rates at a higher temperature during the sintering process of aggregate production [13]. Another researcher claimed that the ACV for sintered fly ash aggregate is 3–4 times greater than the ACV produced through a cold bonded method that has manufactured from the same types of fly ash [14]. This shows that the lightweight artificial aggregate produced through the sintering method produces better properties of aggregate compared to lightweight artificial aggregate produced through the cold bonded method.

5.2 *Aggregate Impact Value (AIV)*

The AIV value present in the percentage form. The high-value percentage for AIV value denotes a low performance of lightweight artificial aggregate and thus show the low strength of concretes. Fly ash sintered artificial aggregates denotes up to 28% [8]. However, in this study, the lowest AIV for SGVA was showed at the sintering temperature of 950 °C of 16.22%, as shown in Fig. 6. This value contributes to the highest strength of SGVA produced in this study. The high strength obtained is due to the formation development of a vitrified shell that strengthens the surface of lightweight artificial aggregate (SGVA) produced.

The AIV value for SGVA obtained at a sintering temperature of 1000 °C is 17.45% which is higher than AIV value produced at 950 °C. This is due to larger pore created at 1000 °C and high-water absorption result produced, thus leading towards weakening the strength of the geopolymer matrix structure. The high-water absorption at 1000 °C of sintering temperature was due to the pore formation at the surface of aggregate and larges porous size produced in the inner part of aggregate produced that allow for high absorption of water. The AIV of lightweight artificial aggregate increased with an increase in the temperature as the chemical reaction of the alkaline solution used increased [15].

6 Techniques Used for Improving the Performance of Sintered Artificial Lightweight Aggregate

Various techniques have been attempted and used to improve the properties of sintered artificial lightweight aggregate to get the excellent quality of aggregates. The incorporation of additives and alkali activators through the technology of geopolymer, surface

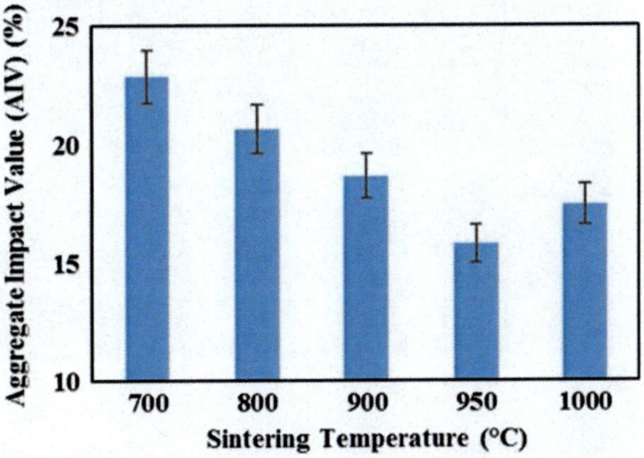

Fig. 6 The AIV value produced at various sintering temperature for sintered geopolymer volcanic ash artificial lightweight aggregate (SGVA)

treatment and curing methods are selected to be the most influencing parameters treatment on producing good properties of sintered artificial lightweight aggregates.

6.1 Incorporation of Additives and Alkali Activators

Incorporation of alkali activator in producing sintered artificial lightweight aggregates has been the focus of many researchers in producing geopolymer artificial lightweight aggregates [16]. The usage of 10 M sodium hydroxide can help activate aggregate production during the sintering process [17]. A study carried out by Geetha and Ramamurthy [18] showed that the alkali activator that consists of a combination of sodium hydroxide and sodium silicate to produce geopolymer aggregate using raw material of bottom ash gives higher strength of 12% with a bulk density of 7–10% higher than the cement-activated aggregate. Volcanic ash activated by the geopolymerization process in this study also showed improving properties (AIV of 16% with low water absorption of 5% at a specific gravity of 1.10) in physical, mechanical and microstructural of the sintered artificial aggregates produced. Figure 7 shows the SEM micrographs of control samples (without the addition of alkali activator) compared with sintered geopolymer volcanic ash artificial lightweight aggregate (SGVA) (with alkali activator of geopolymer).

Sintered geopolymer volcanic ash artificial lightweight aggregate (SGVA) in Fig. 7a shown a completed geopolymer matrix that contributes to the highest strength produced based on ACV and AIV values. The formation of pores was in the range of 2.5–6.9 μm in diameter sizes. The thinner matrix formed is observed for the control

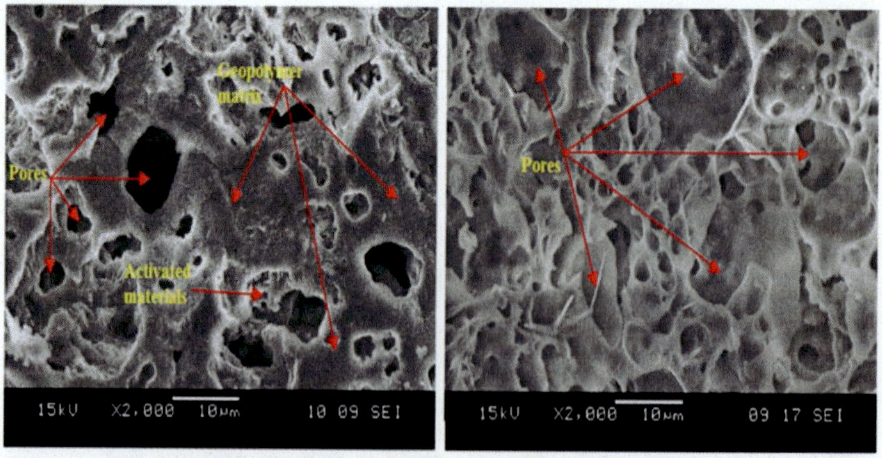

(a) (b)

Fig. 7 SEM images of **a** sintered geopolymer volcanic ash artificial lightweight aggregate (SGVA) compared with; **b** control sample

sample compared to SGVA, as shown in Fig. 7b. The thinner matrix leads to the low strength of lightweight artificial aggregate produced meanwhile the pores existed in control samples are higher than the pore size in SGVA with the sizes range between 2.5 and 18.6 μm in diameter. Thus, it can be concluded that the presence of an alkali activator to activate the geopolymerization process of volcanic ash has contributed to a stable matrix structure. This leads to the high strength of lightweight artificial aggregate of SGVA produced.

Many types of additives also can be added in the production of sintered artificial lightweight aggregates. The addition of polypropylene fiber (PPF) showed increases in the ACV crushing strength of aggregate due to their function as a reinforcement in the structure of hardened sintered artificial lightweight aggregates. Other than that, sodium sulfate (Na_2SO_4) also succeeds in accelerating and activating the pozzolanic reaction of fly ash as it can densify the lightweight artificial aggregate produced and thus strengthen the sintered artificial lightweight aggregates. Sodium salts have been recognized as an adequate flux part to reduce the sintering energy by reducing the sintering temperature [19].

6.2 *Surface Treatment*

To prevent the water absorption problem, one of the alternatives recommended is coating the surface of the lightweight artificial aggregate produced. The sintered artificial lightweight aggregate produced with less or no pores on the aggregate surface will not need to enhance with the surface treatment. Sintered geopolymer volcanic ash artificial lightweight aggregate (SGVA) is naturally produced microscopic pores on the surface, thus contribute to low water absorption. The water glass, which is also known as sodium silicate (Na_2SiO_3), has been used by Güneyisi et al. [14] as the surface treatment of lightweight artificial aggregate. The results showed that the crushing value of ACV of water glass treated lightweight artificial aggregate was 40–100% higher than that of untreated aggregate. This is due to C–S–H phase formation that promotes the hydration reaction process between water glass and calcium oxide that existed in the cementitious system.

Spraying method using alkali activator that consists of sodium hydroxide and sodium silicate (water glass) has been studied by Hwang and Tran [20] on the surface of foamed bottom ash aggregate. The results showed that the crushing strength and unit weight of sintered artificial lightweight aggregate produced was higher than that of untreated spraying aggregate, with low water absorption (33% lower than untreated surface). Spraying the surface of lightweight artificial aggregate with cement-silica fume mixture was recently researched by Tajra et al. [21]. The water absorption of lightweight artificial aggregate treated by using this method was decreased by about 27%. Meanwhile, the density and ACV crushing strength increased by about 13 and 14%, respectively compared with untreated aggregate. Coating the waste paper ash on aggregate can help to produce a smooth surface [22].

6.3 Curing Methods

The right curing method will increase the properties of sintered artificial lightweight aggregate produced. Curing with the condition such as relative humidity of 70% and at a temperature of 20 °C, is the most frequently used method of curing for fly ash aggregate [23]. Another researcher concluded that the water curing methods are needed to enhance the hydration process of the cementitious system before the sintering process and thus improve the properties of aggregate [24]. Another study claimed that the curing at a relative humidity of 99% speeds up to accelerate the reaction in the matrix of fly ash artificial aggregate and resulted in higher crushing strength of 17% [21].

Gomathi and Sivakumar [17] have found out that hardening properties of concrete incorporating fly ash artificial aggregate can be improved under high temperatures of hot water and steam curing. The author observed that heat curing could improve aggregate impact performance, crushing pellets strength, and the aggregate's specific gravity [15]. The geopolymerization process can be speeded up by increasing the curing temperature of artificial aggregate. Geetha and Ramamurthy [18] found that the temperature of 50 to 80 °C gave better properties than water curing. This shows that the optimum curing process also leads to higher performance of sintered artificial lightweight aggregate produced which is controlled by the raw materials used.

7 Performance of Sintered Artificial Lightweight Aggregate in Concrete

7.1 Workability of Concrete

Good quality of concrete should be workable within the acceptable limit, and sufficient target strength produced. The workability value of fresh concrete incorporating sintered geopolymer volcanic ash artificial lightweight aggregate (SGVA) was determined immediately after forming the concrete mixture. The standard slump test has been used in this study under ASTM C143 [25]. The workability work using the slump test is shown in Fig. 8. The slump value of SGVA in concrete was found as 65 mm within the acceptable value of workability. This low value of slump is referred to as the phenomenon of absorption of water from sintered artificial lightweight aggregate due to the existence of pores in the SGVA. However, the slump value obtained in this study is within the acceptable limits. The higher value of slump indicates good workability of concrete produced, thus increasing the flows of concrete and free from segregation.

Fig. 8 The slump test for concrete incorporating sintered geopolymer volcanic ash artificial lightweight aggregate (SGVA)

7.2 Density of Concrete

The average density of concrete produced using sintered geopolymer volcanic ash artificial lightweight aggregate (SGVA) is 1760.1 kg/m^3. The result of density for control concrete produced (control samples with sintered artificial aggregate with-out alkali activator) is 1870.7 kg/m^3. Thus, the percentage difference between these two types of concrete with different artificial lightweight aggregates is 6.0%. However, both types of artificial lightweight aggregates applied in concrete can be classified as lightweight concrete. This statement is referred to ASTM C330 [26], which claimed the density of less than 1920 kg/m^3 could be classified as lightweight concrete.

7.3 Water Absorption of Concrete

The water absorption obtained for concrete incorporating SGVA is 2.8%. Meanwhile, the water absorption of the control sample (sintered artificial aggregate without alkali activator) is 7.3%. This is due to the higher appearance of pores in the control sample, as proved by SEM images in Fig. 7. This value of water absorption produced in this study is still lower compared to concrete incorporating artificial fly ash lightweight aggregate with the addition of bentonite produced by Güneyisi et al. [14] with 11.7% of water absorption.

7.4 Compressive Strength of Concrete

The compressive strength of SGVA at various sintering temperatures applied in concrete is shown in Fig. 9. It is noted that the strength of concrete with SGVA at sintering temperature of 600 °C showed comparable strength with the lightweight commercial aggregate of LECA (Lightweight expanded clay aggregate) in concrete.

Figure 10 shows the result of compressive strength of OPC concrete incorporating SGVA at increasing ages of testing. The result revealed that the compressive strength produced for OPC concrete incorporating SGVA is higher than concrete with OPC concrete incorporating control samples of aggregate (sintered volcanic ash aggregate

Fig. 9 Compressive strength of concrete with sintered geopolymer volcanic ash artificial lightweight aggregate (SGVA) at various sintering temperature

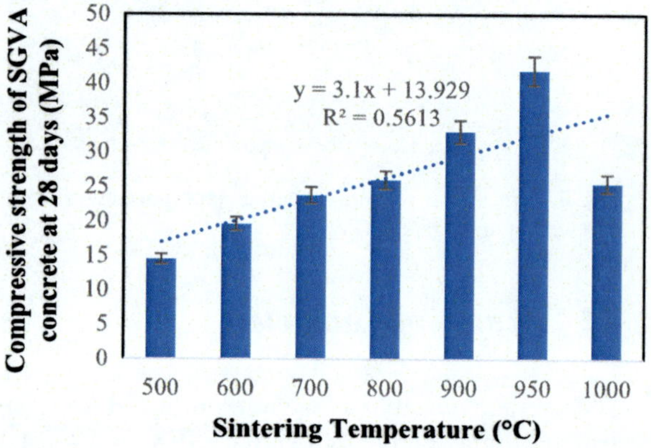

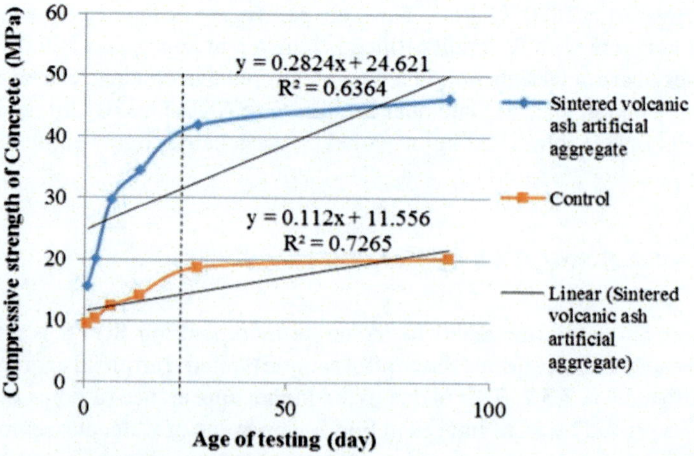

Fig. 10 Compressive strength of sintered geopolymer volcanic ash artificial lightweight aggregate (SGVA) concrete and control concrete (use aggregate without alkali activator) at various ages of testing

without alkali activator). This indicates that geopolymer inclusion in SGVA aggregate plays a significant role in the strength development of concrete produced by ages. The development of strength for SGVA concrete showed an increasing trend of strength produced by ages until 90 days of testing with the R2 value of 0.6364. At 28 days, the compressive strength of SGVA concrete was 41.9 MPa with a low density of 1760 kg/m^3 (considered lightweight concrete). Meanwhile, the compressive strength of concrete with control aggregates (without alkali activator) in concrete was 18.8 MPa. This showed the difference percentage between these two types of concrete is about 123%.

Another study has found the result of compressive strength for fly ash geopolymer concrete with lightweight aggregate was 26.7 MPa at 28 days [27]. This value is lower than SGVA concrete produced in this study (41.9 MPa) that used geopolymer volcanic ash as raw materials. Even though both studies used the technology of geopolymer in producing lightweight artificial aggregate, but different raw pozzolanic materials used may affect the strength of individual and concrete later. However, other researchers found the high strength of artificial lightweight aggregate concrete which can be up to 50 MPa. However, they need a high sintering temperature to produce the aggregate (above than 1100 °C) [14]. Compared to this study, the SGVA just used a lower sintering temperature of 950 °C (less than 1000 °C). Some studies also showed that sintered fly ash artificial aggregate produced was comparable in mechanical properties with sintered geopolymer volcanic ash artificial lightweight aggregate (SGVA) produced in this study. The compressive strength of concrete with sintered fly ash artificial aggregate can be achieved up to 45 MPa at 28 days [4].

8 Interfacial Transition Zone (ITZ) in Concrete

Generally, the good quality of ITZ is highly influenced by the type and properties of the aggregate used. In lightweight concrete, the binding force between artificial lightweight aggregates and cement matrix is very significant, leading to the high strength of concrete produced. The weakest point in the concrete is referred to the interfacial zone between the hydrated cement paste and aggregates. That means the binding force between aggregates and cement matric must be strong to ensure high-quality concrete. Figure 11 showed the image of the cross-section of sintered geopolymer volcanic ash artificial lightweight aggregate (SGVA) surface and cement paste in the concrete that is well interlocked with each other. It can be seen that the cement paste is entering a crushed grain of the surface of SGVA aggregate produced, thus forming interlocking bonding between aggregates and cement paste. This phenomenon will lead to strengthen the position of aggregates when high loading is applied to the concrete, thus increasing the compressive strength values.

Sintered artificial lightweight aggregate usually forms a porous surface as well as in the inner aggregates. When water is added to the mixture, the sintered artificial aggregates start to absorb due to porous structure. At this condition, any build-up of water around the aggregates grains should be prevented, since the water can be

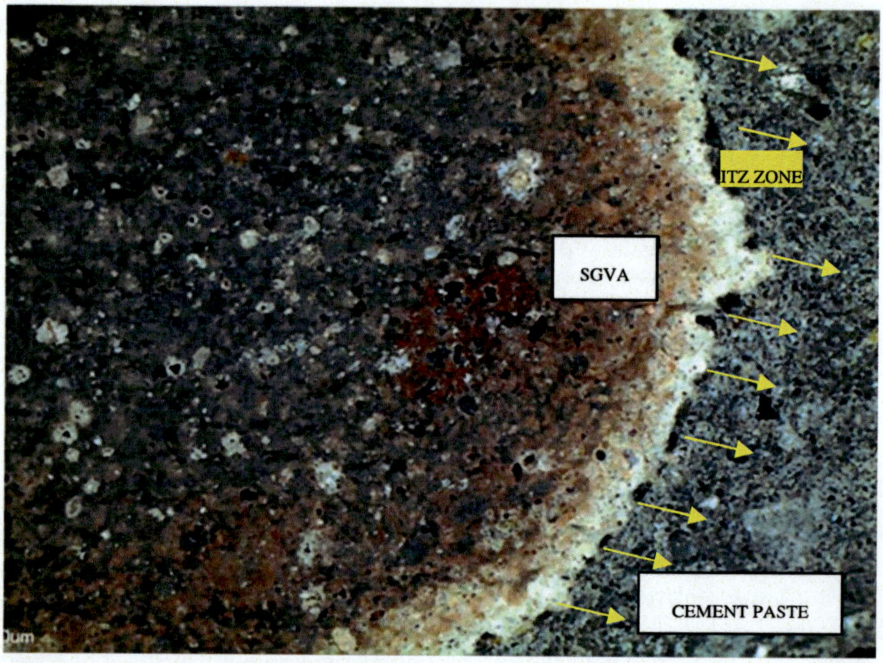

Fig. 11 The ITZ zone of cement paste and sintered geopolymer volcanic ash artificial lightweight aggregate (SGVA) in concrete

absorbed into the grains. The vitrified shell produced in the SGVA acts as the shield to slow the absorption of water from the surface of aggregates. This condition will result in interfacial zones become narrower and increase the density, thus increasing the compressive strength.

A study carried out by Lo et al. [8] and Punlert et al. [28] observed that the sintered fly ash lightweight aggregate was tightly bonded to the cement paste, and that is no clear boundary between the cement paste and aggregates. The appearance of crack also less between two surfaces of cement and aggregate due to enhanced mechanical interlocking. It is noted that the chemical reaction also can occur in aggregate paste interface directly. Punlert et al. [28] mentioned that the chemical reactivity of the sintered fly ash artificial aggregate is more than the reactivity of expanded clay and shale rock [28].

9 Conclusion

The feasibility of sintered geopolymer volcanic ash artificial lightweight aggregate (SGVA) to the development of structural concrete with the addition of comparison between other pozzolanic materials from past literature is presented. The physical, mechanical and interfacial transition zone (ITZ) characteristics of SGVA were also discussed. Based on the results produced, the following conclusions can be made:

1. The sintering temperature for most pozzolanic materials usually varied from 900 to 1200 °C to produce lightweight artificial aggregate. SGVA produced at sintering temperature of 950 °C showed optimum properties as artificial lightweight aggregates.
2. Aggregates of sintered geopolymer volcanic ash (SGVA) are spherical in shape with a rough surface and specific gravity varied from 1.10. The low water absorption obtained was 5.3% which can be classified as the lowest water absorption compared to the other artificial lightweight aggregates produced by literature.
3. It is possible to produce lightweight concrete incorporating SGVA that having compressive strength up to 42 MPa at 28 days with a density of 1760 kg/m^3. The quality of the ITZ of sintered geopolymer volcanic ash artificial lightweight aggregate (SGVA) is superior to commercial lightweight aggregate concrete thus leads to high strength.

References

1. Davidovits J (2017) Geopolymers: Ceramic-like inorganic polymers. J Ceram Sci Technol 8:335–350. https://doi.org/10.4416/JCST2017-00038
2. Razak RA, Abdullah MMAB, Hussin K, Ismail KN, Hardjito D, Yahya Z (2015) Optimization of NaOH molarity, LUSI mud/alkaline activator, and Na_2SiO_3/NaOH ratio to produce lightweight aggregate-based geopolymer. Int J Mol Sci 16(5):11629–11647. https://doi.org/10.3390/ijms160511629
3. Kockal NU, Ozturan T (2011) Durability of lightweight concretes with lightweight fly ash aggregates. Constr Build Mater 25(3):1430–1438. https://doi.org/10.1016/j.conbuildmat.2010.09.022
4. Ramamurthy K, Nair H (2006) Influence of binders on properties of sintered fly ash aggregate. Cem Concr Compos Cem Concr Compos 28:33–38. https://doi.org/10.1016/j.cemconcomp.2005.06.005
5. Cheeseman CR, Makinde A, Bethanis S (2005) Properties of lightweight aggregate produced by rapid sintering of incinerator bottom ash. Resour Conserv Recycl 43(2):147–162. https://doi.org/10.1016/j.resconrec.2004.05.004
6. Cheeseman CR, Virdi GS (2005) Properties and microstructure of lightweight aggregate produced from sintered sewage sludge ash. Resour Conserv Recycl 45(1):18–30. https://doi.org/10.1016/j.resconrec.2004.12.006
7. Bernhardt M, Tellesbø H, Justnes H, Wiik K (2013) Mechanical properties of lightweight aggregates. J Eur Ceram Soc 33(13):2731–2743. https://doi.org/10.1016/j.jeurceramsoc.2013.05.013
8. Lo T, Cui H, Memon S, Noguchi T (2015) Manufacturing of sintered lightweight aggregate using high-carbon fly ash and its effect on the mechanical properties and microstructure of concrete. J Clean Prod 112(1):753–762. https://doi.org/10.1016/j.jclepro.2015.07.001
9. Ayati B, Ferrándiz-Mas V, Newport D, Cheeseman CR (2018) Use of clay in the manufacture of lightweight aggregate. Constr Build Mater 162:124–131. https://doi.org/10.1016/j.conbuildmat.2017.12.018
10. Megat Johari MA, Zeyad A, Bunnori N, Ariffin KS (2012) Engineering and transport properties of high-strength green concrete containing high volume of ultrafine palm oil fuel ash. Constr Build Mater 30:281–288. https://doi.org/10.1016/j.conbuildmat.2011.12.007

11. Nadesan MS, Dinakar P (2017) Structural concrete using sintered fly ash lightweight aggregate: a review. Constr Build Mater 154(15):928–944. https://doi.org/10.1016/j.conbuildmat.2017.08.005
12. Risdanareni P, Ekaputri J, Triwulan T (2017) The effect of sintering temperature on the properties of metakaolin artificial lightweight aggregate. In: Green construction and engineering education for sustainable future: proceedings of the green construction and engineering education (GCEE), p 20045
13. Bogas JA, Real S, Ferrer B (2016) Biphasic carbonation behaviour of structural lightweight aggregate concrete produced with different types of binder. Cem Concr Compos 71:110–121. https://doi.org/10.1016/j.cemconcomp.2016.05.006
14. Güneyisi E, Gesoğlu M, Pürsünlü Ö, Mermerdaş K (2013) Durability aspect of concretes composed of cold bonded and sintered fly ash lightweight aggregates. Compos Part B Eng 53:258–266. https://doi.org/10.1016/j.compositesb.2013.04.070
15. Shivaprasad KN, Das BB (2018) Effect of duration of heat curing on the artificially produced fly ash aggregates. IOP Conf Ser Mater Sci Eng 431:92013. https://doi.org/10.1088/1757-899X/431/9/092013
16. Hemalatha T, Ramaswamy A (2017) A review on fly ash characteristics—towards promoting high volume utilization in developing sustainable concrete. J Clean Prod 147. https://doi.org/10.1016/j.jclepro.2017.01.114
17. Gomathi P, Sivakumar A (2015) Accelerated curing effects on the mechanical performance of cold bonded and sintered fly ash aggregate concrete. Constr Build Mater 77:276–287. https://doi.org/10.1016/j.conbuildmat.2014.12.108
18. Geetha S, Ramamurthy K (2013) Properties of geopolymerised low-calcium bottom ash aggregate cured at ambient temperature. Cem Concr Compos 43:20–30. https://doi.org/10.1016/j.cemconcomp.2013.06.007
19. Wei YL, Cheng SH, Chen WJ, Lu YH, Chen K, Wu PC (2020) Influence of various sodium salt species on formation mechanism of lightweight aggregates made from coal fly ash-based material. Constr Build Mater 239:117890. https://doi.org/10.1016/j.conbuildmat.2019.117890
20. Hwang CL, Tran VA (2015) A study of the properties of foamed lightweight aggregate for self-consolidating concrete. Constr Build Mater 87:78–85. https://doi.org/10.1016/j.conbuildmat.2015.03.108
21. Tajra F, Abd Elrahman M, Lehmann C, Stephan D (2019) Properties of lightweight concrete made with core-shell structured lightweight aggregate. Constr Build Mater 205:39–51. https://doi.org/10.1016/j.conbuildmat.2019.01.194
22. Mohd Sani MSH, Ismail M, Rahman M, Muftah F (2019) Waste paper ash pellets as coarse aggregate replacement in concrete. Int J Eng Adv Technol 8(4):1112–1117
23. Tajra F, Elrahman MA, Stephan D (2019) The production and properties of cold-bonded aggregate and its applications in concrete: a review. Constr Build Mater 225(20):29–43. https://doi.org/10.1016/j.conbuildmat.2019.07.219
24. Manikandan R, Ramamurthy K (2008) Effect of curing method on characteristics of cold bonded fly ash aggregates. Cem Concr Compos 30(9):848–853. https://doi.org/10.1016/j.cemconcomp.2008.06.006
25. American Society for Testing and Materials (ASTM) (2020) Standard test method for slump of hydraulic-cement concrete. ASTM C143, West Conshohocken, The USA
26. American Society for Testing and Materials (ASTM) (2017) Standard specification for lightweight aggregates for structural concrete. West Conshohocken, The USA ASTM C330
27. Jo B, Park S, Park J (2007) Properties of concrete made with alkali-activated fly ash lightweight aggregate (AFLA). Cem Concr Compos 29(2):128–135. https://doi.org/10.1016/j.cemconcomp.2006.09.004
28. Punlert S, Laoratanakul P, Kongdee R, Suntako R (2017) Effect of lightweight aggregates prepared from fly ash on lightweight concrete performances. J Phys Conf Ser 901:12086. https://doi.org/10.1088/1742-6596/901/1/012086

Density, Compressive Strength and Water Absorption Properties of Sand Cement Brick Containing Recycled Concrete Aggregate (RCA) and Crumb Rubber (CR) as Partial Sand Replacement Materials

Faisal Sheikh Khalid, Mohamad Yuzwan Aliff Aminuddin, Nur Fatin Nabila Hissham, Aeslina Abdul Kadir, Mohd Irwan Juki, Shahiron Shahidan, and Syafiqa Ayob

Abstract Commercial construction in Malaysia commonly used natural fine aggregate as the building materials. Concrete production has recently been increased as linear with the economic growth of Malaysia. However, the concrete industry is facing a challenge to find adequate sources of natural fine aggregates and currently considering in utilizing the alternative to substitute fine aggregate with the waste material for concrete production. Thus, a combination of Recycled Concrete Aggregate (RCA) and Crumb Rubber (CR) was used in this research as a partial of sand replacements materials in the sand cement brick. This research aims to study the performance of sand cement brick that contains a combination of RCA and CR as partial sand replacement materials in terms of density, compressive strength and

F. S. Khalid (✉) · M. Y. A. Aminuddin · N. F. N. Hissham · A. A. Kadir · M. I. Juki · S. Shahidan · S. Ayob
Faculty of Civil Engineering and Built Environment, Universiti Tun Hussein Onn Malaysia, 86400 Parit Raja, Batu Pahat, Johor, Malaysia
e-mail: faisalsh@uthm.edu.my

M. Y. A. Aminuddin
e-mail: yuzwanaliff@gmail.com

N. F. N. Hissham
e-mail: nabilahissham95@gmail.com

A. A. Kadir
e-mail: aeslina@uthm.edu.my

M. I. Juki
e-mail: irwan@uthm.edu.my

S. Shahidan
e-mail: mustafa_albakri@unimap.edu.my

S. Ayob
e-mail: nssyafiqa@gmail.com

A. Abdul Kadir et al. (eds.), *Sustainable Waste Utilization in Bricks, Concrete, and Cementitious Materials*, Lecture Notes in Civil Engineering 129,
https://doi.org/10.1007/978-981-33-4918-6_11

water absorption. A total of 144 brick specimens were prepared and mixed with different percentages of RCA and CR as a partial sand replacement material. The design replacement of RCA was set to 0, 15, 30 and 45%, while CR was 0, 2, 4 and 6. Other than, W/C ratio that was used in this research was 0.6. The tests that were conducted included sieve analysis, density, compressive strength and water absorption tests. The brick was tested on day 28 after the brick was prepared. Based on the result, the percentage of water absorption was increased after incrementing of CR while RCA after 30% of replacement. In addition, the percentage of compressive strength and density was decreased after incrementing of CR while RCA after 30% of replacement. Throughout this research, brick that contained 30% of RCA and 2% of CR showed the optimum mix design where it shows the highest percentage of compressive strength and lowest in water absorption compared to another mix design. It is potentially applicable to the RCA and CR as secondary sources of natural fine aggregate and also can reduce waste of RCA and CR in Malaysia.

Keywords Recycled coarse aggregate · Crumb rubber

1 Introduction

Most of the infrastructures in civil engineering industry require a sand cement brick as a construction material. This method has been widely used in Malaysia's building style, where almost every wall structure used sand cement bricks for residential or other buildings. Ismail and Yaacob [1] claimed that most construction developments in Malaysia utilized cement and brick as the building materials due to their availability and lower production cost. A sand cement brick is a block that contains fine aggregate and cement material in masonry construction.

However, the demand for sand cement brick is rising, which contributes to the use of natural resources. This scenario leads to a reduction in natural resources for the production in sand cement brick since natural resources require high levels of energy consumption. Furthermore, due to the limited production of brick as the manufacturers facing the difficulty of obtaining in the limited natural resources, the market price of brick rose significantly. Thus, low-cost buildings could be affected.

Many studies have been conducted to limit natural resource use in the manufacture of sand cement brick. This study has established a new substitute for brick manufacturing, which involves recycling waste materials from the crushed concrete cubes and industrial waste such as rubber tires. The natural fine aggregate replaced by recycled concrete aggregate (RCA) and crumb rubber (CR) has helped the natural resources from depletion and give benefit by making use of the waste generated before disposal. RCA was generated after the renovation and demolition of the old building. The rise of RCA is mainly because new buildings are gradually required to redevelop, renovate, and demolish existing buildings in order to design a new concept that covered more occupants than the old building concept. Figure 1 shows the composition of waste by types of waste generated in Malaysia from conventional

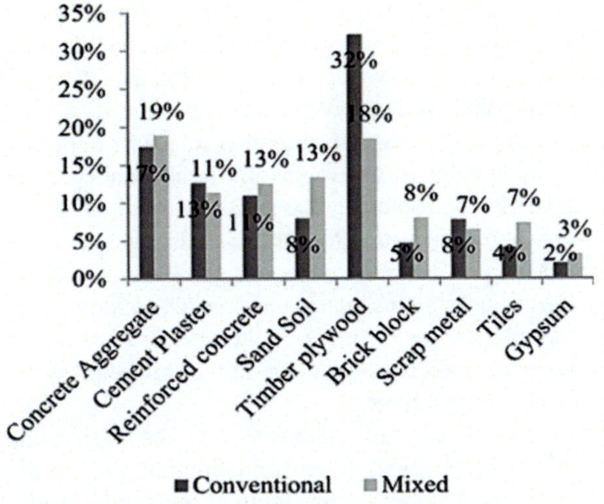

Fig. 1 Waste composition of conventional and mixed development in Malaysia [2]

Table 1 Waste generation rate in Malaysia [2]

Construction method	Project	Volume waste (m^3)	Floor area (m^2)	WGR (t/100 m^2)
Demolition	11	135,790	128,985	130.86
	Average	–	128,985	130.86

and mixed development. Meanwhile, Table 1 shows the rate of waste generation by different methods at various locations in Malaysia [2]. These data indicate that more waste was generated in the construction process of new buildings.

In addition, more waste materials are generated during tire manufacturing. Figure 2 shows the quantity of scrap tire generated in Peninsular Malaysia that has been recorded. This data prove that the increase of rubber tire generated annually reflects more waste materials may be produced.

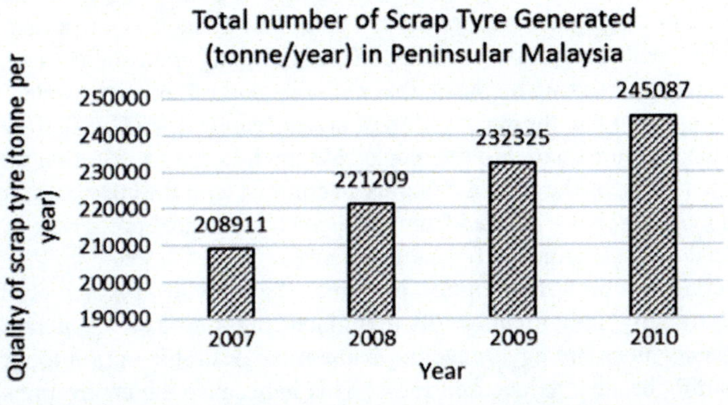

Fig. 2 Quantity of scrap tire generated (ton/year) in peninsular Malaysia [2]

In conjunction with conservation efforts, this study focused on preventing environmental pollution from waste disposal, recognizing the element of sustainable and reduce the construction cost by reusing industrial wastes. The environmental impacts during the building process can be reduced by implementing sustainable construction at every design development stage of a building project [3, 4]. However, the lack of knowledge of environmental issues among contractors is higher and need to be considered. Therefore, the substitution of RCA and CR as fine aggregates become appropriate selection to achieve goals towards sustainable construction development in Malaysia. Moreover, according to Mohammed et al. [5], the mindfulness on sparing the normal sources, reusing of crushed concrete and crumb rubber has given different advantages, for example, make an extra business opportunity, sparing expense of transfer, agreeable with condition and help government to meet the objective of diminishing transfer.

The construction industry was the major consumer of new materials. The increasing demand of new construction, the more amount of natural materials was used. Quick populace development and urbanization dramatically affect the expanded interest in the development of the construction market. Due to this case, many countries were forced to rearrange, renovate and also demolished old building to design a new concept of building that covered more occupants in building than old building concept. Therefore, new construction has produced more waste in the construction project.

Besides that, rubber tire also produced more waste material that need to be disposal. Globally, about 1 billion of used automobile tires were generated each year [6]. It is because, according to the international tire standards, tires can only last for two years after the plant (after which the rubber tire will die (hard) and not safe to be used). As shown, the quantity of scrap tire generated (ton per year) in Peninsular Malaysia in Fig. 2. It is proven that rubber tire was increase more waste year by year in Malaysia.

There was a high mindfulness on the need to give elective utilization to reused materials, particularly in the development of construction field. This resulted in the idea of making use of RCA and CR used to replacement partial of fine aggregates in sand cement brick. RCA and CR can reduce the uses of natural sand cement brick as a fine aggregate. According to Ismail and Yaacob [1], test results showed that sand cement brick that contains RCA is slightly equivalent in physical and mechanical properties with a brick that contains natural fine aggregate. The results also demonstrated that the substitution of normal sand with RCA at 50 and 75% are great on the pressure quality of the blocks. Meanwhile, the substitution with 100% of RCA content lessens the pressure quality in contrast with the control sample. Hence, recycled fine aggregates derived from crushed concrete are appropriate to be used in brick as a decent substitution for natural sand [7, 8].

Recycling is a preferable option to reduce the amount of waste produced from construction activities and rubber tire manufacturing. Based on the hierarchy of solid waste management strategy, recycling is the most desirable option to reduce waste from landfills besides reduce and reuse. This is because the environmentally more desirable and support the growing interest of the public in sustainable development.

In order to achieve the strategy and concept, this research used a recycle material from construction and rubber tire as a partial material of sand in the sand cement brick production.

One of the possibilities is to utilize the RCA and CR as fine aggregate replacement. Therefore, this study aimed to determine the density, compressive strength, and water absorption of sand cement brick containing RCA and CR as partial sand replacement materials. Therefore, the density, compressive strength and water absorption tests were conducted to meet the objective of the study.

2 Experimental

The design mix for cement sand brick was used in accordance with BS EN 998-2 [9] and the mixtures were designed according to the British Standard requirement [10]. Generally, the ratio of cement and sand in sand cement brick is 1:6. The optimum design mix of cement sand brick was 2, 4 and 6% of CR and 0, 15, 30 and 45% of RCA, respectively, as partial sand replacement. The brick was moulded with a mould size of 215 mm in length, 105 mm in width, and 65 mm in depth. Curing is a process that involves the maintenance of a suitable moisture content and temperature in samples for a suitable period of time, immediately after placing and finishing so that the desired properties of samples can be achieved. After pressing, the bricks were stacked on timber palettes and marked according to their aggregate composition percentage.

All the materials were weighed based on each mix design, before casting the brick. The mixed proportion of single brick that contains RCA were 0, 15, 30 and 45% and 0, 2, 4 and 6% for CR, as shown in Table 2.

2.1 Procedure for Making Sand Brick Specimen

The total number of specimens required for this research were 144 blocks of brick. Process of mixture preparation was 0.6% water-cement ratio for all mixture preparations. Figure 3 shows the process of mixing sand cement brick containing RCA and CR wastes.

3 Result and Discussion

The testing results on sand cement brick, including the density test, compressive strength, and water absorption test, are described in this chapter. An analysis was done according to the different percentages of RCA and CR as a partial replacement

Table 2 Mix proportion of single brick

Mix designation	Cement (kg)	Sand (kg)	RCA (%)	RCA (kg)	CR (%)	CR (kg)	Water (kg)	w/c
Control	0.358	2.500	0	0	0	0	0.215	0.6
R_0C_2		2.450		0	2	0.05		
R_0C_4		2.400		0	4	0.10		
R_0C_6		2.350		0	6	0.15		
$R_{15}C_0$	0.358	2.125	15	0.375	0	0	0.215	0.6
$R_{15}C_2$		2.075		0.375	2	0.05		
$R_{15}C_4$		2.025		0.375	4	0.10		
$R_{15}C_6$		1.975		0.375	6	0.15		
$R_{30}C_0$	0.358	1.750	30	0.750	0	0	0.215	0.6
$R_{30}C_2$		1.700		0.750	2	0.05		
$R_{30}C_4$		1.650		0.750	4	0.10		
$R_{30}C_6$		1.600		0.750	6	0.15		
$R_{45}C_0$	0.358	1.375	45	1.125	0	0	0.215	0.6
$R_{45}C_2$		1.325		1.125	2	0.05		
$R_{45}C_4$		1.275		1.125	4	0.10		
$R_{45}C_6$		1.225		1.125	6	0.15		

of NA in sand cement brick under curing conditions of 28 days, respectively. The w/c ratio of 0.6 was constant for each mixture.

All the experimental results were examinations for correlation of compressive strength, density, and water absorption between control and replacement of RCA and CR as an incomplete substitution of sand in sand cement brick.

3.1 *Sieve Analysis*

Sieve analysis was conducted to acquire the size appropriation of granular material by permitting the material to go through a sieve of dynamically littler mesh size and gauging the measure of the material that was held at each sifter as a small amount of the entire mass. The sieve for fine aggregate consists of small size and has different opening sizes at each level, including 10, 5, 2.36, 1.18 mm, 600, 300, 150, and 75 μm. All the grading on this material has been analyzed to ensure that this material is within the overall grading limit BS 882 [11].

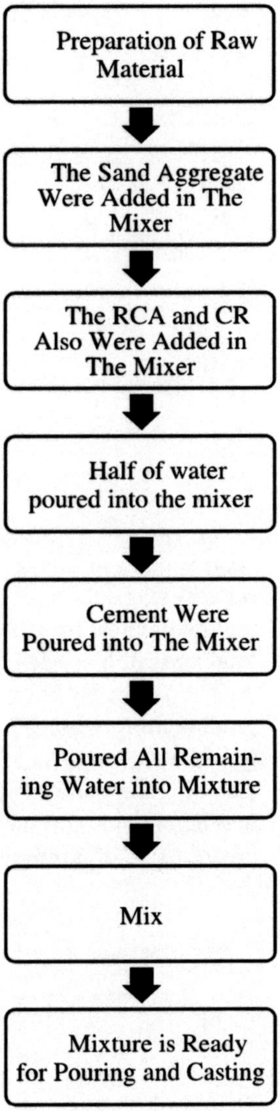

Fig. 3 Procedure of mixing sand and cement brick containing waste RCA and CR wastes

3.1.1 Natural Fine Aggregate

Sieve analysis comprises of soil samples' vibration that passes through a sieve arrangement with continuously smaller openings. This aggregate was obtained from the laboratory and sieved to get less soil less 5 mm. Table 3 shows the result for sieve analysis of natural fine aggregate. From the result of size distribution for natural fine aggregate, the total weight fractions plus the material that reached at the pan comes to 500 g where the weight of the original natural fine aggregate was 500 g. It indicated that the percentage loss is zero and might be due to the precautionary step that has

Table 3 Sieve analysis of natural fine aggregate result

Sieve size	Mass of retained (g)	Percentage of retained (%)	Cumulative percentage retained (%)	Cumulative percentage passing (%)
10 mm	0	0	0	100
5 mm	15	3	3	97
2.36 mm	50	10	13	87
1.18 mm	170	34	47	53
600 μm	140	28	75	25
300 μm	45	9	84	16
150 μm	60	12	96	4
75 μm	15	3	99	1
Pan	5	1	100	0
Total	500			100

been applied in this test, where the sieve was left for one minute after the shaking process was done to avoid loss of dust.

The greatest measure of NA retained at sieve size 1.18 mm with 170 g, 34% of total mass of NA. Meanwhile, the lowest percentage of NA was at the pan layer with 5 g (1%) only. By comparing the sieve analysis result for natural fine aggregate with grading limits from BS 882 [11], the NA at its current state showed that passing rate fell in a region of coarse grading limits. In this manner, it was ordered as coarse class. Figure 4 shows that the grading of natural fine aggregate is in the envelope of maximum and minimum curves. Therefore, it is acceptable to use the NA with the greatest size of 5 mm to the least size of 0.15 mm.

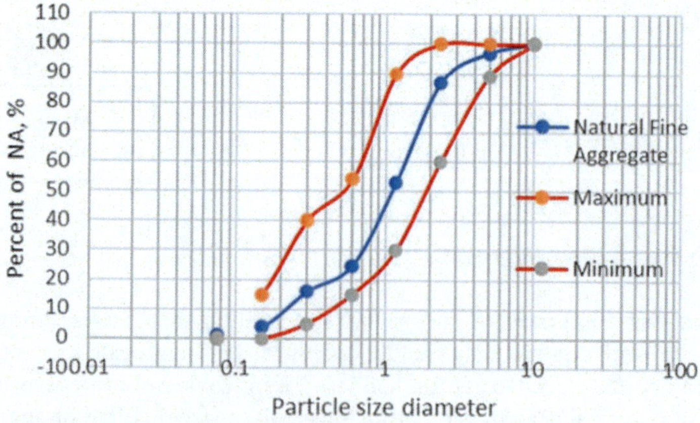

Fig. 4 The grading curve for natural fine aggregate (%)

Table 4 Sieve analysis of recycle concrete aggregate result

Sieve size	Mass of retained (g)	Percentage of retained (%)	Cumulative percentage retained (%)	Cumulative percentage passing (%)
10 mm	0	0	0	100
5 mm	14	1.4	1.4	98.6
2.36 mm	227	22.7	24.1	75.9
1.18 mm	278	27.8	51.9	48.1
600 μm	314	31.4	83.3	16.7
300 μm	75	7.5	90.8	9.2
150 μm	88	8.8	99.6	0.4
75 μm	4	0.4	100	0
Pan	0	0	100	0
Total	1000			100

3.1.2 Recycled Concrete Aggregate

RCA was gotten from crushing the old concrete by using a crusher machine. It was being sieved with a size of less than 5 mm and weighed to 1 kg to determine the grain size distribution of the RCA. The pattern result of sieve analysis was presented in Table 4.

The maximum amount of RCA retained at sieve size 600 μm with 314 g is 31.4% of total mass of RCA. Meanwhile, the lowest percentage of RCA was at the pan layer with 0 g (0%) only. By comparing the sieve analysis result for natural fine aggregate with grading limits from BS 882 [11], the RCA at its current state showed that passing rate fell in a region of coarse grading limits. In this manner, it was ordered as coarse class. Figure 5 shows that the grading of RCA is in the envelope of the maximum and minimum curves. Therefore, it is acceptable to use RCA with the greatest size of 5 mm to the least size of 0.15 mm.

3.1.3 Crumb Rubber

CR was obtained from recycling factory tires at Jalan Bukit Pasir, Muar. The crushed CR were separated according to their size. The size of the sieved waste CR granules with the maximum size of 5 mm was physically similar to the sizes of the fine aggregate. CR was weighed 1 kg to determine the grain size distribution of the CR. The pattern result of sieve analysis is presented in Table 5.

The maximum amount of CR retained at sieve size 600 μm with 340 g is 34.0% of total mass of CR. Meanwhile, the lowest percentage of CR was at the pan layer with 0 g (0%) only. By comparing the sieve analysis result for natural fine aggregate with grading limits from BS 882 [11], the CR at its current state showed that passing rate fell in a region of coarse grading limits. In this manner, it was ordered as coarse

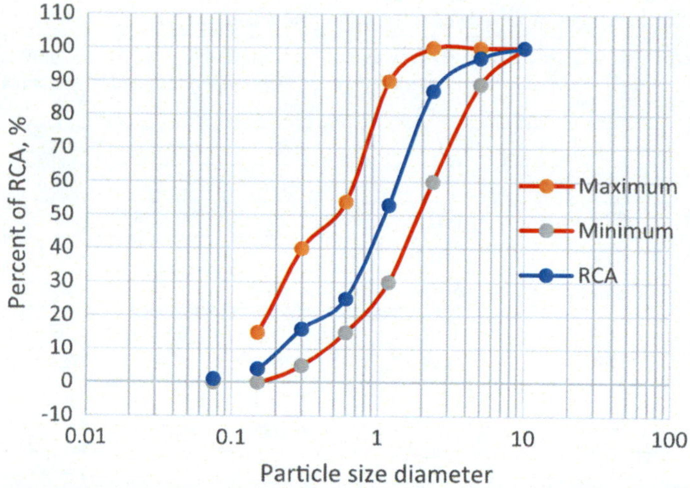

Fig. 5 The grading curve for RCA, %

Table 5 Sieve analysis of CR result

Sieve size	Mass of retained (g)	Percentage of retained (%)	Cumulative percentage retained (%)	Cumulative percentage passing (%)
10 mm	0	0	0	100
5 mm	14	1.4	1.4	98.6
2.36 mm	201	20.1	24.1	78.5
1.18 mm	278	27.8	51.9	50.7
600 μm	340	34.0	83.3	16.7
300 μm	80	8.0	90.8	8.7
150 μm	83	8.3	99.6	0.4
75 μm	4	0.4	100	0
Pan	0	0	100	0
Total	1000			100

class. Figure 6 shows that the grading of CR is in the envelope of the maximum and minimum curves. Therefore, it is acceptable to use the CR with the greatest size of 5 mm to the least size of 0.15 mm.

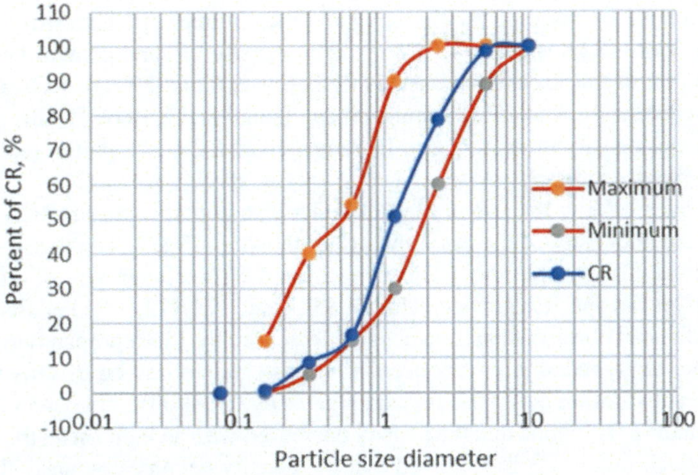

Fig. 6 The grading curve for CR, %

3.2 *Density Test*

This test testing was used in this research to evaluate the density value of sand cement brick that contains the RCA and CR as a partial replacement in sand cement brick. The density test was conducted after the specimen was hardened after the curing process at 28 days after casting the sand cement brick.

Figure 7 shows the pattern of density values for control and replacement mixtures. The result showed that the density of the sand cement brick fluctuated as the percentage of RCA and CR added into the sand cement brick increased. The density

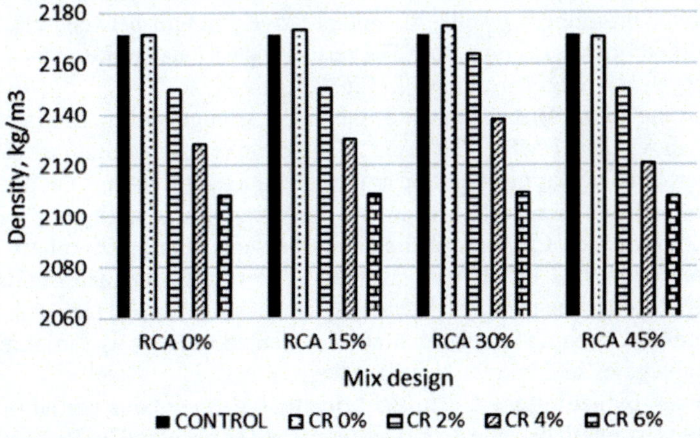

Fig. 7 The density for control and replacement mixture

was evaluated by dividing the mass, kg and the volume, m^3. The density of control sand cement bricks was 2174.5 kg/m^3. The density of sand cement brick without the addition of CR ascended consistently between 0 and 30% of RCA in brick but decreased after RCA at 45% of replacement. Ismail and Yaacob [1] also proved that the value of density increased after incremented of RCA as a partial replacement of fine aggregates in brick.

However, adding the CR in the brick that also contains partial RCA influenced the hardened density of the brick mixes, as shown in Fig. 7. The density decreased after added CR into the brick. This is because the specific gravity of crumb rubber is less than the specific gravity value of RCA and NA [12, 13]. Pelisser et al. [14] reported that rubberized concrete density decreased by 13% compared to the control mix. The steady reduction in the density has also been observed by different authors with the substitution of natural fine aggregates by CR at different percentages. Bisht and Ramana [7] claimed that the density decreased with an increase in the percentage of CR. Each mix of design showed that the density becomes reduced. Thus, it can be concluded that by increasing the percentage combination of replacement of RCA and CR, the density values of sand cement brick mixture will become decrease.

3.3 Compressive Strength Test

The compressive strength of sand cement brick can be presented as the performance of sand cement brick subjected to an ultimate load. The sand cement brick specimens were compressed to obtain the indirect strength of brick. The test was evaluated on brick specimens that reached the age of 28 days and each point presented in graphical plots was taken, obtained from the average of three readings on three specimens. The testing was conducted to find out the strength of the normal sand cement brick (control) and sand cement brick with different replacement percentage of RCA and CR content. From these results, the compressive strength between brick (control) and modified brick was compared. The results for various brick specimens in terms of compressive strength are shown in Fig. 8.

According to the British standard, the target strength of the sand cement brick must be above than 7 MPa. The result of the compressive strength on sand cement brick showed a fluctuating condition as the percentage of RCA and CR added into the sand cement brick increased. Meanwhile, the compressive strength of sand cement brick without added CR increased steadily between 0 and 30% of RCA in brick but decreased at 45% of RCA replacement. The results obtained in this study are relatively closed to the previous research by Ismail and Yaacob [1] in which the value of compressive strength increased after incremented of RCA as partial replacement of fine aggregates in brick.

However, by adding the CR in the brick that also contains partial of RCA, the compressive strength became decreased, similarly as observed by Bisht and Ramana [7]. Each mix design has shown that the compressive strength becomes decreased. It may be inferred that the compressive strength values of sand cement brick mixture

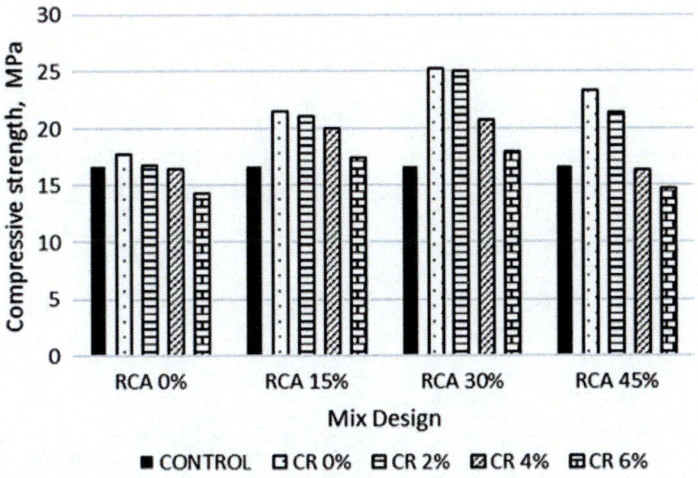

Fig. 8 The average value of compressive strength

are minimized by increasing the percentage combination of replacement of RCA and CR.

The decreased in strength is generally due to the lack of adhesion between the smooth rubber particles and cement paste. Cracks will develop quickly around the rubber particles at the time of loading, which results in rapid rupture of concrete. Another factor is due to the generation of voids which might have developed due to fine nature of crumb rubber. Bisht and Ramana [7] also claimed that as the percentage of crumb rubber in concrete mix increases, the percentage of voids would also increase. The ingredients present in concrete allow finer material to enter into interstitial zones. The filling effect of finer material occurs up to a certain extend of utilization only. Any more incorporation will lead to the generation of voids.

The test results proved it was evident that by added the CR in sand cement brick was reduced the compressive strength. An expanded portion of CR contributed to the absence of grip between the smooth rubber particles and cement paste. As a result, the cracks were quickly developed around the rubber particles when loading was applied. Besides that, the strength was diminished due to the presence of porosity in sand cement brick which may have created because of CR in brick. In general, the rate of decrease in strength was found to diminish with the expansion in the level of CR. Additionally, Bisht and Ramana [7] in their research stated that the strength of this interface is further reduced due to the smooth texture of crumb rubber, which leads to easier propagation of cracks while loading.

The relative compressive strength is defined as the difference of the compressive of the sand cement brick containing RCA and CR as a partial replacement of NA compared to the control brick. Figure 9 shows the difference of average compressive strength of sand cement brick that contains RCA and CR compared to control brick. The range of relative compressive strength was −3.36 (minimum) to 7.6 (maximum). The negative value of the average compressive strength of brick means that the value

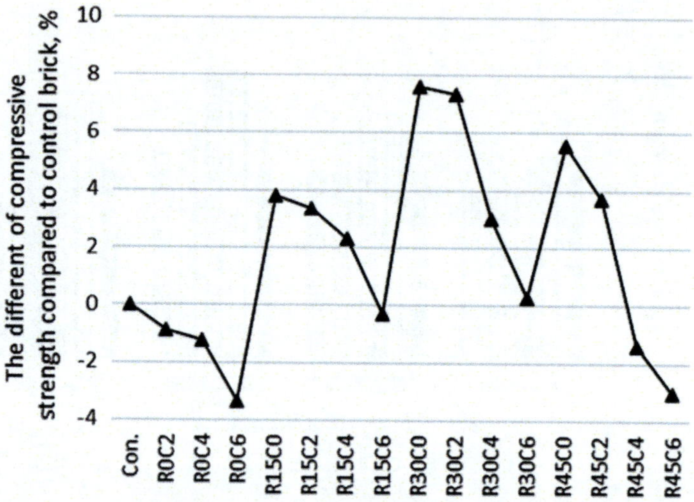

Fig. 9 The different of compressive strength compared to control brick (%)

of average compressive was below the compressive strength of control sand while the positive average was above the compressive strength of control brick. The higher difference in the amount of compressive strength between brick modified (RCA and CR) and control brick indicates the optimum ratio of combination of RCA and CR in sand cement brick. Through the graph result, it can be concluded that the combination of replacement at RCA-30% and CR-2% was the highest compared to the other combinations.

3.4 Water Absorption Test

The water absorption test of sand cement brick can be presented as the performance of sand cement brick in permeability. The testing was conducted to find out the permeability of the normal sand cement brick (control) and sand cement brick with different replacement percentage of RCA and CR content. From these results, the water absorption between brick (control) and modified brick was compared. The results for various brick specimens in terms of water absorption are shown in Fig. 10.

Adding on the CR in the brick that also contains partial of RCA caused the water absorption became increased, parallel with the pattern observed by Bisht and Ramana [7]. The increase in water absorption is due to fine nature of crumb rubber which results in the generation of cracks and voids. It has been concluded that the percentage of voids ratio also increases with the inclusion of crumb rubber in brick mix, which allows the water to penetrate inside the brick. Turatsinze et al. [15] reported an increase in porosity of concrete due to increase in voids with 25% replacement of fine aggregates by waste rubber aggregates. Furthermore, Ganjian et al. [16] illustrated

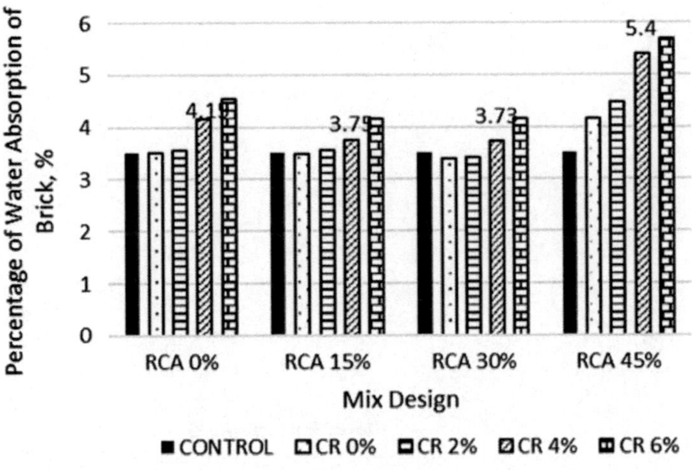

Fig. 10 The average percentage value of water absorption

similar observations for rises of water penetration in concrete mixes, due to weak interlocking and generation of voids between crumb rubber and cement paste.

The relative compressive strength is defined as the percentage difference of the absorption of the sand cement brick that containing RCA and CR as a partial replacement of NA compared to the control brick. Figure 11 shows the percentage difference of average water absorption of sand cement brick that contains RCA and CR compared to control brick. The range of water absorption was −0.12% (minimum) to 2.17% (maximum). The negative percentage of average water absorption of brick means that the value of absorption was very difficult to absorb the water in brick.

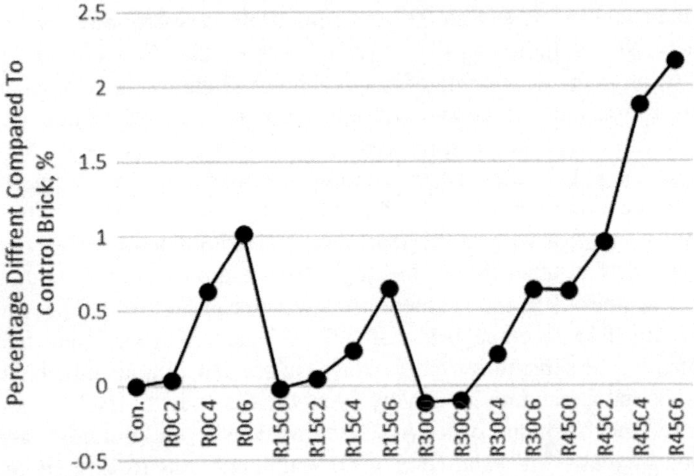

Fig. 11 Percentage different of water absorption compared to control brick (%)

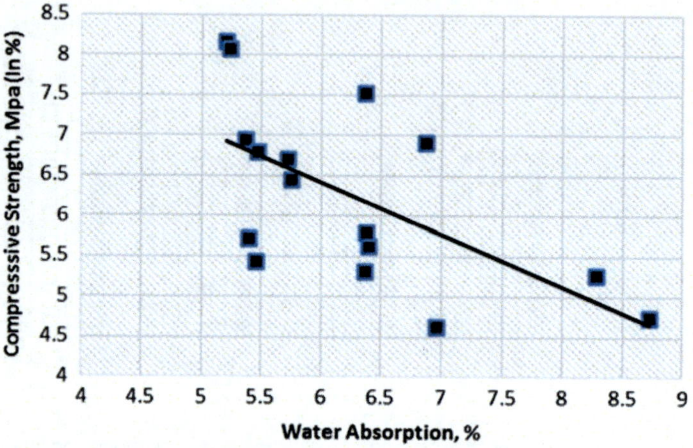

Fig. 12 Comparison between water absorption and compressive strength

In contrast, the positive average was much easier to absorb the water in brick. The lower difference amount of water absorption between brick modified (RCA and CR) and control brick was the optimum ratio of combination of RCA and CR in sand cement brick.

3.5 Relationship Between Density, Compressive Strength and Water Absorption

The relation between density, compressive strength and water absorption were discussed further in this section. Those correlations were defined to evaluate whether the density affects the strength of brick or not and also to determine the effect of water absorption on the strength of bricks. Figure 12 shows the relationship between water absorption and compressive strength. This graph shows the rate of initial water absorption did affect the compressive strength when it reaches its maturity state. Zhang and Zong [17] stated that the compressive strength of brick could be affected if the water absorption rate is high. Consequently, the water is permitted to flow to the sand cement brick without resistance due to the high permeability on the cement brick. Besides that, other molecules might have the possibility to react and disrupt the chemical stability in sand cement brick which caused low compressive strength.

The results in sand cement bricks at 30% replacement of RCA shown least permeability because the substitution caused the tight bonded, diminishing the porosity rate in the brick and expanding its density. Nonetheless, it is inversely to the brick that contains too much amount of RCA, where the density will diminish, and the water absorption increase. The sand cement brick that contains RCA is more permeable than the NA [18]. Their pores are typically broken in a blocks framework and being

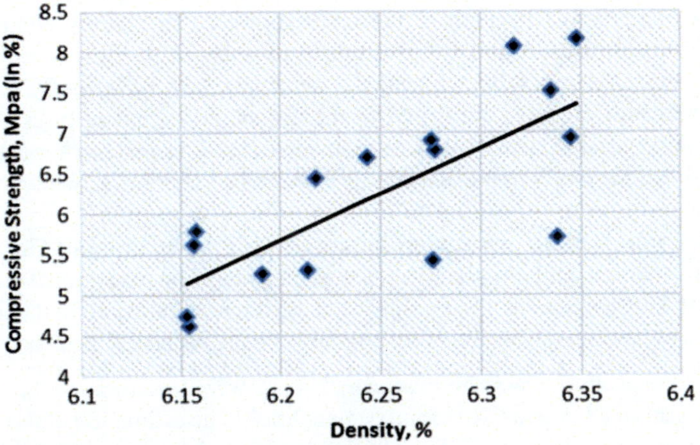

Fig. 13 Comparison between density and compressive strength

totally wrapped by cement paste. This is because, the physical size of RCA is coarser than NA and it make the pressing of grid turns out to be less efficient as the RCA substance was expanded. Thus, the volume of porosity in brick increased and the density is low while the water absorption is higher.

Figure 13 shows that the compressive strength was directly proportional to the density. It defined that the value of density in sand cement brick has affected the compressive strength when it reaches its maturity state. However, according to Zhang and Zong [17], the compressive strength of brick is not affected when the rate of density in brick is high.

4 Conclusion

It clearly showed that the combination of RCA and CR are suitable to be used as partial replacement in sand cement brick. Based on the results of the water absorption, compressive strength test and density test presented above, the conclusions that can be drawn are as follows:

1. The grading curve for fine aggregates was studied for natural fine aggregates, RCA and CR with a sieve size of 5–0.15 mm. As for the sieve analysis test, it was found that the percentage passing retained graph fall between region maximum and minimum limit. All the materials have been categorized in the coarse class followed BS 882:1983,
2. As for the density test, the density value was reduced as the percentage of CR applied to sand cement brick increased while RCA decreased after 30% of replacement. In comparison to normal sand cement brick, the test results for both RCA and CR percentages showed no significant difference,

3. The compressive strength of the brick showed a fluctuating condition as the percentage of RCA and CR was added into the sand cement brick increased. However, the test results for all the percentages of RCA and CR showed no significant difference compared to the normal sand cement brick. The overall results revealed that every mix designation of sand cement brick mix have passed or above the target of normal strength of sand cement brick, which is 7 N/mm^2, and
4. The value of water absorption improved with the addition of CR into a sand cement brick, while for the RCA, the water absorption rose after 30% of replacement. However, there was no significant difference compared to the normal sand cement brick in the results of testing for both RCA and CR percentages applied.

From the analysed data, the main conclusions were to determine the best ratio of replacement of RCA and CR in sand cement brick. The estimation must consider the result of testing towards sand cement brick in terms of density, compressive strength and water absorption. Based on the tests performed, the optimum percentage of replacement of RCA and CR to be added into sand cement brick is 30 and 2%. It is because the performance on compressive strength was better than the control brick and other mix design. Thus, the preferred combination between RCA and CR in sand cement brick is 30 and 2%.

References

1. Ismail S, Yaacob Z (2010) Properties of bricks produced with recycled fine aggregate. World Acad Sci Eng Technol 43(3):878–882
2. Poon CS, Yu A, Jaillon L (2004) Reducing building waste at construction sites in Hong Kong. Constr Manag Econ 22(5):461–470. https://doi.org/10.1080/0144619042000202816
3. Akadiri P, Chinyio E (2012) Design of a sustainable building: a conceptual framework for implementing sustainability in the building sector. Buildings 2(4):126–152. https://doi.org/10.3390/buildings2020126
4. Abid SR, Nahhab AH, Al-aayediHKH, Nuhair AM (2018) Expansion and strength properties of concrete containing contaminated recycled concrete aggregate. Case Stud Constr Mater 9:e00201. https://doi.org/10.1016/j.cscm.2018.e00201
5. Mohammed T, Hasnat A, Awal M, Bosunia S (2014) Recycling of brick aggregate concrete as coarse aggregate. J Mater CivEng 27.https://doi.org/10.1061/(ASCE)MT.1943-5533.0001043
6. Mahla R (2015) Partial replacement of coarse aggregate. 4(2):95–98
7. Bisht K, Ramana PV (2017) Evaluation of mechanical and durability properties of crumb rubber concrete. Constr Build Mater 155:811–817. https://doi.org/10.1016/j.conbuildmat.2017.08.131
8. Ali N, Shahidan S, Radziah S, Abdul Aziz M, Abdul Samad AA, Mohamad N (2018) Compressive strength and initial water absorption rate for cement brick containing high-density polyethylene (HDPE) as a substitutional material for sand. In: The global congress on construction, material and structural engineering (GCoMSE 2017). IOP Conference Series: Materials Science and Engineering
9. British Standards Institution (2016) Specification for mortar for masonry—Part 2: Masonry mortar. London. BS EN 998-2
10. British Standards Institution (2005) Code of practice for the use of masonry—Part 3: Materials and components, design and workmanship. London. BS 5628-3

11. British Standards Institution (1983) Specification for aggregates from natural sources for concrete. London. BS 882
12. Azmi N, Sheikh Khalid F, Irwan JM, Anting N, Mazenan P (2017) A study on the performance of concrete containing recycled aggregates and ceramic as materials replacement. In: IOP conference series: materials science and engineering, vol 271, issue 1, p 012081 https://doi.org/10.1088/1757-899X/271/1/012081
13. Gabr AS (2015) Properties of recycled concrete aggregate under different curing conditions HBRC J 24. https://doi.org/10.1016/j.hbrcj.2015.07.001
14. Pelisser F, Zavarise N, Longo TA, Bernardin AM (2011) Concrete made with recycled tire rubber: effect of alkaline activation and silica fume addition. J Clean Prod 19(6):757–763. https://doi.org/10.1016/j.jclepro.2010.11.014
15. Turatsinze A, Garros M (2008) On the modulus of elasticity and strain capacity of self-compacting concrete incorporating rubber aggregates. Resour Conserv Recycl 52:1209–1215. https://doi.org/10.1016/j.resconrec.2008.06.012
16. Ganjian E, Khorami M, Maghsoudi AA (2009) Scrap-tyre-rubber replacement for aggregate and filler in concrete. Constr Build Mater 23:1828–1836. https://doi.org/10.1016/j.conbuildmat.2008.09.020
17. Zhang S, Zong L (2014) Evaluation of relationship between water absorption and durability of concrete materials. Adv Mater Sci Eng. https://doi.org/10.1155/2014/650373
18. Mat Aris S, Muthusamy K, Uzer A, Ahmad S (2018) Properties of palm oil fuel ash cement sand brick containing pulverized cockle shell as partial sand replacement. In: IOP conference series earth environment science, vol 140, p 12145. https://doi.org/10.1088/1755-1315/140/1/012145

Performance of Concrete Containing Synthetic Wire Waste as Fiber Materials

Faisal Sheikh Khalid, Mohamad Yuzwan Aliff Aminuddin, Azini Amiza Hashim, Aeslina Abdul Kadir, Mohd Irwan Juki, Shahiron Shahidan, and Syafiqa Ayob

Abstract The demand for concrete in the construction field is increasing steadily with the growth of Malaysia economy. This demand is expected to increase over the coming years as a result of technology advancement. This study aims to determine the performance of concrete containing synthetic wire waste as fiber material. The utilization of wire waste as fiber materials could help to reduce the amount of waste dumped into landfills. The excessive amount of wire waste production may lead to various environmental issues. To date, there are no researchers who have conducted studies on the efficiency of concrete containing synthetic wire waste as fiber materials. Hence, an alternative has been presented in this study to overcome the associated environmental issues by using wire waste as fiber materials in the concrete mixture. The proportion of wire waste being incorporated into the concrete mixture is 0, 0.5, 1.0 and 1.5%. Apart from that, the ratio of 0.55 for the water-cement ratio has been used throughout this study. The workability, density, compressive strength and tensile

F. S. Khalid (✉) · M. Y. A. Aminuddin · A. A. Hashim · A. A. Kadir · M. I. Juki · S. Shahidan · S. Ayob
Faculty of Civil Engineering and Built Environment, Universiti Tun Hussein Onn Malaysia, 86400 Parit Raja, Batu Pahat, Johor, Malaysia
e-mail: faisalsh@uthm.edu.my

M. Y. A. Aminuddin
e-mail: yuzwanaliff@gmail.com

A. A. Hashim
e-mail: aziniamizaa@gmail.com

A. A. Kadir
e-mail: aeslina@uthm.edu.my

M. I. Juki
e-mail: irwan@uthm.edu.my

S. Shahidan
e-mail: mustafa_albakri@unimap.edu.my

S. Ayob
e-mail: nssyafiqa@gmail.com

A. Abdul Kadir et al. (eds.), *Sustainable Waste Utilization in Bricks, Concrete, and Cementitious Materials*, Lecture Notes in Civil Engineering 129,
https://doi.org/10.1007/978-981-33-4918-6_12

strength tests have been conducted in this study. The compressive strength test was conducted for 7 days and 28 days. Meanwhile, tensile strength test was conducted for 28 days. The result of compressive and tensile strength test was compared with the control mixture and it was found that the optimum amount of wire waste as fiber materials in concrete is 0.5%.

Keywords Synthetic wire waste · Fiber materials · Workability · Density · Compressive strength · Tensile strength

1 Introduction

In recent years, concrete and synthetic fibers have becomee more common and are often used for construction materials. Concrete has poor tensile strength, ductility as well as energy absorption when it stands alone. Unsatisfactory toughness and presence of defects in the concrete structure have lead to these poor characteristics. Therefore, the concrete toughness is needed to be improved and the size and amount of defects in concrete should be reduced to ensure its good performance in building application. In order to overcome the weakness of concrete, many studies on fiber-reinforced concrete have been performed. Some of the findings suggested that reinforced concrete with the addition of fiber materials drastically improves the performance of concrete and disregard its disadvantages such as poor tensile strength, ductility and energy absorption capacity. Order in magnitude increases in toughness (energy absorption) over plain concrete is commonly observed [1, 2]. Thus, there are many present studies which have been exploring the potential of many different fiber types and shapes with the main focus of investigating the performance of concrete added with fibers.

The idea on using fibers as reinforcement is not relatively new. In particular, asbestos has been used as fiber material in concrete since the year of 1990. However, after it is found that scientifically that the asbestos has the adverse effect on human health, the substitute material for the substance in concrete and other building materials has been critically investigated. For instance, materials such as ppolypropylene (PP), polyethylene (PE), polyvinyl alcohol (PVA), polyvinyl chloride (PVC), nylon, aramid, and polyesters are used as fiber material. Out of these seven materials, PP fibers are the most widely used material in construction applications. Among the applications including shotcrete tunnel linings, blast-resistant concrete, overlays and pavements [3].

In controlling the fiber performance, there are three distinct variables involved which are the aspect of ratio, fiber shape, surface deformation (including anchorage), and surface treatment. Besides, if necessary, the tensile strength of the fiber can be increased to prevent fiber fracture. Generally, the cross-section of an individual fiber can be circular, rectangular, diamond, square, triangular, flat, polygonal, or any substantially polygonal shape (Fig. 1). For the further improvement of the bond characteristics, a fiber can be altered along its length by roughening its surface or by

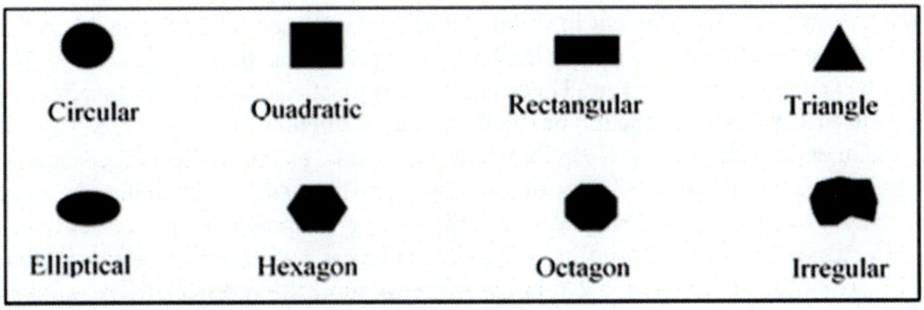

Fig. 1 Examples of cross-sectional geometries of fibers [4]

applying mechanical deformations. It can be smoothed, indented, deformed, crimped, coiled and twisted with end hooks, paddles, buttons, or other anchorages (Fig. 2). One of the main differences between the cross-sections in Fig. 2 is the ratio between their surface area and length. For example, the triangular fiber has the same area with a circular that has 28% larger perimeter [4].

An effective way to increase the concrete performance is by the addition of synthetic fiber to enhance the fracture characteristics and behavior of structures through the capability of the fibers to bridge cracks. Another application of this fiber material is to reduce the shrinkage and shrinkage cracking of concrete associated with hardening and curing. There will be an occurrence of plastic shrinkage if the rate of evaporation on the concrete surface is greater than the rate of water migration

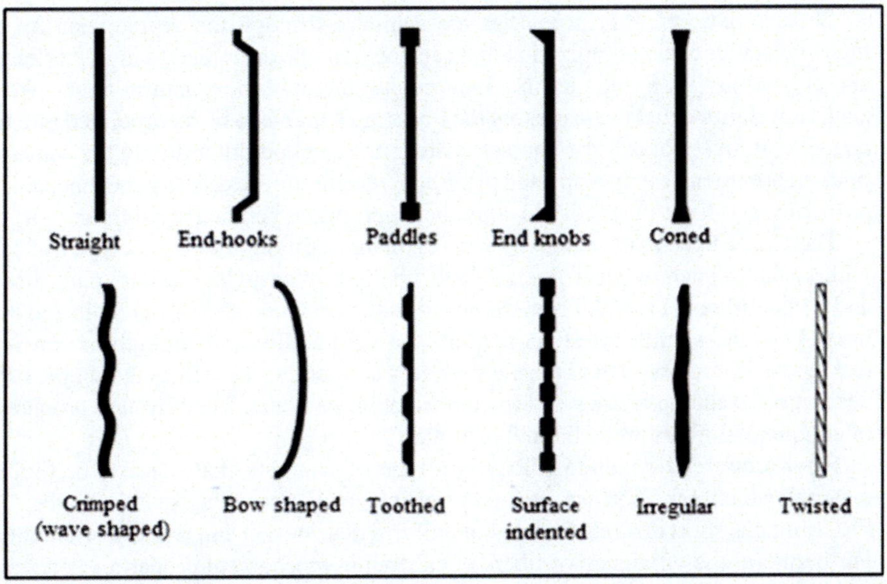

Fig. 2 Examples of some typical fiber [4]

to the surface. This gradient in shrinkage has forced crack to appear on the surface. Plastic shrinkage occurs during the hardening process as the strength and stiffness of wet concrete are very low. The occurrence of plastic shrinkage can be effectively reduced with the introduction of small amount of fibers.

Other benefits of the fiber materials are it can enhance the fatigue strength, wear-resistance and durability of the concrete. By using fiber materials instead of traditional concrete, section thickness can be decreased and cracking can be effectively controlled, resulting in lighter structures with a longer life expectancy. Fiber-reinforced concrete has a broad range of application especially in buildings, highway overlays, bridges, and airport runways. As for load-bearing applications, it is generally used together with traditional steel reinforcement [5]. The use of low-dosage synthetic fiber reinforcement for floor slabs is widely practiced in building construction.

Synthetic fiber is made from chemicals and can be represented as a versatile, macroscopically homogeneous body with a high length to thickness ratio and a small cross-section and can have superior properties to natural fibers such as cotton or silk. Synthetic fiber is manufactured from naturally occurring macromolecules and synthetic polymers. According to Wafa [6], fibers are produced from different materials in various shapes and sizes. Amezugbe [7] also stated that the section of fiber could be square, round, oval, polygonal, triangular or crescent-shaped, relying on the manufacturing process and the raw material used. Fiber is divided into two major categories which are micro and macro. Fiber with the diameters or equivalent diameters less than 0.3 mm (0.012 in.) is known as microfibers whereas fiber with the diameters or equivalent diameters greater than 0.3 mm is known as macrofiber [8].

Characteristic of fiber properties are achieved through the development of an inter-molecular chain organization that can generally be described as highly oriented and crystalline [9]. Synthetic fiber is more durable than most natural fibers. Wang et al. [10] claimed that to ensure that the fibers are suitable to be incorporated into the concrete, it must possess the characteristics such as good durability in the cementitious environment, easily dispersed in the concrete mixture, exhibits good mechanical properties and have the ability to construct appropriate geometric configuration.

There have been many extensive studies on the addition of a various type of fibers with concrete such as steel fibers, wood fibers, polypropylene fibers, glass fibers and carbon fibers [11, 12]. From the result of the previous study, it is indicated that fibers have the significant effect towards the workability and strength of concrete due to several factors such as tensile strength and modulus as well as the shape, stiffness, surface characteristics and the deformation of fibers. The physical properties of commercial fibers are exhibited in Table 1.

Fiber-reinforced concrete (FRC) is a type of concrete that consists of fibrous material which can increase its structural integrity. According to Amezugbe [7], FRC contains short discrete fibers with uniform distribution and random orientation. Furthermore, the efficiency of fibers in enhancing mechanical properties is significantly influenced by their orientation concerning the crack formed in concrete [13]. FRC is made from cement, water, fine and coarse aggregates and fiber materials.

Table 1 Physical properties of commercial fibers [4]

Type of fiber	Diameter (μm)	Specific gravity (g/cm^3)	Tensile strength (MPa)	Elastic modulus (GPa)	Ultimate elongation (%)
Metallic Steel	5–1000	7.85	2000–2600	195–201	0.5–5
Glass E glass	8/15	2.54	2000–4000	72	3.0–4.8
AR glass	8–20	2.70	1500–3700	80	2.5–3.6
Synthetic Acrylic (PAN)	5–17	1.18	200–1000	14.6–19.6	7.5–50.0
Aramid (e.g. Kevlar)	10–12	1.4–1.5	2000–3500	62–130	2.0–4.6
Carbon (low modulus)	7–18	1.6–1.7	800–1100	38–43	2.1-2-5
Carbon (high modulus)	7–18	1.7–1.9	1500–4000	200–800	1.3–1.8
Nylon (polyamide)	20–25	1.16	965	5.17	20.0
Polyester (e.g. PET)	10–8	1.34–1.39	280–1200	10–18	10–50
Polyethylene (PE)	25–1 000	0.96	80–600	5.0	12–100
Polyethylene (HPPE)	–	0.97	4100–3000	80–150	2.9–4.1
Polypropylene (PP)	10–200	0.90–0.91	310–760	3.5–4.9	6–15.0
Natural–organic Cellulose (wood)	15–125	1.50	300–2000	10–50	20
Coconut	100–400	1.12–1.15	120–200	19–25	10–25
Bamboo	50–400	1.50	350–50	33–40	–
Jute	–	1.02–1.04	250–350	26–32	1.5–1.9

Fibers that usually being used for mixing with concrete may be steel, glass, polymeric materials, carbon and cellulose with the length ranging from 3 to 64 mm. Fibers provide the maximum positive effect on FRC efficiency when the fibers are aligned in the direction of stress perpendicular to the crack openings [14]. The FRC's characteristics highly depending on the different concretes, fiber materials, geometries, distribution, orientation and densities.

The Federal Highway Authority report (1989–1994) on High-performance Fiber Reinforced concrete (HPFRC) presents the result of the addition of short discrete fiber or continuous long fibers to the cement matrix. The result conveys that FRC bars and tendons reached their ultimate tensile strength without showing any yielding of the material. As it is known, FRP bar is strong in compression but weaker in tension.

The addition of fibers with the different type and percentage of fibers addition in concrete will influence its mechanical properties. For example, fibers with end anchorage and high aspect ratios were found to be more effective in enhancing the mechanical properties of the concrete. The same length and diameter of crimped-end fiber can give the same result as straight fibers using 40% less fiber [6].

Conversely, some studies conducted by the researcher show that there is no significant improvement in compressive strength when plastic fiber is added in concrete, which is consistent with the outcome reported by Hsie [15], Campione [16], Fraternali [17], and de Oliveira and Castro-Gomes [18]. However, this claim can be contended by Rai and Joshi [19], which found out that the presence of fibers may alter the failure mode of cylinders. Nevertheless, the effect of fiber on the improvement of compressive strength values is very little (0–15%). These differences are attributed to changes in the water to cement ratio between control and fiber-reinforced concrete. In addition, during the compression test, the plain concrete failed catastrophically while the macro plastic fiber-reinforced concrete cylinders failed with many minor cracks on the surface [20]. This condition demonstrated that the plastic fibers still held the concrete together at the failure load. The stress–strain curve of a compressive test on concrete cylinders conducted by [21] (Fig. 3).

Lower density value of concrete portrays that the concrete has a high water content which leads to the decline of concrete's strength. According to Ahmed et al. [22], it is stated that PP fiber has an insignificant effect on the concrete density, but the concrete density has mainly affected by steel fiber. This is clearly due to high specific gravity possesses by steel fibers compared to PP fibers that have low specific gravity. Tables 2 and 3 show the fiber ratio in mixes and the bulk densities of hardened concretes which were measured at 28 days, respectively.

Moreover, the fibers can easily absorb the cement paste to enable their adhesion due to the high content and high surface area of the fibers, hence increasing viscosity of the concrete mixture [23]. There are two suggestions from Mazaheripour [24]

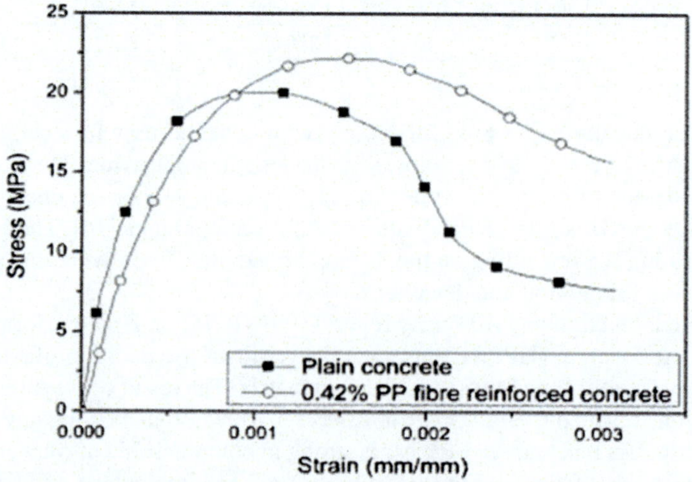

Fig. 3 Average stress–strain curves for concrete with macro plastic fibers [13]

Table 2 Fiber ratio in mixes [22]

Mix No	1	2	3	4	5	6	7	8	9
Steel fiber ratio (%)	0.0	0.5	1.0	0.0	0.5	1.0	0.0	0.5	1.0
PP fiber ratio (%)	0.0	0.0	0.0	0.2	0.2	0.2	0.4	0.4	0.4

Table 3 Bulk density of hardened concrete measured at 28 days [22]

Mix No	Inverted slump test		Bulk density (kg/m^3)
	time(s)	height(cm)	
1	22	–	1760
2	120	12	1860
3	120	8	1870
4	45	–	1780
5	120	13	1800
6	120	10	1880
7	67	–	1740
8	120	4	1840
9	120	3	1880

to improve the workability of fiber-reinforced concrete: (a) to limit the volumetric content of macro plastic fibers to a range of 0.1 to 1% and (b) to increase the moisture content. However, concrete strength will be negatively affected due to the addition of water. On top of that, plasticizer or water-reducing admixtures are often used in fiber-reinforced concrete to improve workability without increasing the water content [21]. Table 4 depicts the slump test results of macro plastic fiber in concrete.

Concrete properties show that it has high compression strength but low in tensile strength. Under tensile loading, cracks propagate rapidly at reduced stress levels, which cause brittle failure in concrete tensions. Developing of fiber materials can

Table 4 Slump test result of macro plastic fiber in concrete [25]

Macroplastic fiber	Fiber dimension	Fiber volumetric content (%)	Slump (%)	Compressive strength (MPa)	Splitting tensile strength (MPa)
Macro PP fiber, wavelength shape [26]	0.9 mm in diameter, 50 mm in length	0	102	35.0	2.2
		1	38	35.4	3.2
		1.5	6.5	30.7	3.2
PP fiber 620 MPa tensile strength and 9.5 GPa Young's Modulus [21]	40 mm × 1.4 mm × 0.11 mm	0	n/a	38.9	3.6
		0.33	n/a	40.5	3.9
		0.42	n/a	41.4	4.1
		0.51	n/a	41.6	4.1

be used to improve the tensile strength of concrete. Fiber-reinforced concrete has superior tensile properties over normal concrete without fiber materials [27]. Choi and Yuan et al. [26] and Hasan et al. [21] claimed that the macro plastic fiber has a positive effect on the tensile strength of concrete. The stress is being propagated to the macro plastic fibers when the stress in concrete reaches the tensile strength of concrete during the split-cylinder test. The fibers have the ability to terminate the propagating macro cracks, thereby enhancing the concrete's tensile strength [15]. Apart from that, macro plastic fiber can retain its shape even after the concrete cracked even it was shown that plain concrete cylinder immediately failed after the concrete cracked.

Concrete is brittle and has low tensile strength which is known to crack under low tensile strains. To increase its tensile strength capacity, concrete has traditionally been reinforced with the steel bar. The usage of FRC has been gradually expanding over the past four decades. In common construction application, FRC uses hydraulic cement, water, fine and coarse aggregates, and short, discrete, randomly distributed fibers. A wide variety of fibers have been developed with varying materials, lengths, and cross-sections. Fibers could increase the mechanical properties of FRC by intersecting the cracks and transferring load across them through the pullout process. Pullout resistance is fundamental to fiber performance in FRC. Steel fibers in FRC use mechanical anchorage such as hooked ends to resist pullout meanwhile synthetic fibers use materials that form a chemical bond with the matrix itself. Fly ash and silica fume which are categorized as mineral admixtures could enhance fiber-matrix bond strength by changing the matrix structure and are usually used in conjunction with fibers to create high-performance concrete.

Fiber-matrix interfacial bond is considered to be the most influential factor in improving the mechanical properties of fiber-reinforced concrete. Bond components can be classified as physical or chemical adhesion, friction, mechanical anchorage or fiber-to fiber interlock. In addition to varying materials or increasing fiber cross-sections to take advantage of adhesion, manufacturing innovations include fibrillation and embossed surfaces to improve the fiber-matrix bond.

In terms of impact and toughness strength, the addition of fibers significantly enhanced engineering properties of concrete. When compared to the plain concrete, the enhanced performance of FRC comes from its improved capacity to absorb energy during fracture. The plain concrete fails in a brittle manner at the occurrence of cracking stresses (first crack load). However, the FRC continues to carry stresses successfully beyond the concrete-matrix cracking. It helps maintain structural integrity and cohesiveness in the material which is denoted as ultimate load. Therefore, this paper presents an examination of concrete containing insulation wire waste as fiber materials in terms of compressive and tensile performances.

2 Experimental

In this study, laboratory testing such as workability, density, compression, and tensile strength tests was conducted to identify the physical and mechanical properties of the concrete containing synthetic wire waste. The raw materials that were being prepared for this study are Ordinary Portland Cement (OPC), coarse aggregates, fine aggregates, synthetic wire waste, and tap water. The maximum size of aggregate that was used is 14 mm to prevent the blockage caused by the aggregates. The aggregate was sieved by using 14 mm sieve pan. Besides, the coarse aggregates were dried under the sun. Figure 4 shows the coarse aggregate.

Natural white sand was used as fine aggregate. The maximum size of the fine aggregate is 5 mm which is sieved by 5 mm sieve pan with a specific gravity of 2.7 and were dried naturally under the sun. Figure 5 shows the fine aggregate.

Ordinary Portland Cement was used in the mixtures. OPC which corresponds to MS 522: Part 1 [28] has chemical composition that is provided in Table 5. The cement was stored in an airtight steel drum in the Material Laboratory at Universiti Tun Hussein Onn Malaysia (UTHM). Figure 6 shows the Ordinary Portland Cement.

In this study, synthetic wire waste was added to the mixture of the concrete. This synthetic wire waste was used as a fiber material. The synthetic wire waste of diameter 1.46 mm with the absence of copper were tested as a fiber material in concrete. The content of synthetic wire waste fibers is 0, 0.5, 1.0 and 1.5%. The length of wire waste was fixed at 40 mm with (L/df) of 34.25. Figure 7 shows the synthetic wire waste without copper.

Fig. 4 Coarse aggregate

Fig. 5 Fine aggregate

Table 5 Chemical composition of the Portland cement and fly ash

Chemical analysis (%)	Portland cement
Loss on ignition	4.40
Silica (SiO_2)	18.08
Iron oxide (Fe_2O_3)	2.43
Alumina (Al_2O_3)	4.72
Calcium Oxide (CaO)	61.94
Magnesium Oxide (MgO)	2.54
Sulphur Trioxide (SO_3)	2.74
Alkali-Sodium Oxide (Na_2O)	0.18
Alkali-Potassium Oxide (K_2O)	0.99

2.1 *Mixtures Proportion of Concrete*

The mixture was designed in accordance to Irwan et al. [29] and Khalid et al. [30]. The absolute volume method was used to measure the absolute volume of each standard concrete component that would occupy 1 m^3 of concrete. The water-cement ratio of the construction concrete mixture is 0.55. The proportion of fiber was measured by the volume of total mix concrete. Table 6 indicates the proportion of the mixture of materials prepared for this analysis.

Fig. 6 Ordinary Portland cement

Fig. 7 Synthetic wire waste

Table 6 Mix proportion for $1m^3$ of concrete

Cement (kg/m^3)	Coarse Aggregate (kg/m^3)	Fine Aggregate (kg/m^3)	Water (kg/m^3)	Synthetic wire waste proportions (%)	Synthetic wire waste (kg/m^3)
345	805	980	189.75	0	0
345	805	980	189.75	0.5	10.65
345	805	980	189.75	1.0	21.3
345	805	980	189.75	1.5	31.95

2.2 Experiment Test

The workability of the concrete mixture was determined through a slump test. This test was carried out before concrete will be poured into the moulds. For each combination, it was conducted for each mixture to ensure that the quantity of concrete materials used and constructed in conjunction with the ruins was calculated. In this experiment, steel cone with height of 300 mm, which has a top opening with 100 mm diameter and 200 mm diameter of the bottom cone was used as an instrument to conduct this test. The mixture placed in a cone in three layers and each layer was compacted by 25 times with a 16 mm diameter rod.

The density value is obtained from a hardened concrete test specimen by dividing the mass with volume. The scale of test specimens that were used is similar to the test specimens in the compressive strength test. The density test is significant as the result of this test will be used for the test of compressive strength and tensile strength. Density value was calculated by using Eq. 1.

$$\rho = \frac{m}{V} \tag{1}$$

where

ρ = Density.

m = Mass.

v = Volume.

The compressive strength test was used to decide whether the concrete mixture meets the criteria for the strength stated in the job specification. The compressive test of concrete was also used to assess the overall strength ability of the material to withstand the load put on it prior to failure. Maximum compressive strength of concrete cubes was obtained when the sample had failed. All of the concrete cubes for each specimen were carried out through these experiments and the average compressive strength value was recorded and determined by using Eq. 2.

$$\sigma = \frac{\mathrm{F}}{\mathrm{A}} \tag{2}$$

where

F = Force applied (N).

A = Area of the specimen (mm^2).

The tensile strength test is an important property of concrete as concrete structure is highly exposed to tensile cracking under different types of effects and applied loads. The force was applied along the vertical diameter through a pair of metal strip interposed between the cylinder and compression testing machine.

3 Result and Discussion

In this study, the objective was to identify the best ratio of the wire waste addition as fiber materials into the concrete mixture which has a great property and performance of the concrete fiber. An analysis was done according to the parameters used in controlling the effect of the percentage of wire waste in concrete mixture under curing conditions of 7 and 28 days, respectively. The water-cement ratio of 0.55 was constant for each mixture. In addition, it also expected that this research could reduce the production of waste wire and thus reducing the pollution caused by plastic materials.

3.1 *Slump Test*

The slump test has been carried out in order to evaluate the fresh state slump of the concrete to find the suitable workability value for concrete incorporating with wire waste as fiber materials. Fiber materials on workability have different effects on the slump and among the three types of fiber including polypropylene, steel and carbon [31]. The comparison slump between normal concrete and concrete containing wire waste as fiber materials is shown in Fig. 8.

For the slump value, the medium range is between 50 and 100 mm. Concrete containing wire waste shows a drop on slump value compared to control concrete. Concrete containing 0.5% wire waste has a slump value of 72 mm which is 4 mm lower than the control slump value. In the meantime, concrete containing 1.0% has a slump value of 69 mm which is 7 mm lower than the control slump. Concrete containing 1.5% wire waste has a slump value of 68 mm which is decreased by 8 mm from control slump.

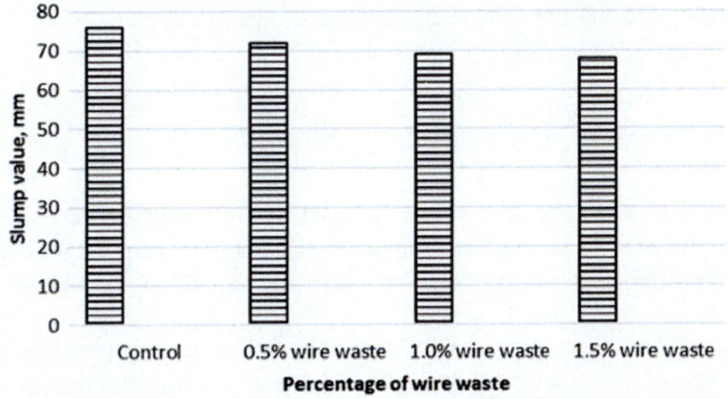

Fig. 8 Slump value of concrete containing wire waste as fiber materials

Table 7 Average density of all specimen and percentage different compared to control

Wire waste fiber content	7 days (cube)		28 days (cube)	
	Average density (kg/m^3)	Percentage different compared to control sample (%)	Average density (kg/m^3)	Percentage different compared to control sample (%)
0% (control)	2360	–	2212	–
0.5%	2369	0.38	2223	0.49
1.0%	2354	−0.25	2245	1.49
1.5%	2363	−0.38	2236	1.08

It shows that the increase in fiber materials can reduce the value of slump. This is because the addition of fibers will create a network structure in the concrete matrix which further restraining mixture from the segregation and flow of the fresh concrete [25].

3.2 Density Test

In order to achieve the objective of study, hardened concrete was used to obtain the average density for cube concrete and cylinder concrete at 7 and 28 days. The result of the average density for all specimens is shown in Table 7. There are no changes in density concrete containing wire waste fiber for 0.5, 1.0 and 1.5% compared to the control specimen. There was a small decline in the curing day density of 28 days compared to 7 days. This trend was confirmed by Libre et al. [32] who stated that PP fiber has insignificant impact to concrete density but the concrete density has mainly effected by steel fiber. It is clear that the pattern of density of concrete has not been influenced by the amount of fiber material as shown in Fig. 9.

3.3 Performance of Compressive Strength

The compressive strength test was calculated by following BS 6073–1 [33]. The result of compressive strength test on concrete cube, 100 mm height and 100 mm width with different percentage of wire waste at 7 and 28 days as shown in Table 8 and Fig. 10.

Based on Table 8, the percentage difference of compressive strength for 7 days of 0.5% wire waste was increased by 17.4% relative to control concrete. However, the compressive strength of concrete containing 1.0 and 1.5% wire waste were decreased by 0.4 and 1.3% when compared to control specimen. The percentage difference of compressive strength for 28 days shows the same trend (Fig. 10). The compressive

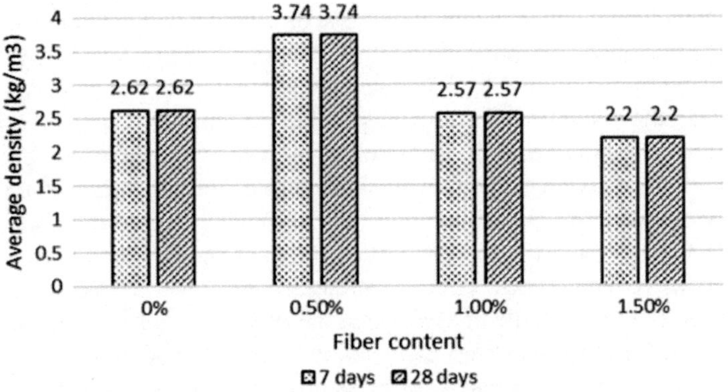

Fig. 9 Average density of concrete containing wire waste as fiber materials

Table 8 Average compressive strength of concrete cube with wire waste as fiber materials

Wire waste fiber content	7 days		28 days	
	Average compressive strength (MPa)	Percentage different compared to control sample (%)	Average compressive strength (MPa)	Percentage different compared to control sample (%)
0% (control)	22.4	–	33.2	–
0.5%	26.3	17.4	41.3	24.4
1.0%	22.3	−0.4	27	−18.7
1.5%	22.1	−1.3	24.7	−25.6

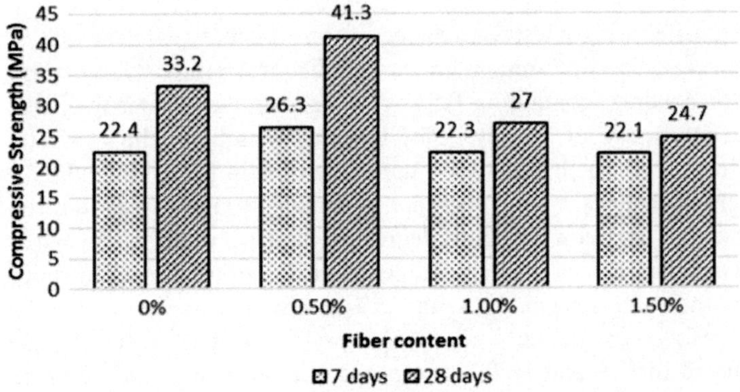

Fig. 10 Average compressive strength of concrete containing wire waste as fiber materials

strength of concrete cubes comprising 0.5% of wire waste was increased by 24.4% relative to the control specimen. Concrete cubes containing 1.0 and 1.5% wire waste were reduced by 18.7% and 25.6%, respectively.

Based on the result of compressive strength at 7 and 28 days, the utilization of wire waste as fiber materials in concrete has been shown to have a substantial effect on the strength of concrete. The result showed that concrete cubes containing 0.5% of wire waste have the highest strength (26.3 MPa) compared to control specimen (22.4 MPa). As a result, the compressive strength of concrete containing wire waste showed an improvement in the concrete strength when 0.5% of wire waste was added.

However, the intensity decreased when the percentage of wire waste rose by up to 1.0 and 1.5%. It was obvious that concrete specimen with 0.5% of wire waste obtained better results compared to the control specimen. The correct amount of wire waste has filled the hole in concrete, making the cube more solid which helped concrete strengthening. The decrease in strength began at 1.0% of the wire waste. The decrease happens due to higher doses of fiber that contribute to excessive distribution resulting in balling of fibers and air pockets which adversely affect the compressive strength of concrete [34].

The overall finding shows that the concrete that contains right amount of wire waste can increase the compressive strength of the concrete. The suitable percentage of wire waste is 0. The overall finding indicates that the concrete containing the correct amount of wire waste will increase the compressive strength of the concrete. The right proportion of wire waste is 0.5%.

3.4 *Performance of Tensile Strength*

Splitting tensile strength test was carried out on concrete cylinder to assess the tensile strength of concrete containing wire waste as fiber materials. It is clear that the tensile strength of concrete containing 0.5% of wire waste is greater than that of control specimen at the age of 28 days. The tensile strength for the concrete containing 0.5% wire waste is the highest with the value of 3.74 MPa. From the data analysis, it is observed that there is an increase of tensile strength with additional of 0.5% wire waste, but the value began to decrease when 1.0 and 1.5% of wire waste were added. Based on Table 9, the percentage difference of tensile strength at age 28 days relative to the control specimen with 0.5% of wire waste was with an improvement of 42.7%. However, the 1.0% wire waste and 1.5% wire waste of tensile strength were reduced for 1.9 and 16.0% compared to control specimen. Figure 11 depicts that the tensile strength was marginally improved by 0.5% relative to the control specimen. The addition of fiber appears to reduce the tensile strength of concrete compared to 0.5% fiber concrete. According to Song et al. [35], the use of 0.6 kg/m^3 nylon and polypropylene fiber in concrete increased from up to 17.1 and 9.7% of the tensile strength value, respectively, relative to control concrete. This observed trend of rising intensity was close to the trend of this study.

Table 9 Percentage comparison of tensile strength test result at 28 days

Wire waste fiber content	28 days	
	Average tensile strength (MPa)	Percentage different compared to control sample (%)
0% (control)	2.62	–
0.5%	3.74	42.7
1.0%	2.57	−1.9
1.5%	2.20	−16.0

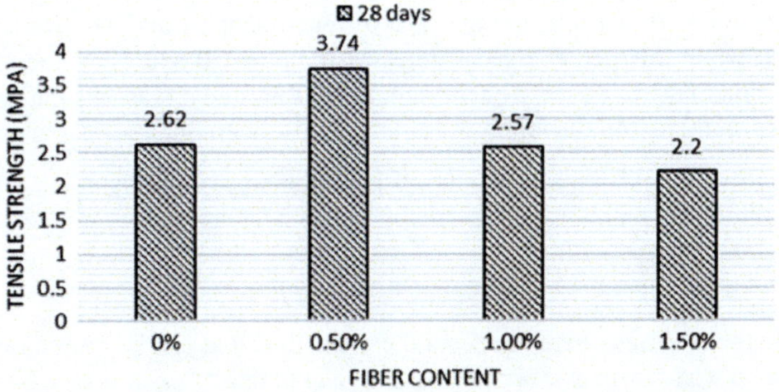

Fig. 11 Average tensile strength of concrete containing wire waste as fiber materials

3.5 *Relationship Between Compressive Strength and Tensile Strength of Concrete Containing Synthetic Wire Waste as Fiber Materials*

Previous studies have shown that the compressive-tensile strength relationship is similar to Eq. 3 [26, 36–38].

$$fct = A\ (fc)B \tag{3}$$

where

A and B: coefficients regression.

Based on Eq. 3, the relationship for compressive-tensile strength for wire waste as fiber concrete is derived from Eq. 4. The equation is derived from mathematical analysis. Table 10 demonstrates the contrast of compressive-tensile relationship between experiments and the use of predicted Eq. 4.

$$fct, wire\ waste = 0.0761\ (fc)^{1.0856} \tag{4}$$

Table 10 Compressive-tensile relationship between experiments and predicted at age 28 days

Wire waste fiber content	fc (Mpa)	fct (MPa) Exp	Predicted fct (MPa)
0% (control)	33.20	2.62	3.41
0.5%	41.70	3.74	4.37
1.0%	27.00	2.57	2.72
1.5%	24.70	2.20	2.47

The study of the relationship between the compressive strength and the tensile strength of concrete containing wire waste as a fiber material shows that the optimal percentage of wire waste is set at 0.5% for each of the tests. The study also indicates that when this percentage of wire waste up to 1.0 and 1.5% the strength for compression and tensile were decreased. From this study, it shows that the force of compression increased the strength for tensile also were increased and vice versa.

4 Conclusion

Based on the findings, there are several conclusions that can be drawn on the usage of synthetic wire waste as fiber materials in a concrete.

There is a major difference in workability between control specimen and specimen that contain wire waste. Wire waste added indicates the poor workability as the concrete was difficult to be mixed compared to the normal specimen. At the age of 7 days and 28 days, the wire waste which serves as additional material in concrete has been proven to enhance the compressive strength. It also increased the tensile strength of the concrete at age 28 days. For 0.5% of wire waste, the strength increased but when 1.0% and 1.5% of wire waste were added to the concrete, the strength has declined. This result demonstrated that it is very important to utilize the right percentage of wire waste that could be added to the concrete.

This study showed that concrete containing the right proportion of wire waste shows a good performance both for compressive and tensile strength. Wire waste used has increased the strength properties of concrete relative to the control specimen. Concrete with an extra 0.5% of wire waste provides the highest strength compared to other concrete samples. It was, therefore, the optimal percentage that was efficient to be used in concrete. Additional wire waste above 0.5% is found to be inefficient, as can be seen by the 1.0 and 1.5% decreases in wire waste content based on data analysis.

References

1. Stingley WM, ACI Committee 316R (1982) Recommendations for construction of concrete pavements and concrete bases. ACI committee report.
2. Bentur A, Mindess S (1990) Fibre reinforced cementitious composites. Elsevier Applied Science, London, New York
3. Kim SB, Yi NH, Kim HY, Kim JHJ, Song YC (2010) Material and structural performance evaluation of recycled PET fiber reinforced concrete. Cem Concr Compos 32(3):232–240. https://doi.org/10.1016/j.cemconcomp.2009.11.002
4. Khalid FS (2015) Engineering properties of ring-shaped polyethylene terephthalate (RPET) fiber self-compacting concrete. Dissertation: Universiti Tun Hussein Onn Malaysia
5. ACI Committee 315 (1988) ACI detailing manual. ACI committee report
6. Wafa FF (1990) Properties and applications of fiber reinforced concrete. JKAU Eng Sci 2:49–63
7. Amezugbe FA (2013) The performance of natural and synthetic fibers in low strength mortar: a pilot study of six selected fibers. Dissertation: University of Florida
8. Banthia N, Bindiganavile V, Jones J, Novak J (2012) Fiber-reinforced concrete in precast concrete application. Research leads to innovative products. PCI J 57(3):33–46.https://doi.org/10.15554/pcij.06012012.33.46
9. Zheng Z, Feldman D (1995) Synthetic fibre-reinforced concrete. Prog Polym Sci 20(2):185–210. https://doi.org/10.1016/0079-6700(94)00030-6
10. Wang Y, Wu H, Li V (2000) Concrete reinforcement with recycled fibers: a review. J Mater Civ Eng 12. https://doi.org/10.1061/(ASCE)0899-1561(2000)12:4(314)
11. Barluenga G, Hernández-Olivares F (2007) Cracking control of concretes modified with short AR-glass fibers at early age. Experimental results on standard concrete and SCC. Cem Concr Res 37(12), 1624–1638. https://doi.org/10.1016/j.cemconres.2007.08.019
12. Cunha V, Barros J, Sena-Cruz J (2010) Pullout behavior of steel fibers in self-compacting concrete. J Mater Civ Eng 22(1). https://doi.org/10.1061/(ASCE)MT.1943-5533.0000001
13. Afroughsabet V, Biolzi L, Ozbakkaloglu T (2016) High-performance fiber-reinforced concrete: a review. J Mater Sci 51(14):6517–6551. https://doi.org/10.1007/s10853-016-9917-4
14. Tejchman J, Kozicki J (2010) Experimental and theoretical investigations of steel-fibrous concrete. Springer, Berlin
15. Hsie M, Tu C, Song PS (2008) Mechanical properties of polypropylene hybrid fiber-reinforced concrete. Mater Sci Eng A 494(1):153–157. https://doi.org/10.1016/j.msea.2008.05.037
16. Campione G, Mendola L, Papia M (2006) Shear strength of steel fiber reinforced concrete beams with stirrups. Struct Eng Mech 24(1), 107–136.https://doi.org/10.12989/sem.2006.24.1.107
17. Fraternali F, Ciancia V, Chechile R, Rizzano G, Feo L, Incarnato L (2011) Experimental study of the thermo-mechanical properties of recycled PET fiber-reinforced concrete. Compos Struct 93(9):2368–2374. https://doi.org/10.1016/j.compstruct.2011.03.025
18. de Oliveira LAP, Castro-Gomes JP (2011) Physical and mechanical behaviour of recycled PET fibre reinforced mortar. Constr Build Mater 25(4):1712–1717. https://doi.org/10.1016/j.conbuildmat.2010.11.044
19. Rai A, Joshi YP (2014) Applications and properties of fibre reinforced concrete. J Eng Res Appl 4(1):123–131. www.ijera.com
20. Brandt AM (2008) Fibre reinforced cement-based (FRC) composites after over 40 years of development in building and civil engineering. Compos Struct 86(1):3–9. https://doi.org/10.1016/j.compstruct.2008.03.006
21. Hasan MJ, Afroz M, Mahmud H (2011) An experimental investigation on mechanical behavior of macro synthetic reinforced concrete. Int J Civ Env Eng 11
22. Ahmed SFU, Maalej M, Paramasivam P (2007) Flexural responses of hybrid steel–polyethylene fiber reinforced cement composites containing high volume fly ash. Constr Build Mater 21(5):1088–1097. https://doi.org/10.1016/j.conbuildmat.2006.01.002
23. Soroushian P, Lee CD (2003) Distribution and orientation of fibers in steel fiber reinforced concrete. ACI Mater J 87(5):433–439

24. Mazaheripour H, Ghanbarpour S, Mirmoradi SH, Hosseinpour I (2011) The effect of polypropylene fibers on the properties of fresh and hardened lightweight self-compacting concrete. Constr Build Mater 25(1):351–358. https://doi.org/10.1016/j.conbuildmat.2010.06.018
25. Yin S, Tuladhar R, Shi F, Combe M, Collister T, Sivakugan N (2015) Use of macro plastic fibres in concrete: a review. Constr Build Mater 93:180–188. https://doi.org/10.1016/j.conbuildmat.2015.05.105
26. Choi Y, Yuan RL (2005) Experimental relationship between splitting tensile strength and compressive strength of GFRC and PFRC. Cem Concr Res 35(8):1587–1591. https://doi.org/10.1016/j.cemconres.2004.09.010
27. Wang Y, Backer S, Li V (1987) An experimental study of synthetic fibre reinforced cementitious composites. J Mater Sci 22:4281–4291. https://doi.org/10.1007/BF01132019
28. Malaysian Standard (2007) Specification for portland cement—part 1 compositions, specifications and conformity criteria for common cements. Department of Standard Malaysia. MS 522–1
29. Irwan JM, Sheikh Khalid F, Othman N, Ibrahim MH, Asyraf R, Annas M (2013) Performance of concrete using light waste PET fibre. Adv Mater Res 795:352–355. https://doi.org/10.4028/www.scientific.net/AMR.795.352
30. Khalid FS, Irwan JM, Ibrahim MH, Othman N, Shahidan S (2018) Splitting tensile and pullout behavior of synthetic wastes as fiber-reinforced concrete. Constr Build Mater 171:54–64. https://doi.org/10.1016/j.conbuildmat.2018.03.122
31. Marzouk H, Chen Z (1995) Fracture energy and tension properties of high-strength concrete. J Mater Civ Eng 7(2):108. https://doi.org/10.1061/(ASCE)0899-1561
32. Libre NA, Shekarchi M, Mahoutian M, Soroushian P (2011) Mechanical properties of hybrid fiber reinforced lightweight aggregate concrete made with natural pumice. Constr Build Mater Constr Build Mater 25(5):2458–2464. https://doi.org/10.1016/j.conbuildmat.2010.11.058
33. British Standards Institution (1981) Precast concrete masonry units—part 1: specification for precast concrete masonry units. London. BS 6073–1
34. Ochi T, Okubo S, Fukui K (2007) Development of recycled PET fiber and its application as concrete-reinforcing fiber. Cem Concr Compos 29(6):448–455. https://doi.org/10.1016/j.cemconcomp.2007.02.002
35. Song P, Hwang S, Sheu B (2005) Strength properties of nylon- and polypropylene-fiber-reinforced concretes. Cem Concr Res Cem Concr Res 35(8):1546–1550. https://doi.org/10.1016/j.cemconres.2004.06.033
36. Xu BW, Shi HS (2009) Correlations among mechanical properties of steel fiber reinforced concrete. Constr Build Mater 23(12):3468–3474. https://doi.org/10.1016/j.conbuildmat.2009.08.017
37. Arioglu N, Girgin ZC, Arioglu E (2006) Evaluation of ratio between splitting tensile strength and compressive strength for concretes up to 120 MPa and its application in strength criterion. ACI Mater J 103(1):18–24
38. Raphael JM (1984) Tensile strength of concrete. ACI J Proc 81:158–165

Environment Acceptability as Building Material from Organic Contaminants Leaching Behavior in Sludge

Nor Amani Filzah Mohd Kamil, Noor Faiza Roslee, and Sadeq Abdullah Abdo Al-Khadher

Abstract Sludge is one of the major hazardous waste generated from industrial activities. Cost-effective remediation method and disposal has become the main issues worldwide. Many types of contaminants are produced from sludge, in which organic contaminants and heavy metals are the most toxic and concerned. Leaching heavy metals in solidification and stabilization (S/S) methods were mostly discussed in the literature and very limited to organic contaminants. This chapter focuses on building material's environmental acceptability based on organic contaminants' leaching behaviour in treated sludge by S/S method. The leaching during the use and the end of life phase is one of the issues in reusing the treated sludge. Studies on the potential sludge measurement as a raw material in the building material production are discussed further in this chapter. Specifically, types of organic contaminants in sludge, solidification and stabilization of organic contaminants from sludge and environment acceptability of sludge as building material are the subtopics in this chapter.

Keywords Geopolymersluge · Organic contaminants · Sustainable building materials · Artificial lightweight aggregate · Volcanic ash · Palletizing · Aggregate impact value (AIV) · Aggregate crushing value (ACV)

N. A. F. M. Kamil (✉) · N. F. Roslee · S. A. A. Al-Khadher
Faculty of Civil Engineering and Built Environment, Universiti Tun Hussein Onn Malaysia, 86400 Parit Raja, Batu Pahat, Johor, Malaysia
e-mail: noramani@uthm.edu.my

N. F. Roslee
e-mail: faizaroslee@gmail.com

S. A. A. Al-Khadher
e-mail: sadeq@uthm.edu.my

A. Abdul Kadir et al. (eds.), *Sustainable Waste Utilization in Bricks, Concrete, and Cementitious Materials*, Lecture Notes in Civil Engineering 129,
https://doi.org/10.1007/978-981-33-4918-6_13

1 Introduction

Due to increasing treatment and landfill cost, waste is reused to find an alternative solution for waste disposal. Besides, the regulatory frame encourages the recovery of waste as much as possible under safe health and environmental conditions. The community's social pressure and environmental commitment in a continuous improvement cycle are significant factors for reusing waste. On the other side of raw material resource, the limited and increasing cost of building material resources, especially quarry closure without opening new capacities, is also the alternative solution of waste reuse.

Solidification and stabilization (S/S) are one method that is commonly practicing in hazardous waste treatment. Many reasons and factors make the S/S method more effective and attractive for researchers to use in the waste reuse. It provides high strength, low permeability and relatively high durability of products that offer the possibility of reusing solidified waste for construction purposes, such as building materials. Extensive research findings showed that utilizing treated sludge as a building material has significant environmental benefits, including a reduction in the energy consumption and greenhouse gases and a decrease in the amounts of residues to be disposed of in landfills, incineration or dumping into water bodies [1].

The regular practice in S/S method measures the physical and mechanical properties to identify the potential as a building material. The physical and mechanical properties including compressive strength, density, water absorption, porosity, permeability, durability and other properties. Besides, one of the crucial factors in waste reuse is the safety of the S/S product or referred to as environment acceptability. The environment acceptability depends on environmental risk assessment, which relates to composition standards (technical) and limits leaching values. The limit values for leaching are based on health risks and mostly referred to the national quality standards of domestic water quality. Meanwhile, in other countries, environment acceptability is based on ecological risk, by which living organisms interact with each other and their physical surrounding [2].

There are many types of hazardous waste produced by industrial activities. This chapter focuses on sludge from sewage treatment, industrial wastewater treatment and petroleum industries. Sludge is a general term used to describe the residual deposits found at the bottom of tanks and other storage vessels. Sludge is commonly produced through the course of water treatment process in various industrial activities and the sludge characteristics are different from their sources. According to the Resources Conservation and Recovery Act (RCRA), sludge is one of the major hazardous waste produced during industrial activities and needs to be treated for safe disposal into a secured landfill.

Economic growth, which improves the living standard, will increase the population. The arisen also creates many industrial complexes that generate vast quantities of domestic sewage and industrial sludge. The sludge is produced during the treatment process of chemical coagulation, flocculation and liquid/solid separation from the wastewater treatment plant. Industrial sludge gave a higher concentration of

Table 1 Types of common sludge that produce from sewage or industrial wastewater treatment plant

Types of sludge	Appearance	Odors	Decomposition
Sludge from primary settling tanks	Gray-coloured slimy slurry	Extremely offensive	Can be readily decomposition under suitable conditions of operation
Chemical-precipitation sludge	Dark in colour. Its surface may be red if it contains much iron	Less offensive	Slower rate decomposition of sludge from primary settling tanks
Activated sludge	Brown flocculants appearance	If in good condition, it is inoffensive	Digest readily alone or mixed with fresh sewage solids
Trickling filter sludge	Brownish	In fresh—inoffensive	Decomposition more slowly than other undigested sludges

contaminants compared to sludge generated from the sewage treatment plant. Thus, the leaching of contaminants from industrial sludge is highly concern compared to sewage sludge. However, sewage sludge contributed to a much higher quantity of sludge than industrial sludge. Table 1 shows the types of sludge that are commonly produced from sewage or industrial wastewater treatment plant.

In Malaysia, sewage sludge disposal activities are controlled by Indah Water Konsortium, the biggest operator on sewage treatment. Gravity or mechanical thickener is used to thicken the sludge and stabilize in the anaerobic digester. The stabilized sludge is dewatered in a screw press and finally disposed into the landfills. This practice is the current implementation of sewage sludge treatment and disposal in Malaysia and other countries as well [3]. For industrial sludge, only licensed contractors appointed by the Department of Environment can treat the sludge in their treatment facility. Besides, the Department of Environment will also be responsible for appointing contractors as sludge transporter to carry the untreated sludge to Kualiti Alam waste management center, which is the biggest operator of hazardous waste treatment in Malaysia.

The hazardous waste generated may comprise wastewater, drilling fluid, and bottom tank sludge during petroleum production. Wang et al. [4] defined petroleum sludge as a general term used to elucidate the residual deposited at the base of the tanks and other types of storage vessels. The capacity of a refinery will dictate the volume of petroleum sludge generated. The bigger the refining capacity, the larger the petroleum sludge production has been generated [4]. Murshid stated that due to the high viscosity of the sludge, it is difficult to be hydrated. The sludge primarily comprises 55.13% of water, 9.25% of sediments, 1.9% of asphaltenes, 10.5% of wax and 23.1% of light hydrocarbons [5]. Most of the oil sludge elements have been recognized as cytotoxic, mutagenic, and potentially carcinogenic.

Sludge can be categorized into two groups, namely organic sludge and inorganic sludge. Organic sludge is produced by treatment facilities that remediate organic contamination in wastewater, such as sewage treatment plants, paper-making industries, food production and pulp mills. Inorganic sludge contains high volumes of sand and metal components produced from wastewater treatment facilities in water purification plants, civil engineering sites and metal plating plants. The organic sludge will be discussed further in the following sections, focusing on types of organic contaminants.

2 Organic Contaminants in Sludge

The origin of the organic compound is from plant and animal tissues made up of carbon and hydrogen. It is in low concentration that exists naturally in soil, water, air and sometimes in food. In high concentration, this organic compound may result in harmful effects on human and environmental health and eventually becomes the toxic and xenobiotic organic contaminants. The majority of organic contaminants are highly mutagenic and carcinogenic. The organic contaminants of sludge are influenced by the origin and treatment of wastewater resulting from the sludge.

Sewage sludge contains contaminants inclusive of heavy metals, pathogens and organic contaminants. Based on 79 papers published on organic contaminants in sewage sludge, the maximum reported concentration of 86% exceeded the EPA risk-based soil screening limits [6]. Based on this review's analysis, certain chemical contaminants such as Polycyclic Aromatic Hydrocarbons (PAHs) and Polychlorinated biphenyl (PCB) were high collected from the analysis report.

In sewage sludge, the concentration of 16 PAHs was detected at 20.67 ± 4.14 mg/kg (dry weight), with 47% is the carcinogenic PAHs [7]. The study also has detected higher concentration of high molecular weight (81%) compared to low molecular weight (19%). The carcinogenic PAHs are mostly derived from high molecular weight. The result from Khillare et al. [7] was different compared to Chen et al. [8] of which 16 PAHs were detected to be very low, 0.5342–1.0666 mg/kg (dry weight). The study also has found 4-ring PAHs, fluoranthene, pyrene, and chrysene were predominant (47.6%) in sewage sludge. Less than 1 mg/kg (dry weight) of organochlorine pesticides (OCP) and PCB were detected in sewage sludge [9]. Dieldrin, chlordane and DDE were measured at 0.77, 0.29 and 0.27 mg/kg, respectively from the study. Meanwhile, OCP and PCB were reported as low concentrations below the standard limit [9].

PAHs were observed at very low concentrations in sludge from the automotive industry treatment plant, varying from 3.03 to 5.93 μg/g (dry weight) [10]. The paper industry produces high concentration of Polychlorinated dibenzo-p-dioxins and polychlorinated dibenzo-p-furans (PCDD/Fs), 0.189 to 1092 ng/TEQ kg and Co-PCBs, 0.265 to 26.6 ng/TEQ kg in sludge [11]. These high concentrations are due to the bleaching process of papermaking.

Table 2 PAHs concentration in petroleum sludge based on the different ring of PAHs [5]

PAHs compounds	Bojes and Pope [5]	*SSL (industrial) (mg/kg)
Napthalene	254.5	200
Acenaphthylene	24.2	2300
Acenaphthene	34.8	37,164
Fluorene	80.6	22,000
Phenanthrene	19.9	18,582
Anthracene	24.8	170,000
Fluoronthene	15.5	22,000
Pyrene	19.9	17,000
Benzo(a)anthracene	9.4	2.1
Chrysene	24.5	2100
Benzo(b)fluoranthene	8.2	210
Benzo(k)fluoranthene	7.3	210
Benzo(a)pyrene	8.4	2.1
Dibenzo(ah)anthracene	4.1	2.1
Benzo(ghi)perylene	5.9	18,582
Indeno(1,2,3-cd)pyrene	5.8	21

*Contaminated Land Management and Control Guidelines No. 1: Malaysian recommended site screening levels for contaminated land. Department of Environment, Ministry of Natural Resources and Environment, Malaysia. (2009)

Many reports have claimed that oil sludge contains volatile organic carbons and semi-volatile organic carbons that are genotoxic [12, 13]. PAHs are typically found in petroleum sludge, including naphthalene, phenanthrene, anthracene, pyrene, benzo(a)-anthracene, chrysene, and benzo(a)pyrene [14]. Table 2 shows PAHs' concentration at different rings measured in crude oil sludge at the bottom tank compared to Site Screening Levels (SSL) for Contaminated Land. From the study, the carcinogenic PAHs such as Benzo(a)anthracene, Benzo(a)pyrene and Dibenzo(ah)anthracene were detected above the SSL limit. Based on the SSL limit, the carcinogenic PAHs show a low concentration of standard limit, for instance, 2.1 mg/kg compared to non-carcinogenic PAHs. These low values were set based on health risk assessment to reduce the adverse impact on human health and the environment.

PAHs and PCB can be categorized as non-polar (insoluble) compounds, which are favorable to be retained in solid [15]. Thus, these compounds are highly likely to be encapsulated in S/S product, especially for the high molecular weight that has higher potential to be retained in S/S compared to low molecular weight. The S/S method may not be suitable to be implemented for polar (soluble) compound because a high concentration of this compound will be detected in leachate during the leaching test. The treatment of sludge by using S/S method will be discussed further in the following section.

3 Solidification and Stabilization of Organic Contaminants in Sludge

The solidification/Stabilization (S/S) method involves mixing a binding reagent into the contaminated media or waste. Even though the terms of solidification and stabilization seem to be identical, each process portrays distinct impacts that are created by the binding reagents to disable the hazardous constituents. According to Malviya and Chaudhary [16], S/S method refers to a group of cleanup methods that hinder or slow down the release of harmful chemicals from the sludge. Inorganic sludge is easily disposed of using this method. Nevertheless, this method is less compatible with organic sludge that has a higher concentration of toxic organic contaminants [17].

Solidification refers to the physical changes of a waste. The growth of the compressive strength, reduction of permeability, and encapsulation of the hazardous constituents are commonly among the desired changes. The chemical changes of the hazardous constituents in sludge can be referred to as stabilization. The required changes leading to the leaching of the contaminant's contents involve converting the constituents into a less soluble, mobile, or toxic form. Thus, as stated by Wang et al. [4], the S/S method includes mixing a binding reagent into the contaminated media or waste with a proper mix design to ensure the encapsulation of contaminants without leaching the hazardous compound.

This method keeps the contaminants from "leaching" above the permissible levels to the surrounding environment [16]. Leaching occurs when water from the rain or other sources dissolve the contaminants and flow them down into the groundwater or over lands towards nearby lakes and streams.

Mix design has to be carried out on each subject of sludge due to the vast diversity of sludge constituents and media. Organic binding reagents such as asphalt, thermoplastic, and urea–formaldehyde have also been tested. As reported by Caldwell et al. [18], organic binding reagents are rarely utilized on a commercial scale because of their extortionate price compared to inorganic binders.

The main issue of the S/S method is that the toxic or the contaminant is still in the non-destructive matrices, which makes the safety of the solidified samples in great danger if the binder does not absorb or react with the toxicity to formulate a stable component. Furthermore, previous research has also claimed that the portland cement usage as the only binder system in S/S method is not effective for the immobilization of several common contaminants [19].

Towards this concern, an effective alternative material is needed to provide a sustainable treatment method. In recent years, tremendous researches have been carried out using ash from different sources in the S/S method as an environmentally-friendly solution to replace cement, as ash is also known as a good binder material with high pozzolanic characteristics [19, 20].

S/S method was considered less compatible with organic waste due to organic contaminants, which may inhibit binder hydration, which is generally not chemically bound in the binding hydration products [21]. The cement setting in the S/S method is hindered by organic contaminants due to the formation of a protective layer around the cement grain, which retards the formation of calcium hydroxide [22]. Also, the chemical uptake of organic contaminants into the hydration product is potentially very low.

By referring to the result of physical and mechanical properties, high organic contaminants in cement paste may increase the porosity and lower the mechanical strength [21]. In the S/S method preparation, sludge is usually dried to reduce the disruption of organic matter in the cement setting. After the drying process, certain organic contaminants, such as phenol and methanol, will form amorphous structures resulting in the detrimental effects on the strength of the S/S product [20]. Immobilization of organic contaminants relies primarily on the physical entrapment in the binder matrix and sorption onto the surface of the binder hydration products, with the possibility of undesired high concentrations of emissions when exposed to environmental leachants.

Table 3 summarizes of S/S studies at different media, namely, contaminated soil, wastewater sludge, and petroleum sludge. The number of papers is based on searching in January 2020, which may increase in future studies. S/S method has been widely implemented on heavy metals. Cement is a common binder used in treated heavy metals containing a high composition of Calcium Oxide. This chemical composition can limit the chemical interaction between the contaminants and the surface of the stabilized waste [20]. This chemical interaction only occurs to contaminants that interact in ions during S/S method. Organic contaminants which are the compounds that are primarily composed of carbon and hydrogen, do not produce these ions. Thus, using only cement as a binder may not efficiently treat the organic contaminants in S/S method.

By comparison on media, high numbers of S/S studies were conducted in contaminated soil. The S/S method formerly was an established method for remediation of contaminated sites. This method has the ability to be used in treating other waste, such as sludge, due to the advantages of the physical and mechanical properties of S/S products. Sludge produced from sewage or industrial treatment plant is usually gray or dark in a colored and slimy slurry of settleable solids. From the observation during testing of the S/S method, sludge from the petroleum industry with high oil content is having difficulty preparing the S/S matrices [20].

Table 3 Studies on the solidification and stabilization at different media

Type of treatment	Media	Number of Papers	Scope
Solidification/stabilization method	Contaminated soil	≈500	Inorganic contaminants. Mostly heavy metals
	Wastewater sludge	≈150	
	Contaminated soil	≈200	Organic contaminants
	Wastewater sludge	≈60	
	Petroleum sludge	4	Inorganic contaminants Mostly heavy metals
		3	Organic contaminants

Karamalidis and Voudrias [22] used Portland and blended cement as binder in encapsulating total petroleum hydrocarbons, alkanes and, 16 PAHs from oil refinery sludge. This study has found that increased cement content will lead to a higher concentration of the leachates' organic contaminants. The cement addition is contributed to the destabilization of the waste. By using cement only as a binder can be categorized as direct immobilization of organic contaminants. Limited studies have shown the success of the encapsulation of organic contaminants.

Due to the unsuccessful of conventional cement-based solidification, additive of cement mixes such as clays, natural materials (fly ash, zeolite, organic polymers) and activated carbon are added as pretreatment in the S/S method. Traditional binders, namely, cement, fly ash and lime, have successfully encapsulated PCB from municipal sewage sludge [23]. Wet-dry and freeze–thaw tests on the S/S product were conducted to measure the effects of climatic stresses in the study. Caldwell et al [18] have introduced activated carbon with cement to encapsulate organic contaminants. The trace amount of organic contaminants with less than 1 mg/g was spiked into waste that high in inorganic concentration and this trace amount was successfully encapsulated. Activated carbon also shows high potential in encapsulating of phenol. Low amounts of activated carbon, which is less than 2%, showed in rapid adsorption of phenol in the S/S method [24]. Phenol was adsorbed into activated carbon much faster than the hydration of cement and complexation reactions of calcium with phenol.

However, due to the high cost of activated carbon production, Leonard and Stegemann [21] have introduced high carbon power plant fly ash to absorb the organic contaminants. This binder is one of the pozzolanic materials with a high composition of unburn carbon. This binder improved the immobilization of the hydrocarbons (0 to 0.6 mg/L) but not for chloride [21]. The study claimed that the materials with high carbon content could be used to encapsulate organic contaminants.

Several studies on the S/S method using oxidizing agents in the encapsulation of organic contaminants have also been done. Johnson et al. [1] had investigated the encapsulation of petroleum sludge in building blocks by using oxidized vegetable oil. As a result, they concluded that sludge could be encapsulated in building blocks, and the products met the minimum specifications for construction blocks. PAHs concentrations detected in leachate are below the USEPA standard limit. To the best of our knowledge, there is no implementation of reducing agents in the S/S method for encapsulating organic contaminants such as PAHs and PCBs in sludge. Reducing agents such as ferum sulfate were used to stabilize chromium in sewage sludge [25].

In conclusion, using S/S method in treating organic contaminants can be categorized into direct immobilization, immobilization after adsorption by using additive of cement mix, and finally, immobilization using oxidizing agents. Treatment using this method produces the S/S products that are commonly disposed into a secure landfill. Besides safe disposal into landfills, the S/S product is potentially being used as a raw material in the building industry. The following section will be discussed further on this potential.

4 Environment Acceptability of Sludge as Building Material

Many criteria are needed to consider the environmental acceptability of waste as a building material. Most studies on S/S method refer to the potential of treated sludge as building material by conducting testing on physical properties, mechanical properties, and leaching tests. Among the tests, leaching testing always refers to measuring the environmental hazard in S/S products.

The leaching behavior of organic contaminants from treated industrial sludge is still under investigation. Johnson et al. [1] measured the potential of petroleum sludge in building block by conducting testing for unconfined compressive strength, water absorption, the initial rate of absorption, efflorescence, and durability test. These tests measure the potential as building block by referring to the physical and mechanical properties of S/S products. By referring to the environmental aspect, the study conducted on leachability on PAHs. The study showed the minimum concentration, which is less than 0.03 ppb, by considering the USEPA standard of environmentally hazardous material. A chemical analysis conducted on chemical oxygen demand testing was also measured, and the result was within the standard specification in accordance with USEPA.

Acid neutralization capacity was used as a leaching test that has indirectly shown the stability of the chemical environment in solidified S/S samples [21]. In relating to the environmental aspect, Dutch intervention value and USEPA drinking water standards were referred to in the study. The S/S product was suggested to control low strength material, landfill liner and daily landfill cover. Vacenovska et al. [26] have verified the possibility of S/S products made of neutralization sludge use in the building block. In the study, the results of partial leachability tests, unconfined compressive strength and the eco-toxicological test showed that solidification as a technological and reclamation material on depots and setting pits is possible.

In another study, Garg et al. [27] have concluded that automobile sludge up to 35% could be utilized as a construction material for making building components. In the study, organic contaminants included in organic matter decreased to 4.3% by drying at 100 ± 5 °C for 48 h. The purpose of drying was to reduce the interruption of organic matter. The leaching test result was compared to standard limits for discharge of pollutants at surface water.

Based on findings from several studies discussed in previous paragraphs, the leaching test refers to the standard limit in which the treated sludge is safely disposed into a secured landfill or compared to standard limit related to standard for water, either standard for surface or groundwater. These standards are uncertainty and can be applied as indicators of safety in the reuse of treated sludge as a building material. Leaching tests are employed to simulate field conditions. It is a method that simulates a sample when it is disposed of in a landfill and is employed to determine whether the contaminated soil has met the guideline [20]. No regulation required any particular leaching test and the type of contaminants to be tested. Commonly, types of contaminants were selected based on sludge characteristics, location or source of sludge, type of treatment method that will be implemented and subsequent use of the sludge. Table 4 shows the list of leaching test modified from USEPA [28].

To the best of our knowledge, the leaching test of the life phase of treated sludge that needs to be done for many years has never been conducted. Thus, the hazardous of the treated sludge is still unknown for a very long duration. Moreover, this limited information gives rise to public concern about the sanitary safety of the S/S product. Residents of the building constructed with treated sludge might think the toxic and hazardous waste are placed directly into their building.

The S/S product was mostly evaluated as building or other construction material based on the leaching test results. Besides, it is highly recommended to measure the release of gasses that adversely impact human health and the environment, such as volatile organic carbon. As mentioned in the previous section, organic contaminants such as PAHs and PCB are not removed in the S/S method. These contaminants with characteristics of the semi-volatile compound may be released during the life phase of treated sludge. A high number of extensive studies are required to develop the

Table 4 Types of leaching test based on USEPA [28]

Leaching test	Description
TCLP	1. Crushed sample 2. Waste is co-disposed with municipal solid wastes in a landfill
SPLP	1. Crushed sample 2. Intended to simulate the effect of acid rain on land-disposed waste
Semi-dynamic leaching test (ANS 16.1)	1. Intact sample—cylindrical shape 2. Procedure specifies the leachant replacement intervals and the cumulative fraction of as substance released during the test is measured
EP Tox (extraction procedure toxicity test)	1. Crushed sample
Cal WET (California waste extraction test)	1. Crushed sample 2. Small size particle, 2 mm
Multiple extraction procedure (MEP)	1. Crushed sample 2. Synthetic acid rain
Modified waste extraction procedure (MWEP)	1. Crushed or monolith sample
Equilibrium leach test	1. Crushed sample 2. Very small size particle, 150 μm 3. Long duration for time extraction, 7 days
Acid neutralization capacity	1. Crushed sample 2. Very small size particle, 150 μm
Sequential extraction tests	1. Crushed sample 2. High number of extraction, 15
Sequent chemical extraction	1. Crushed sample 2. Very small size particle, 150 μm

regulation that can support the S/S product's safety, which can be related to indoor air pollution.

S/S method in treating organic contaminants depends on physical entrapment. A preferable process is required to remove or transform the toxic organic contaminant into less hazardous hydrocarbons for better effectiveness in long-term duration. A combination of the S/S method and the biological process can be implemented to remove or degrade the organic contaminants. This combination process was implemented by Boutammine et al. [29] in the degradation of total petroleum hydrocarbons (TPH) in petroleum drilling cutting and showed 60% degradation of TPH. Based on the result of the study, this combination process also can be implemented for sludge. Thus, the S/S product's hazardous concern, especially on toxic organic contaminants, can be reduced.

In transforming the organic contaminants, high temperatures in the range of 1000–1600 °C vitrification is commonly implemented in preparing clay brick for building construction [30]. This method is oxidizing the organic contaminants into the less hazardous inorganic compound. This method is successful in reducing the hazard of contaminants and eliminate pathogenic. However, this method is expensive in building materials processing and releasing sulfur oxides, nitrogen oxides and dioxins that increase the public unacceptability. Apart from using high temperature, gas purification can be applied to oxidize the toxic organic contaminants [31].

Organic contaminants are carbon sources for living organisms, such as bacteria and fungi. The high concentration of organic contaminants may increase the number of these living organisms, and it becomes highly concerned if the pathogen is growth in the S/S product. Thus, it is not recommended to reuse sludge directly after treated. The treated sludge can be reused after a certain duration, which means the death phase of the bacterial growth. Therefore, besides the leaching test, bacterial growth testing in the S/S method is strongly advised to be conducted.

Permeability, also known as hydraulic conductivity, is one of the important tests related to S/S product's ability to retain contaminants in the matrices. This test is to show how easy water can pass through the S/S product. It is also highly recommended to conduct a test on gas permeability, which means the property of solids that allows the movement of gas through the solids in the presence of a pressure differential.

Besides measuring the S/S product's hazardous, other factors such as management aspects, economy, and regulation enforcement should be considered. In waste reuse as a building material, the product requires site-specific environmental permit, in which the process of permit application process is time-consuming. This lengthy application due to full-scale recycling activities is needed for the permit application [32].

From an economic standpoint, sludge needs to be treated before reuse as the building material. This process can increase transportation costs if the location of treated sludge in a remote area is far from the construction site. By comparing studies on S/S method recently with a decade earlier, high potential binders from other waste have been introduced to replace cement in recent research. Thus, the cost of making the S/S product can be lessened. Lack of data on environmental properties and health risk assessment in sludge reflects the deficiency in environmental regulation.

The standard limit needs to warrant review to revise the regulatory requirements for the concentration of contaminants level in treated sludge for beneficial reuse as building materials. The S/S method's treated sludge can be used as building material without limitation on the environmental characteristics other than the physical and mechanical requirements that each country must meet.

5 Conclusion

Previously, S/S method was implemented in the remediation of contaminated sites. Due to the increase of physical and mechanical properties such as compressive strength and permeability, this method has the potential to treat and reuse waste such as sludge from industrial activities. One of the potentials of treated sludge is utilized as a raw material in building construction. S/S method does not remove the contaminants inside their matrices, and thus environment acceptability is very high concern. Hazardous of treated sludge was measured based on the leaching test. The main problem is the standards limit that was referred to is unreliable to ensure the safety to be used as building materials. Besides measuring leachant in liquid form, it is highly recommended to measure gas emissions due to some organic contaminants' characteristics being volatile or semi-volatile. Other factors, such as management, economy, and regulation enforcement, should be considered to ensure the effective reuse of the S/S products. The process of reusing treated sludge as the building material is still a long way to go, but it is likely to become a reality.

References

1. Johnson O, Napiah M, Kamaruddin I (2015) Encapsulation of petroleum sludge in building blocks. Constr Build Mater 78:281–288. https://doi.org/10.1016/j.conbuildmat.2014.12.122
2. Sorvari J (2008) Developing environmental legislation to promote recycling of industrial by-products—an endless story? Waste Manag 28(3):489–501. https://doi.org/10.1016/j.wasman.2006.12.021
3. Hanum F, Yuan L, Kamahara H, Aziz H, Atsuta Y, Yamada T, Daimon H (2019) Treatment of sewage sludge using anaerobic digestion in Malaysia: Current state and challenges. Front Energy Res 7:1–7. https://doi.org/10.3389/fenrg.2019.00019
4. Wang X, Wang Q, Wang S, Li F, Guo G (2012) Effect of biostimulation on community level physiological profiles of microorganisms in field-scale biopiles composed of aged oil sludge. Bioresour Technol 111:308–315. https://doi.org/10.1016/j.biortech.2012.01.158
5. Bojes HK, Pope PG (2007) Characterization of EPA's 16 priority pollutant polycyclic aromatic hydrocarbons (PAHs) in tank bottom solids and associated contaminated soils at oil exploration and production sites in Texas. Regul Toxicol Pharmacol 47(3):288–295. https://doi.org/10.1016/j.yrtph.2006.11.007
6. Harrison E, Oakes S, Hysell M, Hay A (2006) Organic chemicals in sewage sludges. Sci Total Environ 367(2–3):481–497. https://doi.org/10.1016/j.scitotenv.2006.04.002

7. Khillare PS, Sattawan V, Jyethi D (2018) Profile of polycyclic aromatic hydrocarbons (PAHs) in digested sewage sludge. Environ Technol 41(7):842–851. https://doi.org/10.1080/09593330.2018.1512654
8. Chen CF, Ju YR, Lim YC, Hsieh SL, Tsai ML, Sun PP, Katiyar R, Chen CW, Dong CD (2019) Determination of polycyclic aromatic hydrocarbons in sludge from water and wastewater treatment plants by GC-MS. Int J Environ Res Public Health 16(14). https://doi.org/10.3390/ijerph16142604
9. Clarke B, Porter N, Marriott P, Blackbeard J (2010) Investigating the levels and trends of organochlorine pesticides and polychlorinated biphenyl in sewage sludge. Environ Int 36(4):323–329. https://doi.org/10.1016/j.envint.2010.01.004
10. Karaca G, Tasdemir Y (2013) Removal of polycyclic aromatic hydrocarbons (PAHs) from industrial sludges in the ambient air conditions: automotive industry. J Environ Sci Heal Part A, Toxic/hazardous Subst Environ Eng 48(8):855–861. https://doi.org/10.1080/10934529.2013.761481
11. Ju JH, Lee I, Sim WJ, Eun H, Oh JE (2008) Analysis and evaluation of chlorinated persistent organic compounds and PAHs in sludge in Korea. Chemosphere 74(3):441–447. https://doi.org/10.1016/j.chemosphere.2008.09.059
12. Mishra S, Jyot J, Kuhad RC, Lal B (2001) In situ bioremediation potential of an oily sludge-degrading bacterial consortium. Curr Microbiol 43(5):328–335. https://doi.org/10.1007/s002840010311
13. Prakash V, Saxena S, Sharma A, Singh S, Singh S (2015) Treatment of oil sludge contamination by composting. J Bioremediation Biodegrad 6(3). https://doi.org/10.4172/2155-6199.1000284
14. Castaldi FJ (2003) Tank-based bioremediation of petroleum waste sludges. Environ Prog 22(1):25–36. https://doi.org/10.1002/ep.670220114
15. Mohd Kamil NAF, Hamzah N, Abdul Talib S, Hussain N (2016) Improving mathematical model in biodegradation of PAHs contaminated soil using gram-positive bacteria. Soil Sediment Contam an Int J 25(4):443–458. https://doi.org/10.1080/15320383.2016.1168356
16. Malviya R, Chaudhary R (2006) Factors affecting hazardous waste solidification/stabilization: a review. J Hazard Mater 137(1):267–276. https://doi.org/10.1016/j.jhazmat.2006.01.065
17. Hunce SY, Akgul DT, Demir G, Mertoglu B (2012) Solidification/stabilization of landfill leachate concentrate using different aggregate materials. Waste Manag 32(7):1394–1400. https://doi.org/10.1016/j.wasman.2012.03.010
18. Caldwell RJ, Cote PL, Chao CC (1990) Investigation of solidification for the immobilization of trace organic contaminants. Hazard Waste Hazard Mater 7(3):273–282. https://doi.org/10.1089/hwm.1990.7.273
19. Anastasiadou K, Christopoulos K, Mousios E, Gidarakos E (2012) Solidification/stabilization of fly and bottom ash from medical waste incineration facility. J Hazard Mater 165–170. https://doi.org/10.1016/j.jhazmat.2011.05.027
20. Murshid NB (2019) Leaching behaviors of heavy metals on petroleum sludge waste treatment by using solidification/stabilization (S/S) method with palm oil fuel ash as partial cement replacement. Dissertation: Universiti Tun Hussein Onn Malaysia
21. Leonard SA, Stegemann JA (2010) Stabilization/solidification of petroleum drill cuttings: Leaching studies. J Hazard Mater 174(1–3):484–491. https://doi.org/10.1016/j.jhazmat.2009.09.078
22. Karamalidis AK, Voudrias EA (2007) Cement-based stabilization/solidification of oil refinery sludge: leaching behavior of alkanes and PAHs. J Hazard Mater 18(1–2):122–135. https://doi.org/10.1016/j.jhazmat.2007.02.032
23. Aydilek A, Edil T (2008) Solidification/stabilization of PCB-contaminated wastewater treatment sludges. In: GeoCongress, pp 724–731
24. Hebatpuria V, Arafat H, Rho H, Bishop P, Pinto N, Buchanan R (1999) Immobilization of phenol in cement-based solidified/stabilized hazardous wastes using regenerated activated carbon: leaching studies. J Hazard Mater 70:117–138. https://doi.org/10.1016/S0304-3894(99)00128-4
25. Lin CF, Rou W, Lo KS (1992) Treatment strategy for Cr(VI)-bearing wastes. Water Sci Technol 26(9–11):2301–2304. https://doi.org/10.2166/wst.1992.0721

26. Vacenovska B, Cerny V, Drochytka R, Urbanek B, Vodickova E, Pavlikova J, Valko V (2013) Verification of the possibility of solidification product made of neutralization sludge use in the building industry. Procedia Eng 57:1192–1197. https://doi.org/10.1016/j.proeng.2013.04.150
27. Garg M, Singh LP, Maiti S, Pundir A (2014) Characterization of automobile effluent treatment plant sludge: its utilization in construction materials. Constr Build Mater 73:603–609. https://doi.org/10.1016/j.conbuildmat.2014.09.115
28. USEPA (1989) Risk assessment: guidance for superfund. In: Human health evaluation manual (Part A), interim final, vol 1. Office of Emergency and Remedial Response, U.S. Environmental Protection Agency, Washington DC
29. Boutammine H, Salem Z, Khodja M (2020) Petroleum drill cuttings treatment using stabilization/solidification and biological process combination. Soil Sediment Contam Int J 29(4):1–15. https://doi.org/10.1080/15320383.2020.1722982
30. Areias IOR, Vieira CMF, Colorado HA, Delaqua GCG, Monteiro SN, Azevedo ARG (2020) Could city sewage sludge be directly used into clay bricks for building construction? A comprehensive case study from Brazil. J Build Eng 31. https://doi.org/10.1016/j.jobe.2020.101374
31. Zhang X, Hu Z, Ngo HH, Zhang J, Guo W, Liang S, Xie H (2018) Simultaneous improvement of waste gas purification and nitrogen removal using a novel aerated vertical flow constructed wetland. Water Res 130:79–87. https://doi.org/10.1016/j.watres.2017.11.061
32. Chateau L (2007) Environmental acceptability of beneficial use of waste as construction material—State of knowledge, current practices and future developments in Europe and in France. J Hazard Mater 139(3):556–562. https://doi.org/10.1016/j.jhazmat.2006.02.064

The Effect of Metakaolin in Production of Low Thermal Conductivity Cement Sand Brick

Rafikullah Deraman, Rabiatul Syahindah Mustaffa, Mohd Hanif Ismail, Muhammad Fikri Hasmori, and Sasitharan Nagapan

Abstract The building sector is arguably one of the most resource-intensive industries in the world, accounting for some 38% of global energy consumption. It contributes significantly to total energy and greenhouse gas emissions. It depends highly on equal energy and carbon-intensive supporting industries such as the building material industry of which cement is one of the major components. Cement production significantly impacts anthropogenic carbon dioxide (CO_2) emission responsible for global warming, climate change and airborne dust pollution. In order to reduce energy consumption, CO_2 emission and environmental pollution, a worthwhile venture would be the production of sustainable green building materials. This study investigates the characteristics and potential of metakaolin as a cement replacement in the production of low thermal cement sand bricks. Different proportions of metakaolin from 0, 5, 10 and 15% were utilized and experimented upon to determine their suitability as a green component while meeting all standard brick requirements. Towards this end, laboratory work involving sieve analysis, compressive strength test, specific gravity test, water absorption, scanning electron microscopy and thermal conductivity test was conducted. Based on the results obtained, the optimum proportion of metakaolin replacement in cement sand bricks making is 15% by virtue of its having the lowest thermal conductivity value of 0.580 W/mK and meeting the

R. Deraman (✉) · R. S. Mustaffa · M. H. Ismail · M. F. Hasmori · S. Nagapan
Jamilus Research Centre, Faculty of Civil Engineering and Built Environment, Universiti Tun Hussein Onn Malaysia, 86400 Parit Raja, Batu Pahat, Johor, Malaysia
e-mail: rafikullah@uthm.edu.my

R. S. Mustaffa
e-mail: syhindhmustffa@gmail.com

M. H. Ismail
e-mail: mohdhanif@uthm.edu.my

M. F. Hasmori
e-mail: mfikri@uthm.edu.my

S. Nagapan
e-mail: sasitharan@uthm.edu.my

A. Abdul Kadir et al. (eds.), *Sustainable Waste Utilization in Bricks, Concrete, and Cementitious Materials*, Lecture Notes in Civil Engineering 129,
https://doi.org/10.1007/978-981-33-4918-6_14

standard requirements of 15.6 MPa > 3.5 MPa (ASTM C129-17 for compressive strength), 17.95% < 18% (ASTM C1403-15 for water absorption), and 0.580 W/mK below 1.23 W/mK (ASTM C177-97 for thermal conductivity). It is thus concluded that metakaolin can be a potential cement replacement material in bricks. Its use enhances the thermal insulation property of building envelopes, reduces cement consumption, contributes to the solving of environmental problems and promotes a better environmentally friendly building sector.

Keywords Metakaolin · Cement sand brick · Replacement material · Low thermal conductivity · Insulation properties

1 Introduction

Rapid urbanization brings into the high demand of main raw building materials such as cement, sand, and aggregates causing a shortage of building materials due to increasing housing demand [1]. As demand for housing grows exponentially, it has a direct impact on the construction industry as well as the production of conventional materials such as cement. The production of conventional building materials produces pollution and high energy consumption, which are continuously increasing. Globally, the construction industry is known as a very complex resource-driven activity and the building sector is one of the most resource-intensive industries. The use of energy in the building sector accounts a significant part of the world's total energy and greenhouse gas emissions. Whereby the construction activities occupy 38% of the total energy consumption of the world [2]. The construction activity is supported by other sources such as building materials likes cement. Generally, cement production is an energy and carbon-intensive process. Cement production is a noteworthy contributor to global anthropogenic CO_2 emissions that consequently responsible for the global warming, climate changes and airborne pollution in the form of dust in the environment [3].

Cement itself has adverse effects on greenhouse gases. Cement is the main ingredient in cement-based building materials which is considered as environmentally unfriendly, where the production of cement consumes a tremendous amount of energy and emits significant amounts of carbon dioxide (CO_2) and other greenhouse gases. Cement plants are the significant sources of sulphur dioxide, nitrogen oxide and carbon monoxide, which are associated with the following health [4]. It is reported that the cement industry contributes almost 6% of the world global warming and anthropogenically conservatory gases emissions [5]. With existing issues, the use of cement in cement product based building materials needs to be minimized and replaced with similar supplementary materials to promote a more environmentally friendly in the building sector and consequently reduces the on-site construction waste [6].

Another study reported that energy consumption in Malaysia for indoor cooling is as high as 70% [7]. This high energy consumption is intended to combat the heat of buildings exposed to the tropical sun radiation, which affects the occupants' comfort. Malaysia is located right on the equator, thus receives higher solar and terrestrial radiation than many other countries. Buildings are overheated during the day due to solar heat gain through the building envelope and radiant solar penetration through the windows [8, 9]. The absorption of heat by the building envelope through conduction transfers heat into the interior spaces and increases the heating effects inside the buildings, causing great discomfort. Reducing heat caused by solar radiation in order to achieve a comfortable temperature level requires a lot of electrical energy.

Studies have shown that decreasing the thermal conductivity of a material is the dominant factor in reducing heat that could be transferred to or from the building. Few researchers mentioned that the utilization of low thermal conductivity building materials is important to decrease heat gain through the building envelope, especially for hot climate countries like Malaysia [10–12]. Specific to bricks as a material of the building envelope, it has been postulated that if they are improved by proper processing, they can have significantly lower thermal conductivity and transmittance properties, which means a lower heat gain through the walls of buildings [13]. As a part of the building envelope, walls contribute a significant proportion of heat loss and gain [14] Therefore, selecting the proper thermal properties of a building envelope plays a major role in determining energy consumption patterns and comfort conditions in enclosed spaces [15, 16].

Thus, to combat the aforementioned problems, the construction industry seeks to investigate sustainable solutions to achieve better thermal insulation, which by using low thermal conductivity building materials. Cement sand brick is one of the building materials that have been used as an alternative for a low-cost house in Malaysia. Nowadays, many researchers have been researching green and sustainable materials to replace conventional building materials. One of the researches is by replacing cement partially with other material that has the same effect and chemical composition of cement such as pozzolan. In this study, metakaolin was used to partially replace the cement in the production on of cement sand bricks. Another type of pozzolanic material such as fly ash [17], silica fume [18] and others has shown that the utilization of pozzolanic material proved to improve the properties of cement sand brick in terms of mechanical properties and performance. The brick specimens were evaluated in terms of the compressive strength, water absorption and thermal properties to determine their feasibility for industrial application.

2 Overview of Metakaolin

Metakaolin (MK) is one type of pozzolanic materials that has the potential to be a partial cement replacement material. Other examples of the type of pozzolanic material are rice husk ash (RHA), calcined shale, calcined clay, fly ash, silica fume and ground granulated blast furnace slag (GGBFS). MK is, however, not an industrial by-product like the other pozzolanic materials. The raw material input in the manufacture of metakaolin ($Al_2Si_2O_7$) is kaolin clay. Kaolin is a fine, white, clay mineral that

Table 1 Chemical and physical properties of metakaolin [33]

Properties	MK	Cement
Silicon dioxide (SiO_2)	53.03	21.28
Aluminium oxide (Al_2O_3)	35.63	5.60
Iron (III) oxide (Fe_2O_3)	1.81	3.36
Calcium oxide (CaO)	0.04	64.64
Magnesium oxide (MgO)	0.57	2.06
Sulphur trioxide (SO_3)	–	2.14
Titanium dioxide (TiO_2)	0.93	–
Manganese(II) oxide (MnO)	0.02	–
Sodium oxide (Na_2O)	0.04	–
Potassium oxide (K_2O)	1.88	–
Phosphorus pentoxide (P_2O_5)	0.06	–
Loss on ignition (%)	1.99	0.64
Specific gravity (g/cm^3)	2.60	3.02
Colour	White	Grey
Specific surface area	0.015	1123.3

has been traditionally used in the manufacture of porcelain [19]. Kaolin is one of the most abundant natural minerals and when it is heat-treating, it will become MK [20]. MK is not a by-product because it is made under carefully controlled conditions which produced by thermal processing of purified kaolinitic clay. It means that MK is obtained by heating kaolinitic clay, one the richest natural clay minerals at a temperature between 650 and 900 °C [21]. According to Poon et al. [22], MK typically contains 50–55% silicon dioxide (SiO_2) and 40–45% aluminium oxide ($A1_2O_3$). Table 1 shows the example of chemical and physical properties of MK.

MK has a size of average particle less than 2 μm, which is about ten times finer than cement, and a specific gravity of about 2.6. MK enhances microstructure mainly as a result from micro aggregate filling and the pozzolanic effect, fine particles bridge the gaps between cement particles, hydration products, matrix and aggregates, thus making the pore structure becomes denser [23]. Based on Table 1, it shows that silica and alumina content in the MK are the major oxide compounds and also greater compared to the ordinary Portland cement. Other oxide compounds found in the compositions are TiO_2, Fe_2O_3, MnO, MgO, CaO, Na_2O, K_2O and P_2O_5, which are presented in small percentages. Pozzolanic materials have a high content of silica or alumina which will give a better effect on hydration of cement and improve the strength as well as the microstructure of brick body. MK has higher pozzolanic properties compared with ordinary Portland cement. In the presence of moisture, this material will react with calcium hydroxide in the process of hardening to form secondary Calcium Silica Hydrate (C–S–H) or known as gel which will improve strength and performance of brick. Fine pozzolanic particles also could act as filler in the mixes.

Several studies have been conducted regarding the use of MK material to replace the cement content. The studies showed that with suitable replacement proportion, between 10 and 15% of MK could be considered a suitable replacement for the economic efficiency, fresh and hardened properties of the MK sample [24]. However, some studies have recommended the optimum replacement of cement using MK is at 10% replacement levels. Higher levels of MK replacement could lead to an increased in water requirements. This has due to MK absorbed more water and the particle size is smaller as compared to OPC. Report by Bai et al. [25] revealed that the addition of MK increased the water demand in cement paste.

Many researchers have done comprehensive studies related to pozzolanic materials and industrial wastes for improving the properties of cement-sand bricks. For example, studies using fly-ash [17], silica fume [18], rice hush ash (RHA) [26], plastic [27], pulverized cockle shell [28], oil palm mesocarp fiber [29] oil palm fruit bunch [30] and sugarcane baggase [31]. However, the utilization of MK in cement-sand mixes as a replacement or admixture materials is still limited in the publication amongst the research community. The utilization of MK material is environmentally friendly since it helps in reducing the CO_2 emission to the environment by the minimization of the Ordinary Portland cement consumption.

3 Material and Method

The raw materials used for production of bricks are Ordinary Portland Cement (OPC), river sand and MK. The OPC was manufactured from Tasek Cement and fulfilling the requirement in the MS 522-1 [32]. River sand with at least 98% passing the 4.75 mm sieve was used. The water used for mixing and curing was the portable water treated by the local utility company with the pH value in the range of 6.5–7, which complies with the requirement of MS 28 [33]. Meanwhile, MK was obtained from the Kaolin (Malaysia) Sdn. Bhd.

The mixture ratio used for the specimen preparation was 1:5, representing a cement to sand and a water-cement ratio of 0.55. The mixed ratio was used based on the trial mix result, which was performed earlier. The percentage of MK used ranged from 0 (as control specimen), 5, 10, and 15% as cement replacement. The design mix was prepared to produce a specified mix with different proportions of MK. A summary of design mix of raw materials used is shown in Table 2. The design mix for calculating raw materials is based on specific gravity, density of water and volume of the mould.

The specimens were mixed, cast and tested according to the procedure stated by BS EN 12390-2 [34] and BS EN 206 [35]. The mixing process was performed using mechanical mixing to acquire a uniform mixture. The procedure follows that the sand was first placed in the mixer, then the cement was poured into the mixer, followed by the MK, and finally the dry mixture was gradually added with water. The mixing continues for a few minutes to achieve the homogeneous mix. Then, the mixed were placed into a steel mould. The mix was compacted with 25 strokes of a rod compacter in three layers. The specimens were de-moulded 1 day after casting in

Table 2 The percentage of MK used in the production of cement sand bricks per specimen

Percentage of replacement (%)	MK (kg)	Cement (kg)	Sand (kg)	Water (kg)	Total weight per sample (kg)
0	0.0	0.450	2.250	0.248	2.7
5	0.118	0.430	2.152	0.237	2.7
10	0.248	0.409	2.043	0.225	2.7
15	0.395	0.384	1.921	0.211	2.7

room temperature before the curing process. The specimens have been cured for 7 and 28 days by using wet blanket curing. The manufactured specimens then underwent a series of tests, including physical, mechanical, and thermal properties.

4 Results and Discussions

The laboratory works were conducted in the Heavy Structures Laboratory, Universiti Tun Hussein Onn Malaysia. The results were presented in a table and graphs to assure a fuller explanation. The discussion of the result was discussed in the following sub-sections.

4.1 Particle Size Distribution

The test was performed to determine the suitable size of sand in the production of cement sand brick and carried out using 500 g for each sample. The sand was equivalent in size, which was passing 4.75 mm. The particle size analysis of the sand was illustrated in Fig. 1. The result showed that the sand specimen is within the range limits of the fine sized-sand particles; fine upper limit (F.U.L.), and fine lower limit (F.L.L.). The graph shows that more than 90% of the sand passed through sieve 4.75 mm, which indicates that the sand as a fine aggregate follows the BS 882 [36] and the assessment of the particle size distribution revealed that the sand specimen is well graded.

4.2 Specific Gravity

The test was conducted on MK, cement, and sand specimen to determine its characteristic, carried out using Le Chatelier's flask. The procedure done complied with ASTM C128 standard test method for relative density (specific gravity) and absorption of fine aggregate [37]. The result obtained for MK, cement, and sand were 2.60,

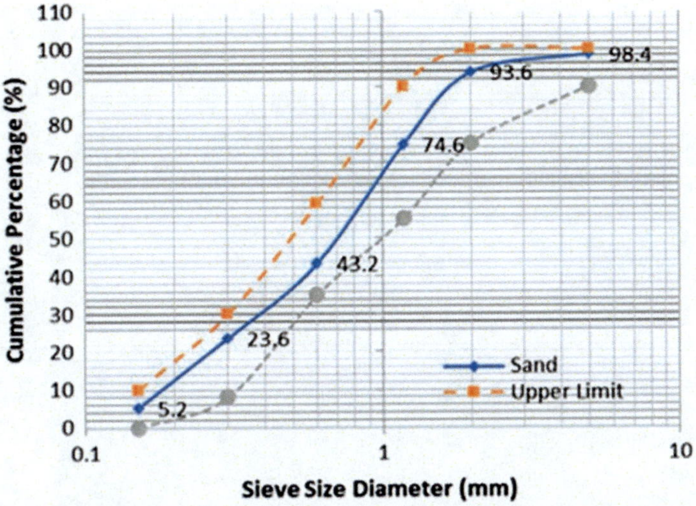

Fig. 1 Particle size distribution (dry sieves analysis) of the sand specimen

3.12, and 2.65, respectively, and aligned with the result obtained by Güneyisi et al. [38]. All the specimens have a specific gravity of more than 2.00, indicates as an inorganic material.

4.3 Scanning Electron Microscope (SEM) Image

A scanning electron microscope (SEM) is a type of electron microscope that produces images of particles of the specimen. The test was conducted to investigate the microstructure of MK, which focuses on porosity on each brick specimen. Figures 2, 3, 4 and 5 show the SEM images of all brick specimens. These images have been magnified 1000 and 2000 to bring out the clear picture of the microstructure of voids and solid phases in cement bricks. Based on the findings, it is clear that the pore sizes are not uniform. However, these figures do signify that there is a dominant pore size, which found in the 15% MK brick specimen. The dominant pore size tends to increase as the brick specimen density reduces due to the higher quantity of MK used.

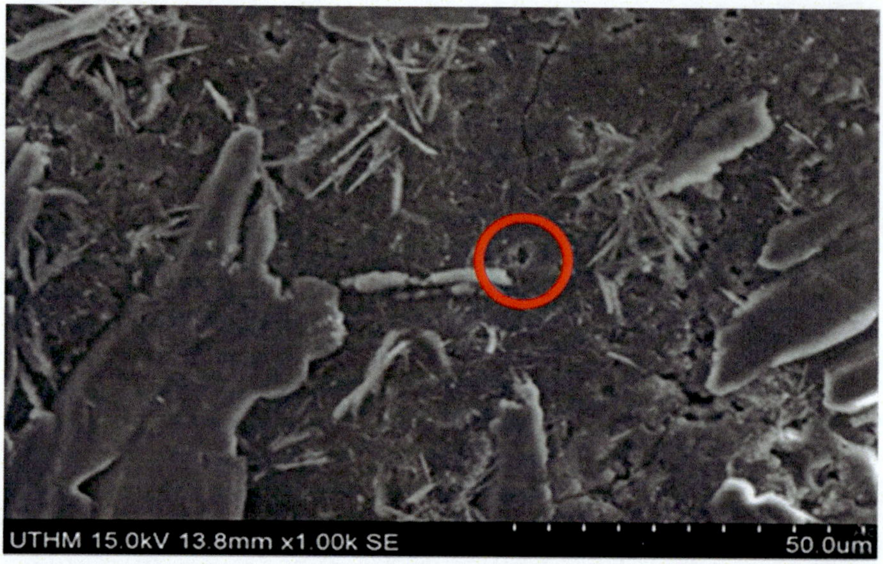

Fig. 2 SEM result for control bricks with 1000 K magnification

Fig. 3 SEM result for 5% metakaolin cement sand brick with 2000 K magnification

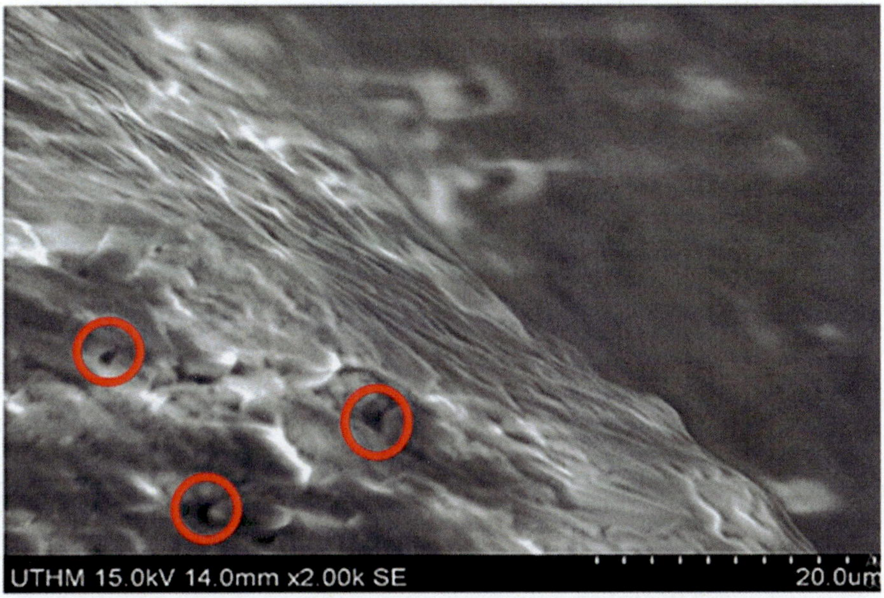

Fig. 4 SEM result for 10% metakaolin cement sand brick with 2000 K magnification

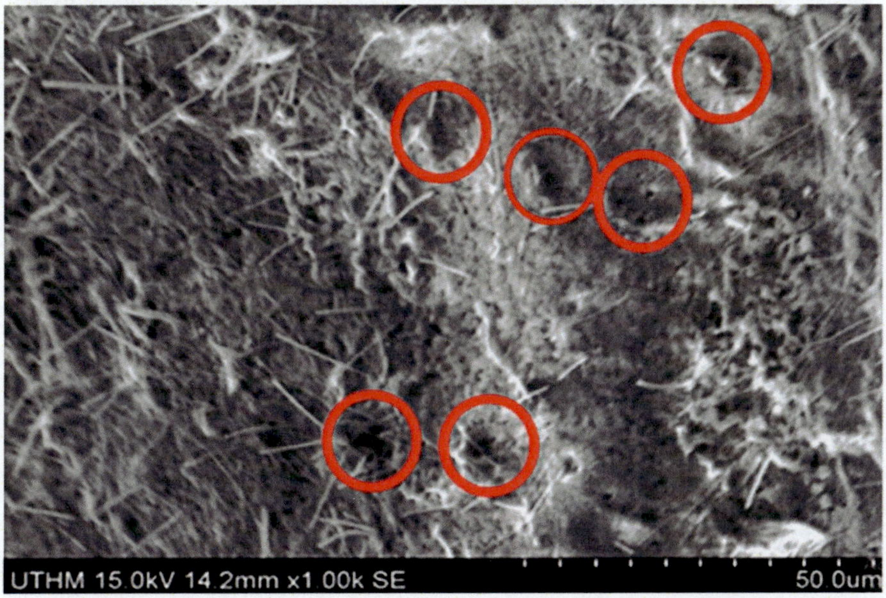

Fig. 5 SEM result for 15% metakaolin cement sand brick with 1000 K magnification

4.4 Compressive Strength Test

The test was performed to obtain the strength value of bricks in order to meet the standard requirement in construction. It was based on the failure stress of the specimens under uniaxial compression. The test was conducted according to BS EN 12390-2 [34]. Based on the ASTM C129-17 [39], the minimum permissible value of compressive strength non-load bearing wall shall be not less than 3.5 N/mm^2. The compressive strength development at 7 and 28 days was given in Fig. 6. Results demonstrated that the compressive strength of cement sand brick with MK increased with the age of curing and decreased as the percentage of MK content increased.

From the figure, the highest compressive strength is 18.9 MPa with 5% of MK replacement at 28 days. Moreover, for 28 days, MK with percentage of 15% replacement has a lower compressive strength compared to others, which is 15.6 MPa. It can be seen that the increase in MK proportion, makes the bricks more brittle and porous. The available silica and aluminous from the 5.0% MK reacted with a portion of C–H released from the hydration process. Thus, the C–S–H released from the pozzolanic reaction was improving the strength [38, 40, 41] and it is shown by the result as 14.9 Mpa (0% MK) to 16.3 Mpa (5% MK) for 7 days and 17.3 Mpa (0% MK) to 18.9 Mpa (5% MK) for 28 days. Furthermore, the incorporation of MK enhanced not only at the early-age strength but also provided higher strength at later ages [38]. In addition, the fine MK particles contributed to the strength development by acting as a microfiller and enhancing the cement paste pore structure. In other word, the effect of the pozzolanic reaction produces more cement gel (i.e. C–S–H and C–A–H) reducing the pore size, blocks the capillary and produces denser brick thus making it stronger and more durable [42].

When the replacement level further increased to 10% and 15%, the increase in the strength values lessened gradually. The decrease in the strength by increasing

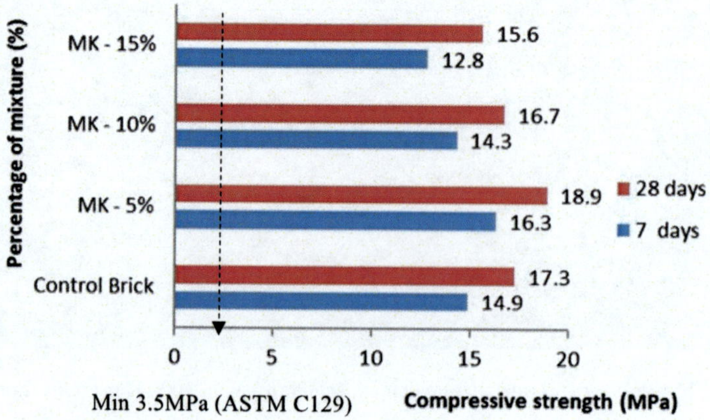

Fig. 6 Compressive strength of specimens with different percentage of MK for 7 and 28 days curing

MK replacement level more than 5% is due to the reduction in the cement amount, and the amount of silica available in the hydrated blended cement matrix is too high. Thus, the amount of the produced C–H is most likely insufficient to react with all the available silica from the additional of MK proportion. As a result, some amount of silica was left without any chemical reaction. However, based on the ASTM C129 [39] requirement, all specimens with different percentages of MK have met and achieved value beyond the standard requirement of compressive strength.

4.5 Water Absorption Test

Water absorption test was performed to measure the percentage of water absorbed by the specimen. It was executed according to the BS 1881-122 [43], method for determination of water absorption at the curing age of 28 days. The standard requirement for the percentage of water absorption should not be more than 18% by weight [44]. Figure 7 shows the percentage of water absorption per different percentage of MK. Based on the figure, the highest water absorption for specimen was 17.95%. Meanwhile, the lowest water absorption for the specimen was 12.59% at 5% MK.

From the figure, it shows that by replacing 5% MK to the control specimen, the water absorption reduced from 14.49 to 12.59%. This was due to the filler effect of ultrafine MK as well as its pozzolanic reaction [38]. Wild et al. [40] reported that the presence of MK, the filling effect is immediate which accelerated by the OPC hydration, and high level of pozzolanic reaction. Moreover, MK significantly influences the pore structure in pastes, besides produces substantial pore refinement leading to significant modifications of water transport properties [24, 42]. Hence, it makes the specimen become dense and less porous.

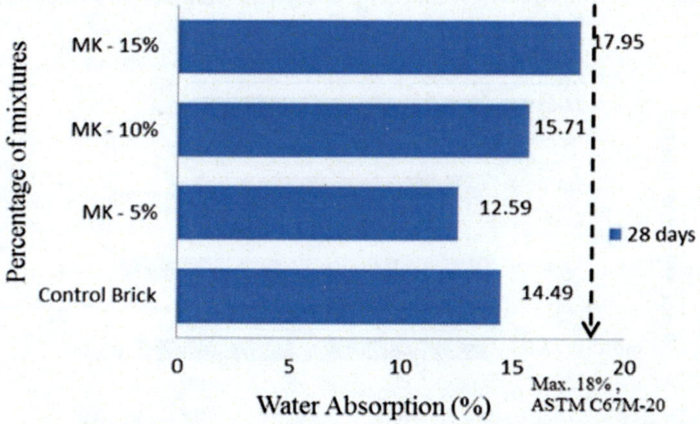

Fig. 7 Water Absorption of specimens with different percentage of MK for 28 days of curing

However, when the replacement level of MK was more than 5%, the water absorption rose significantly. The addition of pozzolan decreased the CH formed by the pozzolanic reaction of the brick specimen [42]. The reduction of the cement amount caused the insufficient of tri-calcium silicate (C_3S) and di-calcium silicate (C_2S) compounds in the mixture which was hydrated to produce calcium hydroxide $Ca(OH)_2$. Therefore, the binding matrix (C–S–H) gel as the main glue that binds between cement paste and sand particles is not strong enough and make the brick specimen less harden and porous [45]. Moreover, MK material contains high silica, which does not cause any binding agent capacity as cement is. MK also absorbs water from the mixture due to its fine pore structure. Therefore, more replacement level of MK required more water in the mixes, makes the harden brick specimen has low workability, porous, brittle and increased water absorption.

4.6 Thermal Conductivity Test

This test was performed to measure the ability of the materials to conduct heat transfer, which presented by its *K*-value. This test was based on the ASTM C177 [46] standard test method for steady-state heat flux measurement and thermal transmission properties by means of the guarded-hot-plate apparatus.

Lyons [47] has mentioned that the conductivity values for cement sand brick are in the range between 1.4 and 1.8 W/mK. Figure 8 shows the thermal conductivity specimens for each different percentage of MK. By referring to the figure, it shows that the higher value of thermal conductivity (*K*-value) within all mix percentage was 5% MK with a reading of 0.744 W/mK. Whereas, the minimum value obtained for thermal conductivity was 0.580 W/mK at 15% of MK. Generally, the brick with a lower value of thermal conductivity is an excellent thermal insulator which provides better insulation for buildings. The drop in thermal conductivity due to the addition

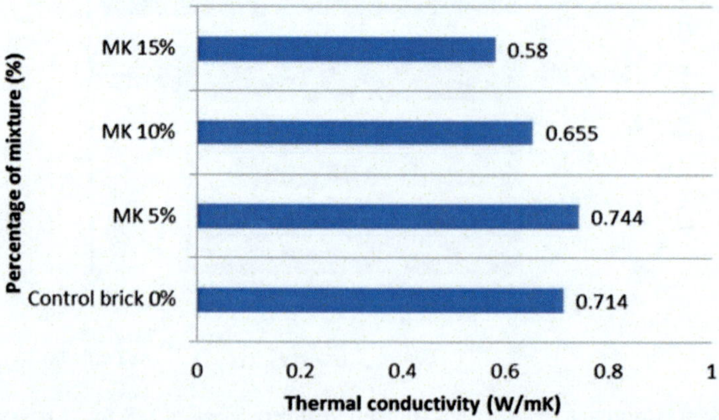

Fig. 8 Thermal conductivity of specimens with different percentage of MK for 28 days of curing

of pozzolanic admixture in mix compositions is significant. The reduction in thermal conductivity in the presence of MK is mainly due to the formation of hydrated products (C–S–H and C–A–H) and porosity which has already been discussed in details in Sects. 4.4 and 4.5. Siddique and Kahn [48] reported that the increased in calcium hydroxide (CH) content due to the Portland cement hydration, while the decrease, later on, is likely due to the dominance of the pozzolanic reaction.

The thermal conductivity is also dependable to the density and compressive strength where thermal conductivity value decreases as the density and compressive strength decrease due to the effect of higher percentages of MK inclusions. Moreover, the drop in thermal conductivity is due to the high air void content in the brick specimen which also has been discussed in Sect. 4.5 [49, 50]. The high porosity of the specimens due to high void content will allow more volume of water entered into the voids, thus increase the percentage of water absorption. The higher the porosity, the lower the thermal conductivity, therefore improving the thermal insulation of the fabrics. Based on the thermal conductivity value obtained, it shows that all the brick specimens have *K*-value below than suggested by Lyons [47].

5 Conclusion

The question of whether cement bricks can be more thermally accommodating has always come to the fore in response to the rising general need for energy conservation and the pressing competition in the building industry for the use of sustainable building materials. Therefore, this study is an attempt to address these issues more thoroughly. The MK was used to partially replace the cement in the production of low thermal cement sand bricks. The decrease in thermal conductivity value is vital to create insulation capability and thermal comfort [51]. Selecting a low thermal conductivity of brick wall is considerable importance in creating low energy consumption buildings that consume less energy to maintain comfortable conditions in enclosed spaces. Thermal insulation improves energy-efficiency of buildings by retarding heat flow through building envelopes and thus reducing the indoor cooling load. It concludes that the mixing of MK improves the insulation property of bricks as compared to the control brick.

Overall, the selection of 15% of MK is the most effective in respect of compressive strength (15.6 MPa) that meets the minimum requirement of the ASTM C129 [39], which is not less than 3.5 MPa. The water absorption was at 17.95%, which is not more than 18% by the ASTM C67M [44]. Whereas, thermal conductivity was 0.58 W/mK is lower than the control brick (0.714 W/mK). The thermal conductivity value was below the standard range of normal cement brick, which ranges from 1.4 to 1.8 W/MK [47].

Due to the aforementioned reasons, it is recommended by the authors that the 15% MK could be used to enhance the thermal viability of the cement sand brick. While some of its properties decrease by incorporating the MK, the overall results suggest that the final brick output complies with the standard requirement and is suitable for non-load bearing purposes. MK is potentially becoming a partial replacement material in the production of low thermal cement sand brick. In addition, the replacement of MK more than 15% is not appropriate due to the high cost of MK and less proportion of cement in mix material will lower the binder between the mixing materials. This study showed that the use of MK cement bricks could reduce heat at the outside of building envelopes and inside of the building. By having a low thermal brick wall in the building, the energy usage could be lessened and thus, increase the energy efficiency of the building. This study may provide a solution to reduce the building's energy demands and curb environmental pollution.

References

1. Madurwar MV, Ralegaonkar RV, Mandavgane SA (2013) Application of agro-waste for sustainable construction materials: a review. Constr Build Mater 38:872–878. https://doi.org/10.1016/j.conbuildmat.2012.09.011
2. Ízzet Y (2015) The evaluation of building materials in terms of energy efficiency. Period Polytech Civ Eng 59:45–48. https://doi.org/10.3311/PPci.7050
3. Bakhtyar B, Kacemi T, Nawaz MDA (2017) A review on carbon emissions in malaysian cement industry. Int J Energy Eco Policy 7(3):282–286
4. Ali N, Jaffar A, Anwer M, Khurram S, Alwi SK, Muhammad AM, Ali N, Muhammad R, Raja HA, Xu, (2015) The greenhouse gas emissions produced by cement production and its impact on environment: a review of global cement processing. Int J Res 2:488–500
5. Zhang J, Liu G, Chen B, Song D, Qi J, Liu X (2014) Analysis of CO_2 emission for the cement manufacturing with alternative raw materials: a LCA-based Framework. Energy Procedia 61:2541–2545. https://doi.org/10.1016/j.egypro.2014.12.041
6. Hasmori M, Zin A, Nagapan S, Deraman R, Abas N, Yunus R, Klufallah M (2020) The on-site waste minimization practices for construction waste. IOP Conf Ser Mater Sci Eng 713(1):12038. https://doi.org/10.1088/1757-899X/713/1/012038
7. Qahtan A, Keumala N, Rao SP (2010) Occupant satisfaction in respect to indoor environmental quality in energy efficient certified buildings in Malaysia. In: 16th international conference of the CIB W104 open building implementation on open and sustainable building, pp 449–458
8. Walter W (2007) Application of infrared thermography in civil engineering. Proc Est Acad Sci Eng 13:436–444
9. Mizal-Azzmi N, Jamaludin N (2014) A review of heat transfer in terraced houses of tropical climate. In: E3S Web Conference, pp 1–6. https://doi.org/10.1051/e3sconf/20140301010
10. Shibib K, Isamael H, Hamza M (2013) Enhancement in thermal and mechanical properties of bricks. Therm Sci 17:1119–1123. https://doi.org/10.2298/TSCI110610043S
11. Mydin O, Azree M (2013) An experimental investigation on thermal conductivity of lightweight foamcrete for thermal insulation. J Teknol Sci Eng 63:43–49. https://doi.org/10.11113/jt.v63.1368
12. Mydin MAO, Nawi MNM, Munaaim MAC, Samad AA, Mohamad N, Johari I (2018) Determination of thermal properties of lightweight cellular mortar (LCM) of different density through guarded hot plate. J Adv Res Fluid Mech Therm Sci 47:127–135
13. Deraman R, Hamzah H, Abdul Kadir A, Abdullah AH, Nagapan S, Yunus R (2017) Assessment of thermal surface temperature of empty fruit bunch (EFB) fired clay brick masonry model using infrared thermography (IRT). Ecol Environ Conserv 23:46–55
14. Rhee-Duverne S, Baker P (2013) Research into the thermal performance of traditional brick walls. English Heritage 47

15. Elias OS, Summers F, Sürmeli A, Yannas S (2006) A comparative study of the thermal performance of building materials. In: 23rd conference on passive and low energy architecture, pp 1–6
16. Edis E, Flores-Colen I, Brito J (2015) Time dependent passive infrared thermographic inspection of facades. In: XIII international conference on durability of building materials and components, pp 842–849
17. Chakraverty S, Panigrahi S, Saini DH (2007) Predicting product parameters of fly ash–cement–sand bricks. Proc Ice Constr Mater 160:65–74. https://doi.org/10.1680/coma.2007.160.2.65
18. Samander S (2013) Effect of silica fume on Fly ash cement bricks. IOSR J Mech Civ Eng 6(4):14–18. https://doi.org/10.9790/1684-641418
19. Nisham K, Sridhar MB, Kumar V (2016) Experimental study on class F fly ash cement bricks using partial replacement of fly ash by metakaolin. Int J Chem Sci 14(S1):227–234
20. Vasudevan G (2019) Effect of alum sludge and metakaolin as a partial replacement cement adding superplastizer. IOP Conf Ser Mater Sci Eng 692:1–6. https://doi.org/10.1088/1757-899X/692/1/012013
21. Owaid H, Hamid R, Taha M (2012) A review of sustainable supplementary cementitious materials as an alternative to all-portland cement mortar and concrete. Aust J Basic Appl Sci 6(9):287–303
22. Poon CS, Kou SC, Lam L (2002) Pore size distribution of high performance metakaolin concrete. J Wuhan Univ Technol Sci Ed 17:42–46. https://doi.org/10.1007/BF02852633
23. Duan P, Shui Z, Chen W, Shen C (2013) Enhancing microstructure and durability of concrete from ground granulated blast furnace slag and metakaolin as cement replacement materials. J Mater Res Technol 2(1):52–59. https://doi.org/10.1016/j.jmrt.2013.03.010
24. Dinakar P, Sahoo P, Sriram G (2013) Effect of metakaolin content on the properties of high strength concrete. Int J Concr Struct Mater 7:215–223. https://doi.org/10.1007/s40069-013-0045-0
25. Bai J, Wild S, Sabir BB, Kinuthia JM (1999) Workability of concrete incorporating pulverized fuel ash and metakaolin. Mag Concr Res 51(3):207–216. https://doi.org/10.1680/macr.1999.51.3.207
26. Habeeb G, Mahmud H (2010) Study on properties of rice husk ash and its use as cement replacement material. Mater Res 13(2):185–190. https://doi.org/10.1590/S1516-14392010000200011
27. Chow MF, Rosidan M (2020) Study on the effects of plastic as admixture on the mechanical properties of cement-sand bricks. In: 2nd global congress on construction, material and structural engineering, pp 12016
28. Mat Aris S, Muthusamy K, Uzer A, Ahmad S (2018) Properties of palm oil fuel ash cement sand brick containing pulverized cockle shell as partial sand replacement. IOP Conf Ser Earth Environ Sci 140:12145. https://doi.org/10.1088/1755-1315/140/1/012145
29. Sali N, Deraman R (2019) The selection of optimum water-cement ratio for production of low thermal conductivity cement sand brick with oil palm mesocarp fibre as admixture. IOP Conf Ser Mater Sci Eng 601:1–9. https://doi.org/10.1088/1757-899X/601/1/012037
30. Ling JH, Lim Y, Leong W, Jusli E, Sia H (2019) Properties of cement brick with partial replacement of sand and cement with oil palm empty fruit bunches and silica fume. J Civ Eng Forum 5(3):289–300. https://doi.org/10.22146/jcef.47982
31. Aminudin E, Abd Khalid NH, Azman N, Bakri K, Md Din MF, Zakaria R, Zainuddin N (2017) Utilization of bagasse waste based materials as improvement for thermal insulation of cement bricks. MATEC Web Conf 103:1–8. https://doi.org/10.1051/matecconf/201710301019
32. Malaysian Standard (2007) Specification for portland cement—part 1 compositions, specifications and conformity criteria for common cements. Department of Standard Malaysia. MS 522-1
33. Standard M (1985) Water for concreting. Department of Standard Malaysia, MS, p 28
34. British Standards Institution (2010) Testing hardened concrete—part 2: making and curing specimens for strength tests. London. BS EN 12390-2

35. British Standards Institution (2013) Concrete—specification, performance, production and conformity. London. BS EN 206
36. British Standards Institution (1991) Specification for aggregates from natural sources for concrete. London. BS 882
37. American Society for Testing and Materials (ASTM) (2015) Standard test method for relative density (specific gravity) and absorption of fine aggregate, pp 23–25. West Conshohocken, The USA. ASTM C128-15
38. Güneyisi E, Gesoğlu M, Mermerdaş K (2008) Improving strength, drying shrinkage, and pore structure of concrete using metakaolin. Mater Struct Constr 41(5):937–949. https://doi.org/10.1617/s11527-007-9296-z
39. American Society for Testing and Materials (ASTM) (2017) Standard specification for nonloadbearing concrete masonry units. West Conshohocken. The USA. ASTM C129-17
40. Wild S, Khatib JM, Jones A (1996) Relative strength, pozzolanic activity and cement hydration in superplasticised metakaolin concrete. Cem Concr Res 26(10):1537–1544. https://doi.org/10.1016/0008-8846(96)00148-2
41. Ismail MH, Rusly NSM, Deraman R (2020) Strength and water absorption of concrete containing metakaolin and nylon fiber. Int J Sustain Constr Eng Technol 11(1):230–242
42. Givi A, Abdul Rashid S, Nora F, Abdul aziz F, Amran M, Salleh A (2010) Contribution of rice husk ash to the properties of mortar and concrete: a review. J Am Sci 6(3):157–165
43. British Standards Institution (2010) Testing concrete—part 122: method for determination of water absorption. London. BS 1881-122
44. American Society for Testing and Materials (ASTM) (2020) Standard test methods for sampling and testing brick and structural clay tile. West Conshohocken. The USA. ASTM C67M-20
45. Jamil M, Kaish ABM, Raman SN, Zain M (2013) Pozzolanic contribution of rice husk ash in cementitious system. Constr Build Mater 47:588–593. https://doi.org/10.1016/j.conbuildmat.2013.05.088
46. American Society for Testing and Materials (ASTM) (2019) Standard test method for steady-state heat flux measurements and thermal transmission properties by means of the guarded-hot-plate apparatus. West Conshohocken. The USA. ASTM C177-19
47. Lyons A (2010) Materials for architects and builders, 4th edn. Elsevier Ltd
48. Siddique R, Khan MI (2011) Supplementary cementing materials. Springer, Berlin Heidelberg
49. Fu X, Chung DDL (1997) Effects of silica fume, latex, methylcellulose, and carbon fibers on the thermal conductivity and specific heat of cement paste. Cem Concr Res 27(12):1799–1804. https://doi.org/10.1016/S0008-8846(97)00174-9
50. Demirboğa R (2003) Influence of mineral admixtures on thermal conductivity and compressive strength of mortar. Energy Build 35(2):189–192. https://doi.org/10.1016/S0378-7788(02)00052-X
51. Moretti E, Belloni E, Agosti F (2016) Innovative mineral fiber insulation panels for buildings: thermal and acoustic characterization. Appl Energy 169:421–432. https://doi.org/10.1016/j.apenergy.2016.02.048